Information and Instructions

This shop manual contains several sections each covering a specific group of wheel type tractors. The Tab Index on the preceding page can be used to locate the section pertaining to each group of tractors. Each section contains the necessary specifications and the brief but terse procedural data needed by a mechanic when repairing a tractor on which he has had no previous actual experience.

Within each section, the material is arranged in a systematic order beginning with an index which is followed immediately by a Table of Condensed Service Specifications. These specifications include dimensions, fits, clearances and timing instructions. Next in order of arrangement is the procedures paragraphs.

In the procedures paragraphs, the order of presentation starts with the front axle system and steering and proceeding toward the rear axle. The last paragraphs are devoted to the power take-off and power lift systems. Interspersed where needed are additional tabular specifications pertaining to wear limits, torquing, etc.

HOW TO USE THE INDEX

Suppose you want to know the procedure for R&R (remove and reinstall) of the engine camshaft. Your first step is to look in the index under the main heading of ENGINE until you find the entry "Camshaft." Now read to the right where under the column covering the tractor you are repairing, you will find a number which indicates the beginning paragraph pertaining to the camshaft. To locate this wanted paragraph in the manual, turn the pages until the running index appearing on the top outside corner of each page contains the number you are seeking. In this paragraph you will find the information concerning the removal of the camshaft.

More information available at Clymer.com
Phone: 805-498-6703

Haynes Publishing Group
Sparkford Nr Yeovil
Somerset BA22 7JJ England

Haynes North America, Inc
859 Lawrence Drive
Newbury Park
California 91320 USA

ISBN-10: 0-87288-422-8
ISBN-13: 978-0-87288-422-9

© Haynes North America, Inc. 1985
With permission from J.H. Haynes & Co. Ltd.

Clymer is a registered trademark of Haynes North America, Inc.

Printed in Malaysia
Cover art by Sean Keenan

All rights reserved. No part of this book may be reproduced or transmitted in any form or by any means, electronic or mechanical, including photocopying, recording or by any information storage or retrieval system, without permission in writing from the copyright holder.

While every attempt is made to ensure that the information in this manual is correct, no liability can be accepted by the authors or publishers for loss, damage or injury caused by any errors in, or omissions from, the information given.

SHOP MANUAL

FORD

5000	6600	6710	7610
5600	6610	7000	7700
5610	6700	7600	7710

Tractor Series identification Plate is located under right hood panel. Tractor Serial Number, along with manufacturing and production code numbers and tractor model number, will appear on implement mounting pad at right front corner of transmission (directly behind engine starter). Numbers will be stamped on top of pad, on mounting face of pad, or partially on top of and partially on mounting face of pad. Refer to following explanation of the numbers that will appear at this location:

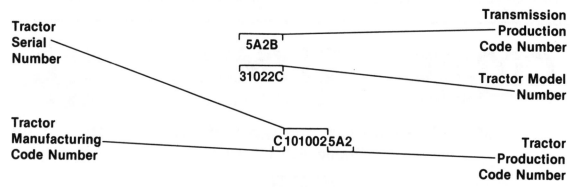

Engine Serial Number, along with tractor size identification, engine type identification and engine production code number, will appear on either the left or right pan rail of cylinder block casting approximately at mid-length of engine. Refer to following explanation of the numbers that will appear at this location:

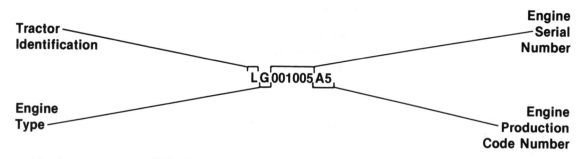

Tractor Identification:
R — 5000 (Before 7-20-68)
E — 5000 (After 7-19-68 through 1975)
L — 5600-5610
E — 6600-6610-6700-6710
F — 7000-7600-7610-7700-7710

Engine Type:

(Early)	(Late)
D — Diesel	1 & 5 — Diesel
G — Gasoline	2 & 6 — Gasoline
P — LP-Gas	

The following tractor models are covered in this manual:

5100 (All-Purpose)	5600 (All-Purpose)	6610 (All-Purpose)	6700
5200 (Rowcrop)	5610 (All-Purpose)	7600 (All-Purpose)	6710
7100 (All-Purpose)	6600 (All-Purpose)	7600C (Rowcrop)	7700
7200 (Rowcrop)	6600C (Rowcrop)	7610 (All-Purpose)	7710

INDEX
(By Starting Paragraph)

DUAL DIMENSIONS

This service manual provides specifications in both the U.S. Customary and Metric (SI) systems of measurement. The first specification is given in the measuring system perceived by us to be the preferred system when servicing a particular component, while the second specification (given in parenthesis) is the converted measurement. For instance, a specification of "0.011 inch (0.28 mm)" would indicate that we feel the preferred measurement, in this instance, is the U.S. system of measurement and the metric equivalent of 0.011 inch is 0.28 mm.

CONDENSED SERVICE DATA

	Series 5000 (Prior to 7-20-68)	Series 5000 (After 7-19-68) 7000	5600	5610	6600 6700	6610 6710	7600 7700	7610 7710
GENERAL								
Engine Make				Own				
No. of Cylinders				4				
Bore, Inches:								
Non-Diesel	4.2	4.4	......	......	4.4	......	......	......
Diesel	4.2	4.4	4.2	4.4	4.4	4.4	4.4	4.4
Bore, mm:								
Non-Diesel	106.7	111.8	......	......	111.8	......	......	......
Diesel	106.7	111.8	106.7	111.8	111.8	111.8	111.8	111.8
Stroke, Inches:								
Non-Diesel	4.2	4.2	......	......	4.2	......	......	......
Diesel	4.2	4.2	4.2	4.2	4.2	4.4	4.2	4.4
Stroke, mm:								
Non-Diesel	106.7	106.7	......	......	106.7	......	......	......
Diesel	106.7	106.7	106.7	106.7	106.7	111.8	106.7	111.8
Displacement, Cubic Inches:								
Non-Diesel	233	256	......	......	256	......	......	......
Diesel	233	256	233	256	256	268	256	268
Displacement, cc:								
Non-Diesel	3818	4195	......	......	4195	......	......	......
Diesel	3818	4195	3818	4195	4195	4385	4195	4385
Compression Ratio:								
Non-Diesel	7.5:1	7.8:1	......	......	7.75:1	......	......	......
Diesel	16.5:1	16.5:1	16.3:1	16.3:1	16.3:1	16.3:1	15.6:1	15.6:1
Compression, Test Gage, (Lbs. at 200 rpm):								
Non-Diesel (Spark plugs removed, WOT)	115-150	115-150	......	......	115-150	......	......	......
Diesel (Stop control out, Throttle closed)	420-510	420-510	290-390	290-390	300-400	300-400	300-400	300-400
Max. Allowable Variation:								
Non-Diesel	25	25	......	......	25	......	......	......
Diesel				50				
Firing Order				1-3-4-2				
Valve Tappet Gap (Warm):								
Intake				0.015 inch (0.381 mm)				
Exhaust				0.018 inch (0.457 mm)				
Valve Face Angle, Degrees:								
Intake	44	29	44.5	44.5	44.5	44.5	29.5	29.5
Exhaust	44	44	44.5	44.5	44.5	44.5	44.5	44.5
Valve Seat Angle, Degrees:								
Intake	45	30	45	45	45	45	30	30
Exhaust				45				
Ignition Timing	Par. 158, 159		......	......	Par. 158	......	......	......
Injection Timing				See Paragraph 133, 134				
Spark Plug Make	Motorcraft	Motorcraft	......	......	Motorcraft	......	......	......
Spark Plug Type	AG-5	AG-5	......	......	AG-5	......	......	......
Engine Low Idle, rpm:	600-650	600-700	600-700	600-850	600-700	600-850	600-700	600-850
Engine High Idle, rpm:								
Non-Diesel	2285-2335	2285-2385	......	......	2285-2385	......	......	......
Diesel	2285-2335	2325-2375	2325-2375	2250-2300	2325-2375	2325-2375	2325-2375	2325-2375
Engine Rated rpm:								
Non-Diesel	2100	2100	......	......	2100	......	......	......
Diesel				2100				
Grounded Battery Terminal				Negative				

SIZES-CAPACITIES-CLEARANCES

Crankshaft Journal Diameter ——————— 3.3713-3.3723 inches ———————
(85.631-85.656 mm)

Crankpin Diameter ——————— 2.7496-2.7504 inches ———————
(69.839-69.860 mm)

CONDENSED SERVICE DATA (CONT.)

	Series 5000 (Prior to 7-20-68)	Series 5000 (After 7-19-68) 7000	5600	5610	6600 6700	6610 6710	7600 7700	7610 7710
SIZES-CAPACITIES-CLEARANCES (Cont).								
Camshaft Journal Diameter...				2.3895-2.3905 inches (60.693-60.718 mm)				
Piston Pin Diameter				See Paragraph 97				
Valve Stem Diameter, Intake .				0.3711-0.3718 inch (9.425-9.443 mm)				
Valve Stem Diameter, Exhaust				0.3701-0.3708 inch (9.400-9.418 mm)				
Main Bearing Diametral Clearance				0.0022-0.0045 inch (0.0559-0.1143 mm)				
Rod Bearing Diametral Clearance:								
Aluminum				0.0021-0.0042 inch (0.053-0.107 mm)				
Copper-Lead				0.0017-0.0038 inch (0.0431-0.0965 mm)				
Camshaft Bearing Diametral Clearance				0.001-0.003 inch (0.0254-0.0762 mm)				
Crankshaft End Play				0.004-0.008 inch (0.1016-0.2032 mm)				
Camshaft End Play..........				0.001-0.007 inch (0.0254-0.1778 mm)				
Piston Skirt-to-Cylinder Clearance				See Paragraph 96				
Cooling System (Less Heater):								
Quarts (U.S.)	15.3	15.3(1)	13.5	13.5	14.5(2)	14.5(2)	18.0(3)	18.0(3)
Liters	14.5	14.5	12.8	12.8	13.8	13.8	17.0	17.0
Crankcase With Filter:								
Quarts (U.S.)	8	8(4)	9	9	9	9	9	9
Liters	7.8	7.8	8.5	8.5	8.5	8.5	8.5	8.5
Transmission, 8-Speed:								
Quarts (U.S.)	19	10.8(5)	12(6)	......	......	......	......	......
Liters	17.9	10.2	11.4	......	......	......	......	......
Select-O-Speed:								
Quarts (U.S.)	11.8	11.8	......	......	......	......	......	......
Liters	11.2	11.2	......	......	......	......	......	......
Transmission & Rear Axle, Dual Power, 540 rpm pto:								
Quarts (U.S.)*	......	......	56	56	56	56	56	56
Liters	......	......	52.9	52.9	52.9	52.9	52.9	52.9
Transmission & Rear Axle, Dual Power With 2-Speed pto:								
Quarts (U.S.)*	......	......	60	60	60	60	60	60
Liters	......	......	56.8	56.8	56.8	56.8	56.8	56.8
Rear Axle, Less Dual Power:								
Quarts (U.S.)	33	34.8(7)	43	......	......	......	......	......
Liters	31.2	32.9	40.7	......	......	......	......	......
Steering Gear, Manual:								
Quarts (U.S.)				0.96				
Liters				0.90				
Power Steering:								
Quarts (U.S.)	1.87	2.35	2.3	2.3	2.3(8)	2.3(8)	2.3(8)	2.3(8)
Liters	1.76	2.22	2.2	2.2	2.2	2.2	2.2	2.2

(1) Model 7000: 14.5 Qts. (13.7 L). (2) Models 6600 and 6610: 13.5 Qts. (12.8 L). (3) Models 7700 and 7710: 19 Qts. (18 L). (4) Model 7000: 9 Qts. (8.5 L). (5) Model 7000: 55 Qts. (52 L). (6) Model 5600 w/o Dual Power. (7) Model 7000: 55 Qts. (52 L). (8) Models 6700, 6710, 7700 and 7710: 2.5 Qts. (2.4 L).
*Models with front-wheel drive refer to paragraph 55N or 69 and 286.

CONDENSED SERVICE DATA (CONT.)

	Series 5000 (Prior to 7-20-68)	Series 5000 (After 7-19-68) 7000	5600	5610	6600 6700	6610 6710	7600 7700	7610 7710
SIZES-CAPACITIES-CLEARANCES (Cont.)								
Front Axle Differential Housing:								
Pints (U.S.)	....	....	11.0	13.0	11.0	13.0	13.7	13.0
Liters	....	....	5.2	6.0	5.2	6.0	6.5	6.0
Front Axle Drive Hubs:								
Pints (U.S.)	....	....	1.6	1.3	1.6	1.3	2.1	1.3
Liters	....	....	0.75	0.6	0.75	0.6	1.0	0.6

FRONT SYSTEM AND STEERING
(All-Purpose Models 5100-7100-5600-5610-6600-6610-7600-7610)

FRONT AXLE ASSEMBLY AND STEERING LINKAGE

All Models

1. **SPINDLE BUSHINGS.** To renew spindle bushings (20 and 23 – Fig. 1), support front of tractor and disconnect steering arms from wheel spindles (25). Slide wheel and spindle assemblies (remove wheels from hubs if desired for clearance) out of front axle extensions (26). Drive old bushings from front axle extensions and install new bushings with piloted drift or bushing driver. New bushings are final sized and should not require reaming if carefully installed. Renew thrust bearing (24) if worn or rough. Refer to Fig. 2 for correct installation of thrust bearing and spindle dust seal.

2. **AXLE CENTER MEMBER, PIVOT PINS AND BUSHINGS.** To remove axle center member, first support front of tractor and remove clamping bolt from tie rod (both tie rods, power steering models). Disconnect drag link front end (17 – Fig. 1) from left steering arm (18) on manual steering models. Remove bolts securing axle extensions (26) to axle center member (4), and slide extensions, spindles and wheels as a unit from center member.

Suitably support center member, remove pivot pin retainers (1 and 9) and pivot pins (2 and 8), then lower center member from front support unit.

Fig. 1—Exploded view of typical front axle, front support and steering linkage for manual steering Models 5100, 7100, 5600, 5610, 6600, 6610, 7600 and 7610.

1. Retainer
2. Front pivot pin
3. Bushing
4. Axle center member & radius rod assy.
5. Cap
6. Thrust washer
7. Bushing
8. Rear pivot pin
9. Retainer
10. Tie rod
11. Tie rod end assy.
12. Front support
13. Steering gear assy.
14. Dust cover
15. Drag link
16. Locknuts
17. Drag link end assy.
18. Steering arm (L.H.)
19. Dust cover
20. Spindle bushing (upper)
21. Dust seal
22. Woodruff key
23. Spindle bushing (lower)
24. Thrust bearing
25. Spindle (L.H.)
26. Axle extension (L.H.)

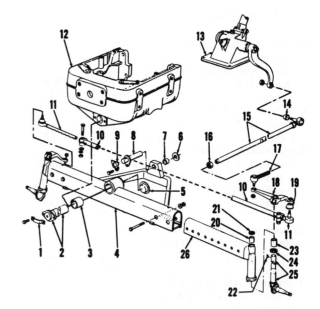

5

Pivot pin bushings (3 and 7) are final sized and may be renewed at this time if scored or worn. Reinstall by reversing the removal procedure.

3. **FRONT SUPPORT.** To remove front support, proceed as follows: Drain cooling system and remove hood, grille and radiator. Support front end of tractor.

On models with manual steering, disconnect rear of drag link (15 – Fig. 1) from steering gear (13) and remove

pivot pins (2 and 8). Raise front of tractor and roll complete front axle unit away from tractor. Attach a hoist to front support and remove bolts securing front support to engine block and oil pan. Be careful not to lose or damage any shims that may be located on bolts securing front support to oil pan.

Reinstall by reversing the removal procedure, observing the following: Install the four bolts securing front support to engine block and tighten to a torque of 250-270 ft.-lbs. (340-367 N·m). Do

Fig. 2 — Top view shows proper installation of dust seal on spindle after spindle is installed in axle extension. Lower view shows proper installation of thrust bearing on spindle prior to installing spindle in axle extension.

not install the two lower front support bolts at this time. Using a feeler gage, measure any existing clearance between front support and oil pan at the two attaching points; then install shims equal to measured clearance when bolts are installed. Shims are available in thicknesses of 0.014, 0.017, 0.021, 0.024 and 0.027 inch (0.357, 0.432, 0.533, 0.610 and 0.686 mm). Tighten the two lower bolts to 250-270 ft.-lbs. (340-367 N·m) after shims are installed.

Front support service on Model 5100 or 7100 with power steering is the same as that for manual steering models except for removal of center steering arm and valve unit (early 5100 models) or center steering arm (late 5100 models and 7100 models) and service on steering arm and pivot pin. Refer to paragraph 19 for service on steering arm and pivot pin and to the preceding paragraph for service on support unit.

4. DRAG LINK, TIE ROD AND TOE-IN. Front wheel spindle arm tie rod ends and drag link ends are of the non-adjustable automotive type and procedure for renewing is self-evident.

Toe-in should be correct for each tread width position when tie rod clamp bolt is placed in corresponding notch of rod end assembly. If toe-in is not within limits of 0 to ¼-inch (6.35 mm), check for bent or excessively worn parts.

Length of drag link between steering gear arm and right front spindle arm should be adjusted for different tread width positions as follows: When right front axle extension is in innermost position or extended 2 inches (5.08 cm) (one adjustment hole), drag link should be in shortest position. Then, lengthen drag link one notch for each additional 2-inch (5.08 cm) extension of right front axle.

STEERING GEAR UNIT

All Manual Steering Models; Early 5100 Models With Power Steering

Refer to Fig. 3 for exploded view of steering gear unit used on all manual steering models and early Model 5100 tractors with power steering.

5. ADJUSTMENT. If there is no perceptible end play of either steering (worm) shaft (6 – Fig. 3) or rocker shaft (22) and a pull at outer edge of steering wheel of 1 to 2¾ lbs. (0.45 to 1.25 kg) is required to turn gear unit past mid-position (with drag link or power steering cylinder disconnected from steering arm), adjustment can be considered correct. Although usually performed only during reassembly of gear unit, adjustments for wormshaft and rocker shaft end play can be made to correct ex-

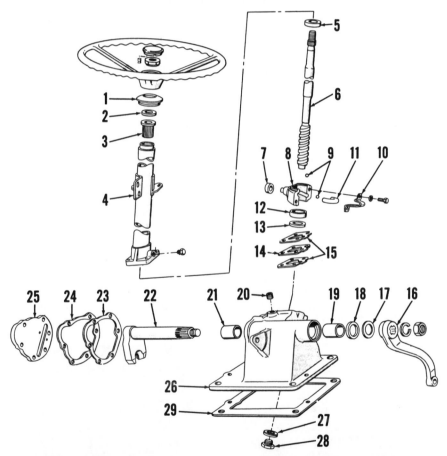

Fig. 3 — Exploded view of steering gear unit of the type used on all manual steering models and on early 5100 models with power steering. Thirty-four ⅜-inch (9.53 mm) diameter steel balls are used in the assembly; 20 are used for steering shaft bearings and 14 are used in the ball nut (8) and tube (11) recirculating groove. Parts 27, 28 and 29 are not used on Models 5600, 5610, 6600, 6610, 7600 and 7610.

1. Grommet	9. Steel balls (⅜-in.)	16. Steering arm
2. Dust seal	10. Retainer	17. Dust seal
3. Bushing	11. Tube	18. Oil seal
4. Steering column	12. Lower bearing race	19. Bushing
5. Upper bearing race	13. Spacer	20. Plug
6. Wormshaft	14. Shims	21. Bushing
7. Roller	15. Gasket	22. Rocker shaft
8. Ball nut		

23. Shims	
24. Gasket	
25. Side cover	
26. Gear housing	
27. Sealing washer	
28. Plug	
29. Gasket (Select-O-Speed only)	

cessive end play or turning effort. With unit removed from tractor as outlined in paragraph 8, proceed as follows:

6. WORMSHAFT END PLAY. Remove side cover (25 – Fig. 3) and inspect condition of unit. If no obvious damage or excessive wear is noted, add or remove shims (14) and gaskets (15) so wormshaft turns freely, but has no perceptible end play. Approximate shim and gasket thickness can be determined by installing steering column without shims or gaskets and measuring resulting gap between steering column and gear housing. Tighten column retaining nuts finger tight and measure gap at several points with feeler gage as shown in Fig. 7. Paper gaskets are 0.010 inch (0.254 mm) thick and steel shims are available in thicknesses of 0.003 and 0.010 inch (0.076 and 0.254 mm). Use one gasket on each side of shim pack and on final assembly, apply a light coat of sealer to gaskets. Tighten steering column retaining nuts to a torque of 25-35 ft.-lbs. (34-48 N·m).

7. ROCKER SHAFT END PLAY. First, adjust wormshaft end play as outlined in paragraph 6, then proceed as follows:

Be sure rocker arm shaft and ball nut are in mid-position and roller is in place as shown in Fig. 4, then install side cover (25 – Fig. 3) without shims (23) or gaskets (24). Tighten retaining nuts and cap screws equally finger tight, then measure gap between side cover and steering housing at several points with feeler gage. Average gap measurement is approximate thickness of shims and gaskets required. Shims are available in thicknesses of 0.005 and 0.010 inch (0.127 and 0.254 mm). Use one 0.010 inch (0.254 mm) gasket on each side of shim pack and, on final assembly, apply a light coat of sealer to gaskets.

8. **R&R STEERING GEAR ASSEMBLY.** To remove steering gear assembly, proceed as follows: Disconnect battery ground cable and remove steering wheel. Remove sheet metal covers at each side of steering gear. Remove screws retaining instrument panel, disconnect ground terminal at left side of panel and rotate panel up out of opening in sheet metal. Disconnect wiring from instrument panel and remove the panel assembly. Remove light switch and choke or diesel shut-off cable. If equipped with Select-O-Speed transmission, refer to paragraph 220 for removal of controls. Remove engine hood and sheet metal surrounding fuel tank. Shut off fuel supply, disconnect fuel supply line and diesel fuel return line from tank, then remove fuel tank assembly. Thoroughly clean steering gear and surrounding area. Disconnect drag link or power steering cylinder from steering arm, then unbolt and remove steering gear assembly from transmission housing. Drain lubricant from steering gear housing if unit is to be disassembled.

NOTE: On models with Select-O-Speed transmission, take care that no dirt or foreign material enters transmission housing while removing steering gear assembly and place a cover over opening in transmission housing while steering gear is removed.

To reinstall steering gear assembly, place a new gasket on transmission housing (Select-O-Speed transmission models only), then reinstall steering gear assembly by reversing removal procedure. Refill gear housing with Ford 134 oil or a suitable equivalent.

9. **OVERHAUL STEERING GEAR UNIT.** With steering gear assembly removed from tractor as outlined in paragraph 8, refer to exploded view of unit in Fig. 3 and proceed as follows:

Remove nut retaining steering arm (16) to rocker shaft (22) and, using suitable pullers, remove arm from shaft. Unbolt and remove side cover (25), shims (23) and gaskets (24). Remove roller (7) from ball nut (8) and slide rocker shaft from housing. Unbolt and remove steering column (4), shims (14) and gaskets (15) from gear housing (26). Remove bushing (3) from upper end of steering column. Pull wormshaft (6) upward, then remove upper bearing race (5) and the ten loose bearing balls (9). Remove wormshaft and ball nut assembly from gear housing as shown in Fig. 5, then remove the ten loose bearing balls from gear housing. Unscrew ball nut assembly from wormshaft and remove the 14 recirculating balls from nut. Tube (11 – Fig. 3) can be removed from nut (8) if necessary. Remove lower bearing race (12), spacer (13), bushings (19 and 21) and oil seal (18) from gear housing (26).

To reassemble, proceed as follows: Install new bushings (19 and 21) using piloted drift or bushing driver, then install new seal (18) with lip to inside of gear housing. Install spacer (13) and lower bearing race (12) in gear housing, then stick the ten bearing balls in race with grease. Assemble tube (11) to ball nut (8) if removed, then stick the 14 recirculating balls in tube and groove of nut with grease. Thread ball nut assembly onto wormshaft, then install shaft and nut assembly in gear housing as in Fig. 5. Carefully insert wormshaft into lower bearing to avoid dislodging bearing balls, then while holding shaft in bearing, place upper bearing race over shaft and invert assembly allowing gear housing to rest against end of shaft. Stick the ten bearing balls in upper race with grease, then push bearing assembly up into housing as shown in Fig. 6. While holding against upper bearing, turn assembly upright. Install new bushings (3 – Fig. 3) in steering column, then refer

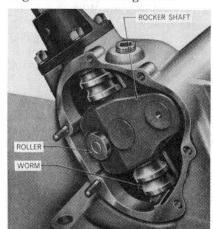

Fig. 4 – View of steering gear assembly with side cover removed. Roller moves in slot in side cover.

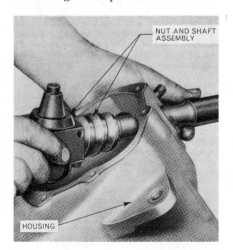

Fig. 5 – Removing ball nut and steering shaft assembly from gear housing.

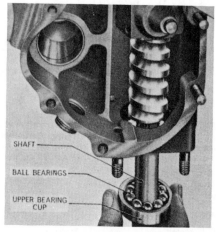

Fig. 6 – Installing upper bearing race and ball bearings.

to paragraph 6 for wormshaft adjustment and column installation. Insert rocker shaft and place dust seal (17) on outer end of shaft. Install steering arm (16) and tighten retaining nut to a torque of 150-190 ft.-lbs. (204-258 N·m). Place roller on end of ball nut (Fig. 4) and install side cover with proper shims and gaskets from rocker shaft adjustment as outlined in paragraph 7.

Late Power Steering Models 5100-7100-5600-5610-6600-6610-7600-7610

10. These later power steering models use the integral power assist unit shown schematically in Fig. 8 and exploded in Fig. 9. A piston is built into shaft ball nut and a cylinder machined into gearcase housing; and entire case unit is pressurized by steering oil. Control is by means of a rotary valve which is built into piston and ball nut unit (12) and is not available separately. Pressure passage to top of piston (P) is internal while lower end is pressurized by external flow through pressure tube (24). Manual operation of steering gear is made possible by a check ball (7) located in valve housing which recirculates oil within gear housing when pump is inoperative.

11. **REMOVE AND REINSTALL.** To remove steering gear assembly, first disconnect battery ground cable and remove steering wheel. Remove sheet metal covers at each side of steering gear. Remove screws retaining instrument panel, disconnect ground terminal (left side) and rotate panel up and out of opening in cowl. Disconnect wiring and remove instrument panel. Remove light switch and choke (or diesel shut-off cable). On Select-O-Speed models, refer to paragraph 220 for removal of controls. Remove engine hood and cowl. Shut off fuel, disconnect and remove fuel tank. Clean steering gear unit.

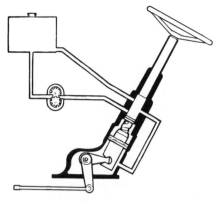

Fig. 8—Schematic view of late power steering gear unit used on some 5100 models and all others except 5200, 6600C, 7600C, 6700, 6710, 7700 and 7710. Steering cylinder is built on ball nut and gear unit is pressurized. Refer to Fig. 9 for exploded view.

Disconnect drag link and pump pressure and return lines, then unbolt and lift off steering gear assembly.

NOTE: On Select-O-Speed models, make sure no dirt or foreign material falls into transmission during removal or while steering gear is off.

To install steering gear, reverse removal procedure. Torque four housing

base bolts to 135-165 ft.-lbs. (184-224 N·m). Use a new transmission housing gasket on Select-O-Speed models. Refill steering gear after complete assembly by cycling power steering, engine running, while keeping pump reservoir filled.

12. **OVERHAUL.** Before disassembling the removed steering gear, temporarily reinstall steering wheel and disconnect external oil feed pipe (24—Fig. 9). Turn steering wheel from lock to lock several times until as much fluid as possible is pumped from housing.

Remove steering arm (15) using Special Tool 1001 or other suitable puller. Remove side cover (23), gasket (20), and rocker shaft end float shim (22). Turn steering shaft until rocker shaft arm is centered in housing opening as shown in Fig. 10, then withdraw rocker shaft.

Remove the four stud nuts securing steering column (1—Fig. 9) and lift off column and shaft (2). Remove and save shim pack (5). Install oil seal protector sleeve (Tool SW 23/1 or FT.3147) over steering shaft spline as shown in Fig. 11, gently tap valve housing (6) away from

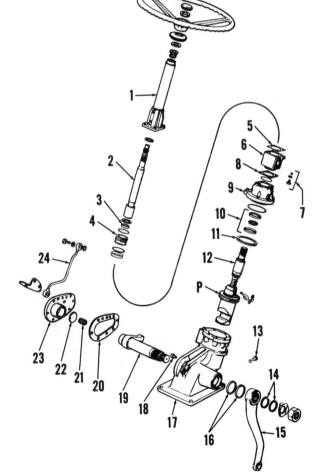

Fig. 9—Exploded view of late power assist steering gear.

1. Steering column
2. Steering shaft
3. Housing seal
4. Bushing sleeve
5. Shim pack
6. Valve housing
7. Check valve
8. Shim pack
9. Bearing housing
10. Bearing
11. Piston ring
12. Ball nut
13. Guide peg
14. Dust seal
15. Steering arm
16. Oil seal
17. Gear housing
18. Wear pin
19. Rocker shaft
20. Gasket
21. Spring
22. Float shim
23. Side cover
24. Pressure tube
P. Piston

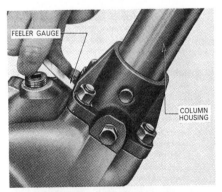

Fig. 7—Measuring clearance between gear housing and steering column housing to determine shim and gasket thicknesses needed. Shim and gasket thickness required between gear housing and side cover is determined in similar manner.

bearing housing (9) and lift off valve housing, saving bearing adjustment shims (8) as valve housing is removed.

Remove cap screws retaining bearing housing (9–Fig. 9); then, turning splined end of steering shaft clockwise, force bearing housing (9) and bearing (10) up and out of ball nut and main gear housing. Shaft bearing (10) contains fifteen 5/16-inch (7.94 mm) diameter loose steel balls which are free to fall as the parts are removed. Bearing balls are interchangeable with the 28 steel balls used in steering nut (N–Fig. 12).

Working through side opening, carefully push ball nut (N), piston and associated parts out of main housing. Note that ball nut is prevented from turning in housing by groove (G) which fits over guide peg (13). Be careful not to damage piston rings (11) as piston is withdrawn. Remove clamp bracket (B), transfer tube and the 28 bearing balls from main nut.

Examine all parts for wear or scoring and make sure parts are thoroughly cleaned. Renew all seals, gaskets and "O" rings when unit is disassembled, as all parts are under system pressure. "O" rings are located on piston guide peg (13–Fig. 9); also between bushing sleeves in control valve housing as shown in Fig. 13. Carefully push bushings out top end of housing. Seal (3) must be installed from underside (chamfered) end of bushing sleeve (4), using Special Tool SW 23/2 which applies pressure in groove between the two seal lips. Coat sealing "O" rings sparingly with a suitable lubricant and install carefully using Fig. 13 as a guide.

Assemble steering gear as follows: Install piston rings (11–Fig. 12), if removed, and rotate rings until end gaps are 180 degrees apart. Align groove (G) with locating pin (13) and carefully install piston using a suitable ring compressor. Position ball nut assembly so rocker shaft arm slot is aligned with main housing side opening (Fig. 10) and install rocker shaft with spring (21) removed. Temporarily install main housing side cover (23–Fig. 9), using a new gasket (20) but omitting end float shim (22) at this time. Install and tighten side cover retaining cap screws to 35-45 ft.-lbs. (48-61 N·m); then using a dial indicator, measure and record rocker shaft end float. Remove side cover then reassemble, installing spring (21) and an end float shim (22) which most nearly equals measured end float minus 0.008 inch (0.203 mm). Shims (22) are available in thicknesses of 0.050, 0.060 and 0.080 inch (1.27, 1.52 and 2.03 mm). Retighten side cover cap screws to 35-45 ft.-lbs. (48-61 N·m).

Turn rocker shaft until ball nut and piston unit is at top of its stroke. Install wormshaft and lower half of transfer tube as shown in Fig. 14, then feed in the 28 bearing balls using clean grease. Install upper half of transfer tube and clamp bracket (B–Fig. 12).

Install bearing housing (9–Fig. 9) and tighten retaining cap screws to a torque of 15-20 ft.-lbs. (20-27 N·m). Slide lower race of bearing (10) into bearing housing bore, grooved side up, then install the 15 steel balls in bearing groove using clean grease.

Position Oil Seal Protector Sleeve (SW 23/1 or FT.3147) over shaft splines as shown in Fig. 15; then, omitting bearing adjusting shim pack (8–Fig. 9), install control valve housing. Tap housing lightly into place until it bottoms and, using a feeler gage, measure clearance between bearing housing and valve housing as shown in Fig. 15. Install a shim pack (8–Fig. 9) equal to measured clearance minus 0.003 inch (0.076 mm). Shims (8) are available in thicknesses of 0.005, 0.010 and 0.025 inch (0.127, 0.254 and 0.635 mm). Shim pack must be accurate to within 0.0015 inch (0.0381 mm). Shim pack (8) controls preload of worm gear bearing (10).

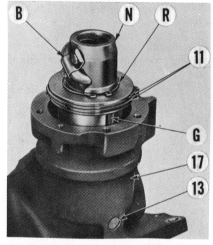

Fig. 12 – Partially disassembled view of steering gear. Refer to Fig. 9 for parts identification except for the following:

B. Clamp bracket	N. Ball nut
G. Locating groove	R. Snap ring

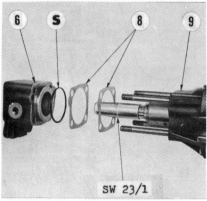

Fig. 10 – Rocker shaft can be withdrawn when arm is centered in housing openings as shown. Refer to Fig. 9 for parts identification.

Fig. 13 – Control valve housing partially disassembled, showing location of bushing sleeves and "O" ring seals.

3. Oil seal	L. Lower sleeve
4. Bushing sleeve	M. Middle sleeve
6. Valve housing	S. "O" ring

Fig. 11 – Oil seal protector sleeve (SW 23/1 or FT.3147) should be used as shown. Refer to Fig. 9 for parts identification except for "O" ring seal (S).

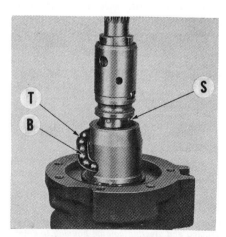

Fig. 14 – With piston at top of stroke, install worm shaft (S) and lower half of transfer tube (T), then feed in the 28 loose balls (B) using clean grease. Refer to text.

With bearing preload correctly adjusted, assemble and install steering column, upper steering shaft and bearing washer. Omit shim pack (5) on trial assembly. Make sure steering column is bottomed on steering shaft splines; then measure clearance between steering column flange and valve housing as shown in Fig. 16. Install shim pack (5 – Fig. 9) equal to measured clearance PLUS 0.005 inch (0.127 mm). Shim (5) is available in 0.005 inch (0.127 mm) thickness only. Tighten steering column stud nuts to a torque of 25-30 ft.-lbs. (34-41 N·m). Complete assembly by reversing disassembly procedure. Tighten external feed line banjo bolts to 25-30 ft.-lbs. (34-41 N·m), steering arm nut to 300-350 ft.-lbs. (408-476 N·m) and steering wheel nut to 60-80 ft.-lbs. (82-109 N·m). Install steering gear unit as outlined in paragraph 11.

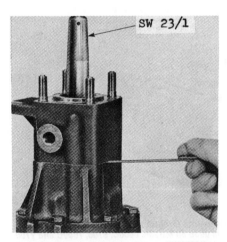

Fig. 15 — Using protector sleeve (SW 23/1 or FT.3147), install and bottom valve housing without shims; then measure shim pack thickness as shown. Install shims equal to measured clearance minus 0.003 inch (0.0076 mm).

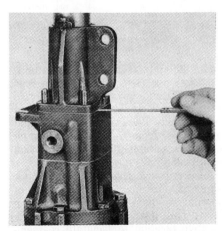

Fig. 16 — Make sure steering column is bottomed in steering shaft splines, then measure clearance for upper shim pack as shown. Install shims equal to measured clearance PLUS 0.005 inch (0.125 mm).

FLUID AND BLEEDING

All Models

CAUTION: The maintenance of absolute cleanliness of all parts is of the utmost importance in the operation and servicing of the hydraulic power steering system. Of equal importance is the avoidance of nicks or burrs on any of the working parts.

13. Recommended power steering fluid is Ford M-2C41A oil. Maintain fluid level to full mark on dipstick on early models with separate reservoir; or bottom of filler neck on late models with integral pump and reservoir. After each 600 hours of operation, it is recommended that filter element and fluid be changed and reservoir cleaned.

The power steering system is self-bleeding. When unit has been disassembled, refill reservoir to full level, start and idle engine, and refill if level lowers. Cycle steering gear by turning steering wheel at least five times from lock to lock, maintaining fluid level at or near full mark. System is fully bled when no more air bubbles appear in reservoir and fluid level ceases to lower.

14. **SYSTEM PRESSURE AND FLOW.** Power steering system pressure should be 1050-1150 psi (7.25-7.94 Mpa) for all models.

Early models with separate steering system reservoir were equipped with a flow control valve which maintained a regulated fluid flow of 3.5 gpm (13.25 liters/min.) at 1000 engine rpm. On late model pumps with integral reservoir used on Models 5600, 6600 and 7600, a flow control valve is not used and normal pump flow at 1000 engine rpm is 2.74 gpm (10.37 liters/min.). Normal pump flow at 1000 engine rpm on Models 5610, 6610 and 7610 is 3.6 gpm (13.6 liters/min.).

On all models, pressure and flow can be checked by teeing into pump pressure line. On early models, pressure relief valve cap plug (7 – Fig. 17) is externally located and pressure can be adjusted

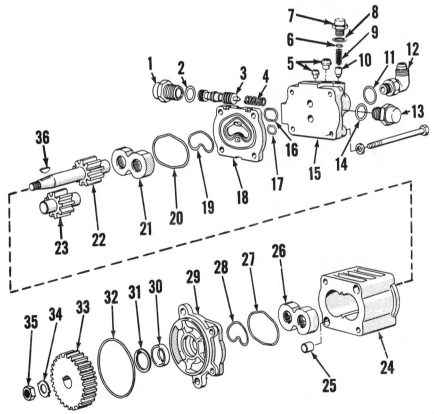

Fig. 17 — Exploded view of early power steering pump. Note that flow control valve spring (4) and small tip end of valve (3) is towards side of rear cover (15) containing relief valve assembly (items 6 through 10).

1. Cap plug	11. Seal ring	20. Outer seal ring	29. Front cover
2. "O" ring	12. Outlet elbow	21. Bearing block	30. Seal (except rowcrop
3. Flow control valve	13. Cap plug	22. Drive gear & shaft	models)
4. Spring	14. "O" ring	23. Driven gear & shaft	31. Locating ring
5. Tubing seats	15. Rear cover	24. Pump body	32. "O" ring
6. Shims	16. "O" ring	25. Bolt (dowel) rings (2)	33. Drive gear
7. Cap plug	17. "O" ring	26. Bearing block	34. Tab washer
8. "O" ring	18. Rear plate	27. Outer seal ring	35. Nut
9. Spring	19. Inner seal ring	28. Inner seal ring	36. Woodruff key
10. Pressure relief valve			

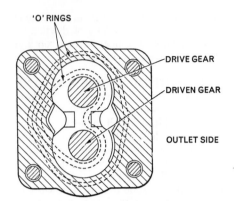

Fig. 18 — Drawing should correct installation of gears and bearing blocks in pump body. View is from rear (flanged) end of pump body.

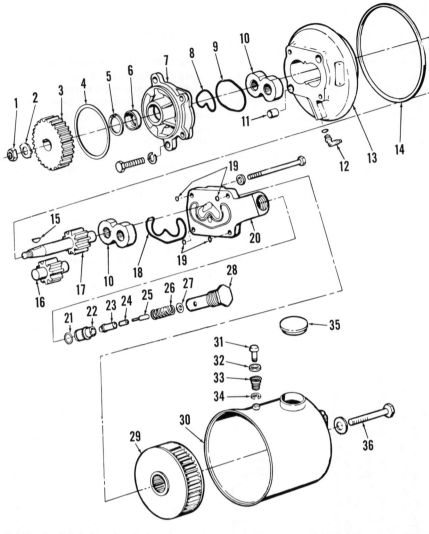

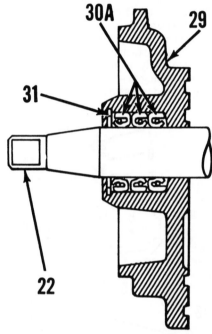

Fig. 19 — On Model 5200 and 7200 tractors, power steering pump is fitted with three shaft seals (30A) instead of single seal (30 — Fig. 17) used on other models. Lips of two inner seals face inward, outer seal lip is toward retaining snap ring (31). Pump front cover is (29); drive shaft is (22).

Fig. 20 — Exploded view of typical power steering pump with integral reservoir used on later Model 5100 and on all other models. A one-piece cover seal is used on Models 6600C, 6700, 7600C and 7700 in place of "O" rings (19).

1. Nut, 7/16-20	10. Bearing/bushing (2)	20. Pump cover	28. Valve body
2. Tab washer	11. Ring dowel (2)	21. Relief valve "O" ring	29. Reservoir filter
3. Drive gear	13. Pump body	22. Relief valve seat	30. Reservoir
4. "O" ring	14. Reservoir seal	23. Relief valve head	31. Vent valve
5. Snap ring	15. Woodruff key	24. Relief valve seal	32. Valve seal
6. Drive gear seal	16. Driven gear	25. Spring guide	33. Vent spring
7. Pump flange	17. Drive gear & shaft	26. Relief valve spring	34. "E" ring retainer
8. Inner seal ring	18. Cover seal	27. Shim*, 0.010 & 0.015	35. Filler cap
9. Outer seal ring	19. "O" rings (4)	in. (0.254 & 0.381 mm)	36. Bolt, 5/16-18 x 2.650

without disassembly. On late models with integral pump reservoir, pump must be removed for relief valve adjustment, refer to paragraph 15.

HYDRAULIC PUMP

All Models

15. **R&R AND OVERHAUL PUMP.** Thoroughly clean pump, lines and surrounding area. Disconnect lines from pump and allow fluid to drain. Cap all openings to prevent dirt from entering pump or lines, then unbolt and remove

pump assembly from engine front plate. When reinstalling pump, use new sealing "O" ring and tighten retaining bolts to a torque of 23-29 ft.-lbs. (31-39 N·m). Reconnect lines, fill and bleed system as in paragraph 13.

On early models, refer to exploded view of pump in Fig. 17 and disassemble pump as follows: Scribe an assembly mark across pump covers and body. Straighten tab on washer (34) and remove nut (35). Pull drive gear (33) from pump shaft and remove key (36). Remove the four through-bolts and separate rear cover assembly (15), plate (18), body (24) and front cover (29). Remove bearing blocks (21 and 26) and

gears (22 and 23) from pump as a unit. Remove caps (1, 7 and 13) from rear cover (15) and withdraw flow control valve (3), pressure relief valve (10) and related parts. Remove snap ring (31) and oil seal (30) from front cover. Clean all parts in a suitable solvent, air dry, then lightly oil all machined surfaces.

Inspect bearing blocks (21 and 26) for signs of seizure or scoring on face of journals. (When disassembling bearing block and gear unit, keep parts in relative position to facilitate reassembly.) Light score marks on faces of bearing blocks can be removed by lapping bearing block on a surface plate using grade "0" emery paper and kerosene.

Examine body for wear in gear running track. If track is worn deeper than 0.0025 inch (0.064 mm) on inlet side, body must be renewed. Examine pump gears for excessive wear or damage on journals, faces or teeth. Runout across gear face to tooth edge should not exceed 0.001 inch (0.0254 mm). If necessary, gear journals may be lightly polished with grade "O" emery paper to remove wear marks. The gear faces may be polished by sandwiching grade "O" emery paper between gear and face of scrap bearing block, then rotating gear. New gears are available in matched sets only.

When reassembling pump, install all new seals, "O" rings and sealing rings.

NOTE: On models (5200 and 7200) with three shaft seals (Fig. 19), the two inner seals are installed lips first and outer seal lip out to serve as a dust seal.

Install flow control valve (3–Fig. 17), spring (4) and plugs (1 and 13) with new "O" rings (2 and 14). Install pressure relief valve (10), spring (9) and plug (7), being sure that all shims (6) are in plug and using new "O" ring (8). Assemble pump gears to bearing blocks (use Fig. 18 as a guide, if necessary) and insert unit into pump body. Be sure the two bolt rings (hollow dowels) are in place in pump body, then position front cover on body. Place rear plate (18–Fig. 17) at rear of body and install rear cover. Tighten the four cap screws (throughbolts) to a torque of 13-17 ft.-lbs. (18-23 N·m). Install pump drive gear key, drive gear, tab washer and nut. Tighten nut to a torque of 55-60 ft.-lbs. (75-82 N·m) and bend tab of washer against flat on nut.

On models with integral pump and reservoir, refer to Fig. 20. Clean pump and surrounding area and disconnect pump pressure and return lines. Remove the two cap screws securing pump to engine front cover and lift off pump and reservoir as a unit. Drain reservoir and remove through-bolt (36), reservoir (30) and filter (29).

Relief valve cartridge can now be removed if service is indicated. For access to shims (27), grasp seat (22) lightly in a protected vise and unscrew body (28). Shims (27) are available in thicknesses of 0.010 and 0.015 inch (0.254-0.38 mm). Starting with the removed shim pack, substitute shims thus varying total pack thickness, to adjust opening pressure. Available shims permit thickness adjustment in increments of 0.005 inch (0.127 mm) and each 0.005 inch (0.127 mm) shim pack thickness will change opening pressure about 35 psi (241.5 kPa). If parts are renewed, the correct thickness can only be determined by trial and error, using the removed shim pack as a guide.

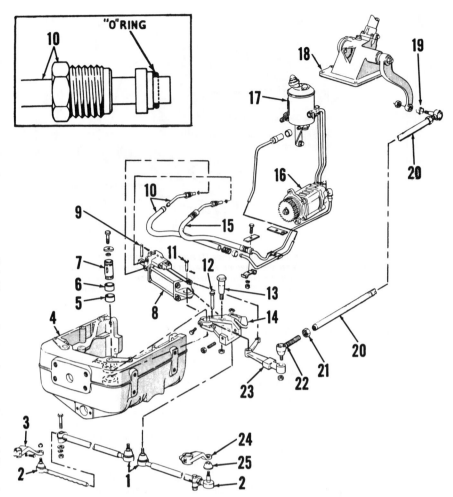

Fig. 22 – Exploded view of early Model 5100 power steering and linkage. Front axle and spindle units are same as those shown in Fig. 1 for manual steering models. A longer steering arm pivot shaft (7) and upper bushing (6) are used on late production units; refer to text. Note proper installation of "O" rings on tubes (10 and 15) where they enter cylinder (8); see inset at upper left.

1. Tie rods	8. Power cylinder assy.	13. Cylinder to steering arm bolt
2. Tie rod ends	9. Piston rod to front support pin	14. Steering arm
3. Steering arm (R.H.)	10. Cylinder to reservoir return line	15. Pump to cylinder pressure line
4. Front support	11. Link to actuating arm pin	16. Power steering pump
5. Bushing, lower	12. Actuating arm pin	17. Fluid reservoir
6. Bushing, upper		18. Steering gear assy.
7. Steering arm pivot shaft		19. Dust cover
		20. Drag link
		21. Locknuts
		22. Drag link end assy.
		23. Actuating arm
		24. Steering arm (L.H.)
		25. Dust cover

To disassemble pump, bend back tab washer and remove shaft nut (1), drive gear (3) and key (15). Mark or note relative positions of flange (7), pump body (13) and cover (20); then remove pump through-bolts. Keep parts in their proper relative position when disassembling pump unit. Pump gears (16 and 17) are available in a matched set only. Bearing blocks (10) are available separately, but should be renewed in pairs if renewal is because of wear. Bearing blocks should also be renewed with gear set if any shaft or bore wear is evident.

When reassembling the pump, align reservoir and tighten through-bolts to a torque of 13-17 ft.-lbs. (18-23 N·m) and drive gear nut (1) to a torque of 55-60 ft.-lbs. (75-82 N·m).

POWER CYLINDER AND CONTROL VALVE

Early Model 5100

16. **REMOVE AND REINSTALL.** To remove power cylinder and control valve assembly (8–Fig. 22), first drain cooling system and remove hood, grille and radiator. Remove any cover plates from top of front support (4) and proceed as follows:

Disconnect power steering fluid lines (10 and 15) from unit and immediately cap all openings. Remove cotter pins from pins (9, 11 and 12) and pull pins from above. Remove cotter pin and nut from lower end of anchor bolt (13), move actuating arm forward and remove bolt.

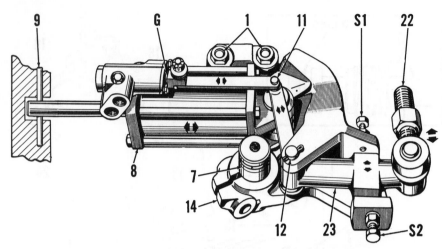

Fig. 23 — Drawing showing power cylinder (8), tie rods (1) and rear drag link (22) end connected to steering arm and actuating arm. Clearance between actuating arm (23) and ends of screws (S1 and S2) should be adjusted as outlined in text. Clearance between control valve retainer (14 — Fig. 24) and yoke (16 — Fig. 24) is shown as (G). Refer to Fig. 22 for parts identification.

Remove cylinder and control valve assembly from front support.

Reinstall power cylinder and control valve assembly by reversing removal procedure. Refill and bleed power steering system as outlined in paragraph 13, then check actuating arm travel adjustment as outlined in paragraph 17.

17. ADJUST ACTUATING ARM STOPS. Stop screws (S1 and S2 – Fig. 23) are provided to limit travel of control valve linkage, thus preventing damage to valve or linkage due to overtravel. To adjust stop screws, loosen locknut on each screw and turn screws in until no power assist can be obtained when turning steering wheel in either direction. Then, back screws out just far enough until power assist is obtained through full range of travel in either direction. Tighten locknuts while holding stop screws in this position.

18. OVERHAUL POWER CYLINDER AND CONTROL VALVE ASSEMBLY. Remove unit as outlined in paragraph 16, refer to exploded view in Fig. 24 and proceed as follows:

To disassemble unit, proceed as follows: Unscrew nuts (38) at each end of transfer tube (39). Remove nuts from through-bolts (32) and unscrew bolts from valve housing. Remove cylinder end (34) from cylinder tube (31) and remove tube from piston and valve housing. Clamp flat of piston rod in a vise, remove cotter pin and unscrew piston retaining nut (30). Remove washer (29) and piston (27), then withdraw piston rod from outer side of valve housing. Remove outer screws holding retaining plate (14) to valve housing and withdraw control valve spool, centering spring and yoke assembly. To disassemble, insert steel rod through hole in valve spool,

clamp in soft-jawed vise as shown in Fig. 25 and unscrew yoke. Remove end cover (1 – Fig. 24), felt washer (2), retainer washer (3) and "O" ring (4) from valve housing. Unscrew plug (18) and remove check ball (20). Remove piston rod scraper (22), back-up ring (23) and "O" ring (24) from valve housing. Remove "O" rings (26 and 33) from housing and end assembly and remove "O" ring (7) from valve spool.

Thoroughly clean all parts in solvent, air dry and inspect for excessive wear,

scoring or other damage. Renew any parts not suitable for further use. Control valve spool (6) and housing (25) are available only as a valve assembly. Bushing (35) is renewable in cylinder end (34); bushing is final sized and should not require reaming if carefully installed.

To reassemble, proceed as follows: Insert new "O" ring (24) in valve housing cylinder rod bore and insert new back-up ring (23) at outer side of "O" ring. Lubricate rod (36) and place new scraper (22) on rod with scraper lip to outer end. Insert rod through bore in valve housing. Place new "O" ring (26) in groove on housing and new piston ring (28) in groove on piston. Clamp flat end of rod in rise, then install piston (27), washer (29) and piston retaining nut (30). Tighten nut securely and retain with cotter pin. Using a fabricated piston ring compressor similar to that shown in Fig. 26, install cylinder tube (31 – Fig. 24) over piston and shoulder on housing. Place transfer tube (39), with ferrules (37) and nuts (38) installed in position and install new "O" ring (33) on cylinder end (34), then install end of cylinder tube and transfer tube and install the four through-bolts. Tighten nuts securely, then tighten transfer tube nuts. With a sleeve type driver, drive scraper (22) into valve housing. Install new "O" ring (4) in groove in outer end of valve housing and new "O" ring (7) in groove on inner end of valve spool. Assemble washers, "O"

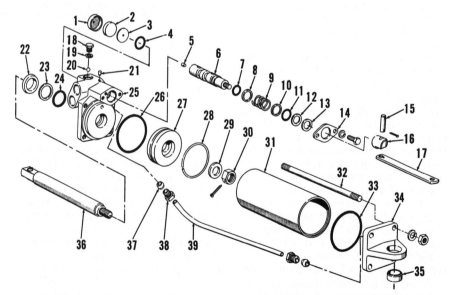

Fig. 24 — Exploded view of power cylinder and control valve assembly for early Model 5100 tractor.

1. Cover	11. "O" ring	21. Plug	31. Cylinder tube
2. Felt washer	12. Back-up ring	22. Piston rod wiper	32. Through-bolt
3. Washer	13. Washer	23. Back-up ring	33. "O" ring
4. "O" ring	14. Retainer	24. "O" ring	34. Cylinder end
5. Plug	15. Pin	25. Valve housing	35. Bushing
6. Valve spool	16. Yoke	26. "O" ring	36. Piston rod
7. "O" ring	17. Link	27. Piston	37. Ferrules
8. Washer	18. Plug	28. Piston ring	38. Nuts
9. Centering spring	19. Washer	29. Washer	39. Transfer tube
10. Washer	20. Check valve ball	30. Castellated nut	

ring (11) and centering spring (9) on valve spool, then install yoke (16) with retainer plate (14). Lubricate control valve spool and insert assembly into valve housing; tighten retainer plate cap screws securely. Insert retainer washer (3) and felt washer (2) in opposite end of valve spool bore, then check valve ball (20) and plug (18) with new gasket (19). If removed from yoke, reinstall control link (17) and pin (15). Reinstall plug (21) ·if removed from valve housing for cleaning purposes.

19. R&R CENTER STEERING ARM AND ACTUATING ARM. With power steering cylinder and control valve assembly removed as outlined in paragraph 16, proceed as follows:

Remove nuts from tie rod (1 – Fig. 22) ends and disconnect end assemblies from center steering arm. Turn steering wheel until nut on steering gear drag link front end (22) is above hole in bottom of front support, remove nut and disconnect end from actuating arm. Remove clamp bolt from center steering arm and remove cap screw and the two washers from top end of pivot shaft (7). Pull shaft using a ⅜-24 bolt, large washer and a 3-inch (7.62 cm) length of pipe.

NOTE: The cap screw and two washers may not be present on some early Model 5100 production units; if not, procure a kit, Ford part 309733, and· install on reassembly.

Remove center steering arm and actuating arm from front support. Renew bushings (5 and 6) if excessively worn. On early units, both bushings are 1 inch (2.54 cm) in length; later units have upper bushing (6) that is 1.38 inches (3.5 cm) long and lower bushing (5) is 1 inch (2.54 cm) long. Be sure correct parts are obtained and are installed in correct location.

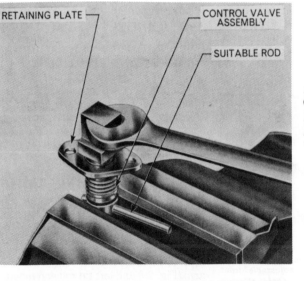

RETAINING PLATE — CONTROL VALVE ASSEMBLY — SUITABLE ROD

Fig. 25 – To remove yoke (16 – Fig. 24) from control valve spool, insert rod through hole in valve and hold unit in soft jawed vise, then unscrew yoke from valve spool.

When reinstalling unit, the following procedure should be observed in adjusting stop screws (S1 and S2 – Fig. 23) as over-travel of actuating arm may damage control valve unit. Reinstall center steering arm and actuating arm, but leave steering gear drag link end (22 – Fig. 22) disconnected from actuating arm. Then reinstall power cylinder and control valve unit as outlined in paragraph 16, but make initial adjustment of stop screws before starting engine as follows: With control valve spool in centered position, adjust stop screws so there is 0.045 inch (1.143 mm) clearance between end of each screw and actuating arm. Fill power steering reservoir, start engine and bleed system as outlined in paragraph 13, but take care that steering gear is not turned tight against stops until final adjustment of stop screws is made as outlined in paragraph 17.

Late Models 5100-7100 and Models 5600-5610-6600-6610-7600-7610

20. INTEGRAL POWER ASSIST STEERING GEAR UNIT. All late models use the integral power steering gear unit shown exploded in Fig. 9. Remove or reinstall unit as outlined in paragraph 11 and overhaul removed unit as in paragraph 12.

21. CENTER STEERING ARM. Refer to Fig. 27 for an exploded view of

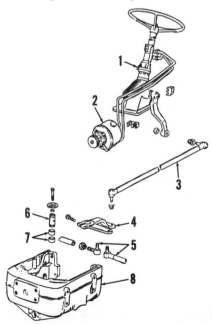

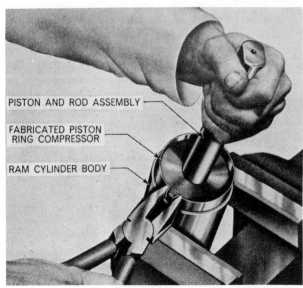

PISTON AND ROD ASSEMBLY — FABRICATED PISTON RING COMPRESSOR — RAM CYLINDER BODY

Fig. 26 – Fabricate a piston ring compressor similar to one shown to aid in installing piston and rod assembly in cylinder tube.

Fig. 27 – Exploded view of late Model 5100, 7100, 5600, 5610, 6600, 6610, 7600 and 7610 steering gear, front support and associated parts.

1. Steering gear	5. Tie rods
2. Pump	6. Pivot shaft
3. Drag link	7. Bushings
4. Center steering arm	8. Front support

front support, center steering arm and associated parts. To remove center steering arm, first drain cooling system and remove hood, grille and radiator. Remove front support cover (baffle). Disconnect both tie rods and turn steering wheel if necessary, until drag link front end nut is accessible through front support, remove nut and disconnect drag link at front end. Remove clamp bolt from center steering arm and cap screw and washers from upper end of pivot shaft. Pull pivot shaft using a 3/8-24 inch bolt, large washer and 3-inch (7.62 cm) length of pipe. Renew bushings (7) in front support if excessively worn. Longer bushing is installed in upper bore. Reinstall by reversing the removal procedure.

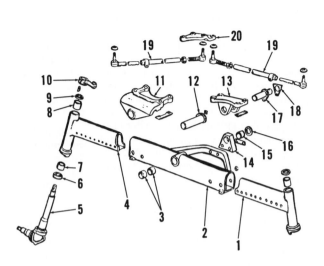

Fig. 28—Exploded view of "Wide Adjustable" type front axle used on Rowcrop Models 5200, 7200, 6600C and 7600C. "All Purpose" type is similar.

1. Extension, L.H.
2. Center member
3. Pivot bushings
4. Extension, L.H.
5. Spindle
6. Thrust bearing
7. Bushing
8. Bushing
9. Dust seal
10. Steering arm
11. Front pivot bracket
12. Pivot pin
13. Rear pivot bracket
14. Trunnion
15. Bushing
16. Thrust washer
17. Pivot pin
18. Retainer
19. Tie rod
20. Center steering arm

FRONT SYSTEM AND STEERING (Models 5200-7200-6600C-7600C)

22. These Rowcrop model tractors are equipped with a front pedestal adaptable to a wide adjustment front axle, a single front wheel or a dual wheel tricycle type front end. All versions are equipped with a tilt steering wheel and hydrostatic power steering system which consists of an engine driven pump with fluid reservoir, a Saginaw Hydramotor steering unit or a Ross-type steering motor with a steering drive motor mounted in front pedestal.

WIDE ADJUSTABLE FRONT AXLE

Models 5200-7200-6600C-7600C

23. **REMOVE AND REINSTALL.** To remove wide front axle assembly as a unit, first disconnect center steering arm (20—Fig. 28) from steering motor. Support front of tractor and unbolt front and rear pivot pin supports (11 and 13) from front pedestal and roll complete axle assembly away from tractor. Reinstall by reversing the removal procedure. Tighten center steering arm cap screws to a torque of 180-220 ft.-lbs. (245-299 N·m).

NOTE: These tractors may be optionally equipped with an "All Purpose" type front axle unit which is similar to "Wide Adjustable" front axle shown in Fig. 28. Major differences in construction are in wheelbase. On models with "Wide Adjustable" front axle, spindles center on front of front support giving a long wheelbase best suited for mounted equipment. On "All Purpose" axle models spindles approximately center on front of

engine block giving a short wheelbase best suited for drawbar work. Removal is similar except that rear pivot pin bracket on "All Purpose" models is cast into oil pan.

Fig. 29—Installed view of "All Purpose" front end showing method of attaching tie rods to center steering arm. On "Wide Adjustable" front end, both tie rods mount below center steering arm as shown in Fig. 27.

24. **SPINDLE BUSHINGS.** Spindle bushings (7 and 8—Fig. 28) are pre-sized and can be renewed without removing axle extensions from tractor. Pull old bushings and install new ones using a piloted puller to prevent damage to bushings or axle bores.

25. **FRONT AXLE SUPPORT BRACKET AND PIVOT PINS.** Front axle support brackets (11 and 13—Fig. 28) can be removed after removing axle assembly as outlined in paragraph 23. Bushings (3 and 15) are pre-sized. Tighten support bracket cap screws to a torque of 180-220 ft.-lbs. (245-299 N·m) when reinstalling.

26. **TIE RODS AND TOE-IN.** Recommended toe-in is 1/8 to 1/4-inch (3.175-6.350 mm) for all models. Both tie-rods must be adjusted to equal length. Note that on "All Purpose" (Short Wheelbase) models tie rods cross over to farthest hole in center steering arm (Fig. 29) and left tie rod end mounts above steering arm. On "Wide Adjustable" models, tie rods attach to

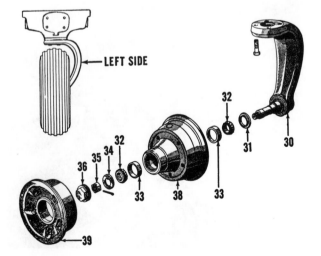

Fig. 30—Exploded view of tricycle single front wheel spindle and wheel assembly. Unit is interchangeable on model with wide adjustable front axle or dual front wheel and pedestal assembly.

30. Spindle
31. Retainer
32. Bearing cone
33. Bearing cup
34. Washer
35. Nut
36. Hub cap
37. Hub
38. Wheel disc
39. Wheel flange

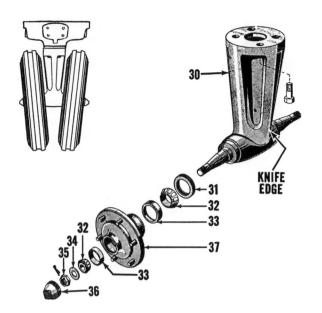

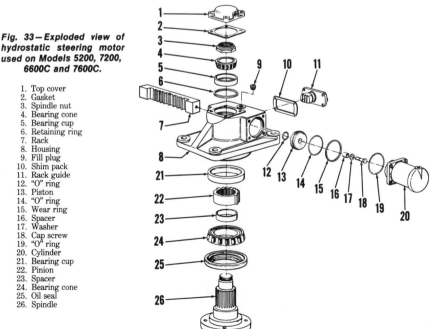

Fig. 31 — Exploded view of dual wheel tricycle spindle and wheel hub unit. Note that knife edge of spindle must be towards front of tractor. Unit is interchangeable with wide adjustable front axle or single front wheel.

30. Spindle
31. Retainer
32. Bearing cone
33. Bearing cup
34. Washer
35. Nut
36. Hub cap
38. Wheel disc
39. Wheel flange

Fig. 34 — Inset shows proper installation of piston rings. Refer to paragraph 29 for procedure.

nearest hole in center steering arm and both rods mount below arm as shown in Fig. 28.

TRICYCLE FRONT SPINDLES

All Models So Equipped

27. Refer to Figs. 30 and 31 for exploded views of single and dual front spindle assemblies. Either unit can be bolted to steering motor flange. Single wheel spindle must be installed with offset to left side of tractor as shown in Fig. 30 and dual wheel spindle must be installed with knife edge to front as

shown in Fig. 31 so proper caster will be obtained.

NOTE: Be sure to deflate tire before attempting to disassemble single wheel disc (38 — Fig. 30) and flange (39).

FRONT SUPPORT

All Models

28. **R&R STEERING MOTOR.** To remove steering motor (4 — Fig. 32), first drain cooling system and remove grille, radiator shell and radiator. On tricycle models, support front of tractor and unbolt spindle (pedestal) assembly from steering motor flange. On wide axle models, unbolt center steering arm.

Disconnect power steering lines (1 and 3) and bleed line (2), capping all connections to prevent dirt entry. Remove the four bolts securing steering motor to front support casting, and lift motor straight up out of support unit.

Install by reversing the removal procedure. Tighten retaining cap screws to 180-220 ft.-lbs. (245-299 N·m); then fill and bleed steering system as outlined in paragraph 31.

29. **OVERHAUL STEERING MOTOR.** To disassemble removed steering motor, refer to Fig. 33. Remove top cover (1) and cylinders (20). Pistons can be removed after cylinders are off, by removing cap screws (18). If spindle (26), spindle bearings or pinion (22) are to be removed, unstake and remove spindle nut (3) and use a suitable press or large puller to push spindle (26), lower bearing (24), pinion (22) and seal (25) downward out of housing (8) and upper bearing (4).

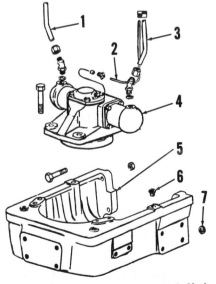

Fig. 32 — Exploded view of typical Model 5200, 7200, 6600C and 7600C front support showing steering motor and attached parts.

1. Steering line	5. Front support
2. Bleed line	6. Dust plug
3. Steering line	7. Spacer shims
4. Steering motor	

Fig. 33 — Exploded view of hydrostatic steering motor used on Models 5200, 7200, 6600C and 7600C.

1. Top cover
2. Gasket
3. Spindle nut
4. Bearing cone
5. Bearing cup
6. Retaining ring
7. Rack
8. Housing
9. Fill plug
10. Shim pack
11. Rack guide
12. "O" ring
13. Piston
14. "O" ring
15. Wear ring
16. Spacer
17. Washer
18. Cap screw
19. "O" ring
20. Cylinder
21. Bearing cup
22. Pinion
23. Spacer
24. Bearing cone
25. Oil seal
26. Spindle

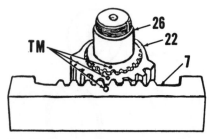

Fig. 35 — Rack and pinion must be timed when steering motor is assembled. Refer to Fig. 33 for parts identification.

Fig. 36 — Rolling torque can be checked by installing a cap screw in threaded center hole in spindle flange then using a torque wrench as shown.

Piston seals consist of an outer Teflon wear ring (15) and an inner sealing "O" ring (14). Refer to inset, Fig. 34, for assembled view. If Teflon wear rings are to be removed for inspection, heat piston and rings in boiling water to soften and expand Teflon wear ring, and remove with a blunt pointed tool while hot. If rings are being renewed, have both pistons ready for service and heat both Teflon wear rings in boiling water. Install "O" rings in clean grooves, making sure "O" rings are straight. Remove Teflon wear rings from water and install while hot. Immediately clamp both pistons in a ring compressor and allow to cool. Compression while hot allows Teflon wear ring to contract in piston groove, thus permitting cylinder to be installed. When reinstalling pistons on rack (7 – Fig. 33), make sure counterbore for "O" ring (12) is toward rack and that "O" ring is installed. Spacer (16) permits some movement of piston when not under pressure and "O" ring (12) forms piston inner seal when cylinder is pressurized. Tighten cap screw (18) to a torque of 15-20 ft.-lbs. (20-27 N·m) when piston is installed.

If spindle has been removed, press shaft out of lower bearing (24), and install lower bearing and seal (25) in housing bore, renewing any parts which are questionable. Note that alignment marks are affixed on spindle, pinion and rack (TM – Fig. 35). Make sure marks are aligned when parts are installed.

To check spindle bearing preload, rack guide (11 – Fig. 33) must be installed and retaining screws tightened **Finger Tight**. Thread a ¾-inch cap screw into center hole in shaft flange, then use a torque wrench to check preload as shown in Fig. 36. Tighten spindle nut until a rolling torque of 52-63 inch-pounds (5.9-7.1 N·m) is required to turn spindle shaft. With preload correctly adjusted, stake spindle nut into notch in shaft.

Rack guide (11 – Fig. 33) must maintain firm contact with rack when turned past center position, but should not bind. With steering motor completely assem-

bled, breakaway torque required to turn spindle from center position should be 20-25 ft.-lbs. (27-34 N·m). If adjustment is incorrect, add or remove shims (10) as required. Shims are available in thicknesses of 0.007, 0.009, 0.010, 0.012, 0.015 and 0.020 inch (0.178, 0.229, 0.254, 0.305, 0.381 and 0.508 mm). Tighten guide retaining cap screws to a torque of 20-25 ft.-lbs. (27-34 N·m).

After steering motor is installed, remove plug (9) and completely fill housing using power steering fluid M-2C41A or a suitable equivalent.

30. R&R FRONT SUPPORT. To remove front support casting (5 – Fig. 32), first remove wide front axle assembly (if so equipped) as outlined in paragraph 23 and steering motor as in paragraph 28. Attach a hoist to front support and remove attaching bolts, then swing unit away from engine, being careful not to lose shims (7), if present.

When installing front support casting, first be sure oil pan cap screws are properly tightened. Install front support using only the four cap screws securing support to engine block. Tighten these bolts to a torque of 250-270 ft.-lbs. (340-367 N·m), then using a feeler gage measure any existing clearance between front support and oil pan at attaching bolt holes. Loosen the four installed cap screws, then install the two lower screws using shims (7) equal in thickness to the measured clearance. Shims (7) are available in thicknesses of 0.014, 0.017, 0.021, 0.024 and 0.027 inch (0.356, 0.432, 0.533, 0.610 and 0.686 mm).

POWER STEERING SYSTEM

Models 5200-7200 (Early)

CAUTION: The maintenance of absolute cleanliness of all parts is of the ut-

most importance in the operation and servicing of the hydraulic power steering system. Of equal importance is the avoidance of nicks or burrs on any of the working parts.

31. FLUID AND BLEEDING. Recommended power steering fluid is Ford M-2C41A oil. Maintain fluid level in pump reservoir at bottom of filler neck. After each 600 hours of operation, it is recommended that filter element and fluid be changed and reservoir cleaned.

The power steering system is self-bleeding. When unit has been disassembled, refill reservoir to full level, start and idle engine, and refill if level lowers. Cycle steering gear by turning steering wheel fully in each direction at least five times, maintain fluid level at or near full mark. System is fully bled when no more air bubbles appear in reservoir, steering action is steady and solid, and fluid level ceases to lower.

32. SYSTEM PRESSURE AND FLOW. The power steering pump assembly incorporates a pressure relief valve and a flow control valve. System relief pressure should be 1050 to 1150 psi (7.25-7.94 MPa). All models maintain a regulated flow of 3.5 gpm (13.25 liters/min.) at 1000 engine rpm.

33. STEERING SYSTEM TROUBLESHOOTING. Refer to the following paragraphs for checking causes of steering system malfunction:

HARD STEERING. Check column bearings and bearings in Hydramotor unit; renew if rough or damaged. Check ring, rotor and vanes for wear and renew assembly if necessary. Check for sticking control valve spool or blocking spool in Hydramotor; clean valves or renew Hydramotor unit as required.

EXCESSIVE WHEEL DRIFT. Check blocking spool spring and guide assembly and renew if spring is broken. Check for leakage past blocking valve; if excessive, renew Hydramotor unit. Check seals on steering cylinder pistons and renew pistons and/or cylinders as required.

STEERING WHEEL TURNING UNAIDED. Check Hydramotor unit for sticking control valve spool, broken valve spool spring, actuator shaft binding or torque shaft (inside actuator shaft) broken. Clean spool and bore or renew Hydramotor unit as required.

STEERING WHEEL SLIPPAGE. Hydramotor control valve spool scored (renew Hydramotor unit) or rotor seals leaking (renew seals).

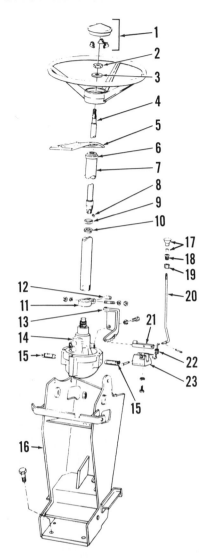

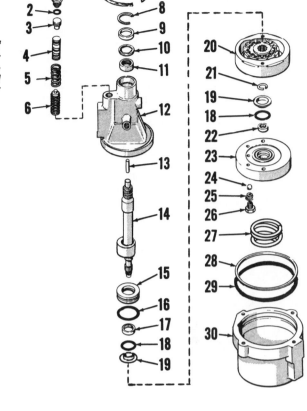

Fig. 38—Exploded view of Saginaw Hydramotor steering unit. Also refer to Figs. 30 to 52 for photos showing disassembly and reassembly techniques.

1. Block valve adjuster
2. "O" ring
3. Plug
4. Blocking valve
5. Spring
6. Spring & guide assy.
7. Snap ring
8. Snap ring
9. Dust seal
10. Oil seal
11. Needle bearing
12. Housing
13. Dowel pins (2)
14. Actuator shaft assy.
15. Bearing support
16. "O" ring
17. Needle bearing
18. "O" ring
19. Rotor seal ring
20. Ring, rotor & vane assy.
21. Snap ring
22. Needle bearing
23. Pressure plate assy.
24. Check valve balls (2)
25. Check valve springs (2)
26. Retaining plugs (2)
27. Pressure plate spring
28. Back-up ring
29. "O" ring
30. Cover

Fig. 37—Exploded view of steering shaft assembly, tilt mechanism and steering motor support used on earlier 5200 and 7200 models. See Fig. 38 for exploded view of steering motor (14).

1. Cap & retainers
2. Steering wheel nut
3. Flat washer
4. Steering shaft
5. Column seal
6. Dust cover
7. Jacket assy.
8. Motor shaft pin
9. Taper ring
10. Shaft connector nut
11. Tube clamp
12. Spacer
13. Tilt sector
14. Steering motor
15. Motor pivot pin (2)
16. Support assy.
17. Knob & locknut
18. Rod guide
19. Guide nut
20. Tilt control rod
21. Pawl
22. Pawl spring
23. Pawl pivot

EXCESSIVE NOISE. Hydraulic lines vibrating against tractor frame or broken control valve spool spring; insulate lines from tractor or renew Hydramotor unit if spring is broken.

ERRATIC MOVEMENT OF FRONT WHEELS. Check Hydramotor ring, rotor or vanes for scoring, wear or binding condition and renew ring and rotor assembly if necessary.

WILL NOT STEER IN EITHER DI-

RECTION. Check fluid level in reservoir. The manual steer check ball between pump return and pressure passages in Hydramotor unit may not be seating. Disassemble unit and clean passage with solvent and dry with compressed air. Renew Hydramotor unit if check ball cannot be made to seat.

FRONT WHEELS JERK OR TURN WITHOUT MOVING STEERING WHEEL. Check for sticking rotor vanes, rotor springs out of place or broken, scored pressure plate, scored rotor ring, scored housing, ball check valves in pressure plate leaking, improper assembly causing gap between rotor components. Disassemble Hydramotor unit, carefully clean and inspect all parts and renew components or complete Hydramotor unit if necessary.

34. R&R AND OVERHAUL PUMP. Pumps used are the same as those used on Models 5100 and 7100. Refer to paragraph 15 for overhaul procedure.

35. HYDRAMOTOR STEERING UNIT. Refer to following paragraphs for information on removal, overhaul and reinstalling Saginaw Hydramotor power steering unit. Refer to paragraph 33 for troubleshooting information.

36. R&R HYDRAMOTOR UNIT. To remove steering unit, proceed as follows:

Disconnect battery ground cable. Remove steering wheel cap (1 – Fig. 37) and nut (2), then remove steering wheel. Remove retainer screws from steering column seal (5) and lift seal from instrument panel. Disconnect fuel shut-off cable at forward control (stop control lever on diesel injection pump) and remove cable retainer clamps. On nondiesel Model 5200, disconnect choke cable. Remove steering tilt control knob, hand throttle control knob from its lever and knob from dual power control lever on tractors so equipped. Remove six instrument panel screws and reach below to disconnect proof meter drive cable from behind meter head. Remove electrical connections for engine coolant temperature gage or warning light, oil pressure warning light, air cleaner restriction warning light (if so equipped), fuel gage leads and charge indicator warning light. Remove interfering leads to voltage stabilizer (constant voltage unit). Identify leads to help in reassembly. When all interfering parts are disconnected, lift out instrument panel with gage assembly.

Working through panel opening within hood rear panel assembly, uncouple pressure and fluid return lines from steering motor (14). Back out motor pivot pins (15) and lift out steering motor and steering column assembly.

Reverse removal procedure to reinstall, then bleed power steering system as outlined in paragraph 31.

37. R&R BLOCKING SPOOL (RE-ACTION) VALVE. Blocking spool valve and related parts can be removed and reinstalled after Hydramotor steering unit has been removed as outlined in paragraph 36. Refer to Fig. 39 and proceed as follows:

Remove lockout adjuster nut (1). Plug (3) and spool valve (4) may now be removed by pushing plug into bore against

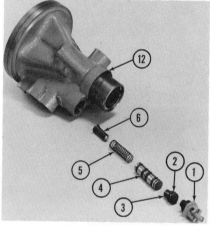

Fig. 39—Exploded view of Hydramotor housing and blocking valve components. Blocking valve can be removed without disassembly of Hydramotor.

1. Lockout
2. "O" ring
3. Plug
4. Blocking valve
5. Spring
6. Spring & guide assy.
12. Housing

spring pressure with screwdriver, then quickly releasing plug to allow spring to pop it out of bore. Remove plug and, if spool sticks in bore, invert unit and tap housing (12) with soft faced mallet to jar spool out. Invert unit and allow spring (5) and spring and guide assembly (6) to drop from bore.

Spool is not serviced separately from complete Hydramotor unit. Renew other parts as necessary, or install new Hydramotor unit if spool and/or spool bore in housing are not serviceable.

To reassemble, install parts in bore of housing (12) as shown in exploded view, renewing "O" ring (2) on plug (3) and tightening adjuster nut to a torque of 10-15 ft.-lbs. (14-20 N·m).

38. R&R COVER RETAINING SNAP RING. To remove snap ring (7–Fig. 38) used to retain cover (30) to housing (12), proceed as follows:

With unit removed from tractor as outlined in paragraph 36, check to see that end gap of snap ring is near hole in cover as shown in Fig. 40; if not, bump snap ring into this position with hammer and punch. Insert a pin punch into hole and drive punch inward to dislodge snap ring from groove. Hold punch under snap ring and pry ring from cover with screwdriver. Usually, coil spring (27–Fig. 38) will push housing from cover; if not, bump cover loose by tapping around edge with mallet as shown in Fig. 41.

To reinstall cover retaining snap ring, housing must be held in cover against pressure from spring. It is recommended that unit be placed in an arbor press and housing be pushed into cover with a sleeve as shown in Fig. 42.

CAUTION: Do not push against end of shaft (14 – Fig. 38).

Place snap ring over housing before placing unit in press. Carefully apply force on housing with sleeve until flange on housing is below snap ring groove in cover. Note that lug on housing must enter slot in cover. If housing binds in cover, do not apply heavy pressure; remove unit from press and bump cover loose with mallet as shown in Fig. 41. When housing has been pushed far enough into cover, install snap ring in groove with end gap near hole in cover as shown in Fig. 40.

39. OVERHAUL HYDRAMOTOR STEERING UNIT. With unit removed from tractor as outlined in paragraph 36 and cover retaining snap ring removed as outlined in paragraph 38, proceed as follows:

Clamp flat portion of Hydramotor housing in a vise and remove cover (30 – Fig. 38) by pulling upward with a twist-

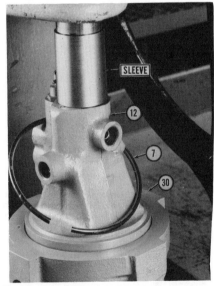

Fig. 42—Using sleeve and arbor press to push housing into cover to allow installation of the cover retaining snap ring (7).

Fig. 40—To remove cover retaining snap ring, drive punch through hole in cover to disengage snap ring from groove.

Fig. 41—In event coil pressure plate spring does not push cover from housing, tap alternate mounting lugs on cover with soft face mallet as shown above.

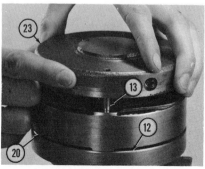

Fig. 43—Lifting pressure plate (23) from dowel pins (13).

ing motion. Remove pressure plate spring (27). Remove the two cap screws, then lift off pressure plate (23) as shown in Fig. 43. Remove dowel pins (Fig. 44), then remove snap ring (21) from shaft (14) with suitable snap ring pliers and screwdriver; discard snap ring. Pull pump ring and rotor assembly (20) up off shaft as shown in Fig. 45. Tap outer end of shaft with soft faced mallet as shown in Fig. 46 until bearing support (15) can be removed, then carefully remove actuator shaft assembly from housing as shown in Fig. 48.

NOTE: As actuator shaft and control valve spool assembly is a factory balanced unit and is not serviceable except by renewing complete Hydramotor steering unit, it is recommended that this unit not be disassembled.

Carefully clean and inspect removed units. Refer to paragraph 30 for information on blocking valve assembly. If housing control valve bore or blocking valve bore are deeply scored or worn, or if blocking valve spool or actuator shaft

and control valve spool assembly are damaged in any way making the unit unfit for further service, a complete new Hydramotor steering unit must be installed. If these components (housing, blocking valve and actuator assembly) are serviceable, proceed with overhaul as follows:

Remove check valve retainers (26 – Fig. 38), springs (25) and check valve balls (24) from pressure plate (23) and blow passages clear with compressed air. Renew pressure plate assembly if check valve seats or face of plate are deeply scored or damaged. Renew needle bearing (22), springs (25) and/or check valve balls (24) if damaged and pressure plate is otherwise serviceable.

NOTE: Drive or press on lettered (trademark) end of bearing cage when installing new needle bearing.

If bearing support (15) is otherwise serviceable, a new needle bearing (17) may be installed; drive or press only on lettered end of bearing cage.

Remove snap ring (8), dust seal (9) and oil seal (10) and inspect needle bearing (11); renew needle bearing if worn or damaged. Press only on lettered (trademark) end of bearing cage when install-

Fig. 48 – Removing the actuator shaft assembly from housing. Be careful not to cock control valve spool in bore of housing.

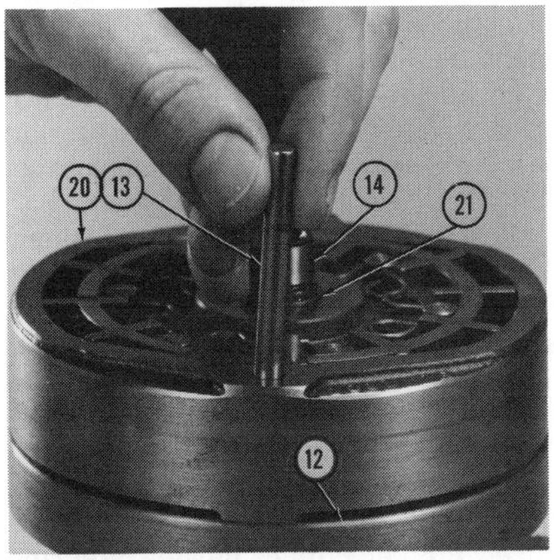

Fig. 44 – Remove dowel pins (13) from motor ring and housing. Then, remove snap ring (21) retaining rotor to actuator shaft (14).

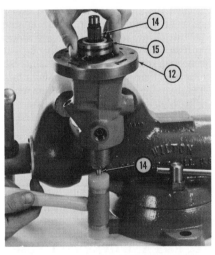

Fig. 46 – Tap on outer end of actuator shaft (14) to bump bearing support (15) from housing (12).

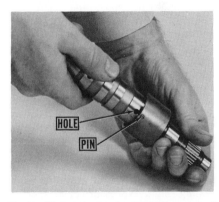

Fig. 49 – Pin in actuator sleeve must be engaged in hole in end of control valve spool before actuator assembly is installed. If spool cannot be pulled out of sleeve, pin is engaged.

Fig. 45 – Lifting motor ring, rotor and vane assembly (20) from actuator shaft (14) and housing (12).

Fig. 47 – Lifting Teflon rotor seal (19) and "O" ring (18) from bearing support (15). Needle bearing (17) is serviced separately from bearing support. Groove (G) is for support sealing "O" ring (16 – Fig. 38). Identical seals (18 and 19) are used in pressure plate.

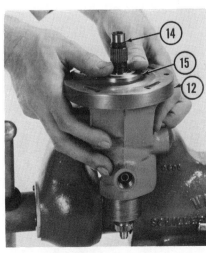

Fig. 50 – When pushing bearing support (15) into housing (12), be careful not to damage "O" ring on outside groove of support.

ing new bearing. Install oil seal with lip towards inside (needle bearing), then install dust seal and retaining snap ring.

If rotor ring, rotor or vanes are worn, scored or damaged beyond further use, or if any vane springs are broken renew unit as a complete assembly (20). If unit was disassembled and is usable, reassemble as follows: Place ring on flat surface and place rotor inside ring. Insert vanes with rounded edges out in the rotor slots aligned with large inside diameter of ring, turn rotor 90 degrees and insert remaining vanes. Hook springs behind vanes with a screwdriver as shown in Fig. 51, then turn assembly over and hook springs behind vanes on opposite side of rotor.

To reassemble Hydramotor unit, place housing (with needle bearing, seals and snap ring installed) in a vise with flat (bottom) side up. Check to be sure that pin in actuator is engaged with hole in valve spool; if spool can be pulled away from actuator as shown in Fig. 49, push spool back into actuator and be sure pin is engaged in one of the holes in spool. Then, lubricate spool and shaft and carefully insert assembly into bore of housing. Place bearing support, with outside "O" ring and needle bearing installed, on shaft and carefully push support into housing as shown in Fig. 50. Insert rotor sealing "O" ring and motor seal into bearing support. Place ring and rotor assembly on shaft and housing with chamfered outer edge of ring up (away from housing). Install a new rotor retaining snap ring and insert dowel pins through ring into housing. Using heavy grease, stick "O" ring and rotor sealing ring into pressure plate, then install pressure plate assembly over shaft, pump ring and rotor assembly and the two dowel pins. Install the two retaining cap screws, tighten to 120 inch-pounds (13.6 N·m) then back off ¼-turn. Be sure cap screws are within circumfer-

ence of pressure plate. Place coil spring on top of pressure plate. Install new "O" ring and back-up ring in second groove of cover (Fig. 52), lubricate rings and push cover down over pressure plate and ring. While holding cover on assembly, place unit in an arbor press and insert cover retaining snap ring as outlined in paragraph 38.

Later Models 5200-7200 and Models 6600C-7600C

These Rowcrop models are equipped with tilt steering gear mechanism and hydrostatic power assist Ross type steering motor shown in Fig. 53.

40. **FLUID AND BLEEDING.** Use Ford M-2C41A fluid in this system. Fluid level should be kept at bottom of reservoir filler neck. System capacity is 4.6 U.S. pints (2.2 liters). Reservoir (30 – Fig. 20) should be disassembled for cleaning and new filter (29) installed after each 600 hours of operation.

System is self-bleeding. Whenever power steering system has been disassembled for any reason, after reassembly, refill fluid reservoir. Start engine and run at low idle rpm. Operate steering from full right to full left through at least five complete cycles. Replenish reservoir fluid as needed to maintain level. System is free of trapped air when no bubbles appear, steering is firm and reservoir level remains steady at full.

41. **PRESSURE AND FLOW.** System pressure is regulated at 1050 to 1150 psi (7.25-7.94 MPa) by action of pressure relief valve incorporated in pump. Rate of flow is maintained at 3.5 gpm (13.25 liters/min.) at 1000 engine rpm.

42. **R&R AND OVERHAUL PUMP.** Refer to paragraph 15 and Fig. 20 for procedure and exploded view of pump.

43. **R&R ROSS TYPE STEERING UNIT.** To remove unit from these Rowcrop tractors, proceed as follows:

Disconnect battery ground cable. Remove steering wheel cap (1 – Fig. 53) and nut (2), then remove steering wheel. Use a puller. Remove screws from steering column flexible seal assembly and lift seal from instrument panel. Disconnect fuel shut-off cable at engine end and cable retainer clamps, then loosen mounting nut under panel so cable can be removed. On non-diesel models, choke knob and cable correspond to this fuel shut-off control. Remove steering tilt control knob and nut (23). Remove hand throttle control knob from its lever, and on tractors so equipped, remove knob from dual power control lever. Remove six screws which attach instrument panel to hood rear panel assembly and reach below to disconnect proof meter drive cable from behind meter head. Remove electrical connections from engine coolant temperature gage, oil pressure warning light, air cleaner restriction warning light (Model 7600C only), fuel gage leads and charge indicator warning light. Disconnect leads to voltage stabilizer (constant voltage unit) which will interfere with panel removal. Identify leads for convenience in reassembly. When all interfering parts are clear, lift out instrument panel and gage assembly.

Working through panel opening and from below rear hood panel, disconnect fluid pressure and return lines from steering motor (11). Release locknuts (15), back out two tilt pivot pins (14) and lift out steering motor (11) and steering column (10).

Reverse this procedure to reinstall, then refer to paragraph 40 for procedure for bleeding power steering system.

44. **OVERHAUL STEERING MOTOR.** Separate motor unit from removed motor/column assembly by removing four cap screws which retain steering motor (11 – Fig. 53) to base of steering column (10). Slide motor (11) and shaft (6) out of column (10), remove two retainers (9) and dowel pins (8), then separate motor from shaft at coupling (7). Plug or cover hydraulic ports of motor (11) and thoroughly clean motor exterior.

Insert a threaded fitting into one of the hydraulic ports, then clamp this fitting in a vise so motor (11) is held securely with input shaft (47) pointing downward. Remove assembly screws (24) and housing end cover (25).

NOTE: From this point on, great care must be used to protect finely lapped matting surfaces of disassembled components. Proper sealing between these

Fig. 51 – Be sure all vane springs are engaged behind rotor vanes. Springs can be pried into place with screwdriver as shown.

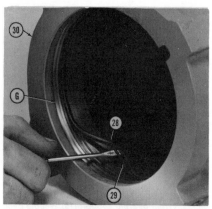

Fig. 52 – "O" ring (28) and back-up ring (29) are installed in cover (30); be sure back-up ring is to outside (open side) of cover. Groove (G) is for cover retaining snap ring.

parts is dependent upon their finish. During disassembly, parts may hang together by an oil film and be accidentally dropped.

When seal (27) and retainer (28) are removed, carefully remove washer (26) and commutator set which comprises commutator (29) and ring (30) followed by manifold (31). Lift off spacer (34), drive actuating link (33) and rotor/stator set (32). Separate link and spacer from rotor/stator set. Do not allow vanes and springs to fall out of rotor/stator (32). This assembly should remain together for the time being.

Release housing (35) from vise and re-clamp with input shaft (47) pointed upward.

Remove dust seal (63) and after placing light match marks on rims of housing (35) and cover assembly (58), unbolt and lift off cover (58), shaft (47) and valve spool assembly (38).

Remove and discard seal ring (57), then slip cover (58) from input shaft and remove spacer (48). Remove shims (54 and 55) and note number used for guidance in reassembly. Remove snap ring (62) and hub seal ring (61). Retain these parts. Remove brass back-up washer (60) and packing seal (59) and discard. Needle bearing (56) should be left in place within cover (58) unless renewal is required.

Remove snap ring (53), thrust washer (52), control valve bearing (51), thrust washer (50) and spring washer (49) from shaft (47).

With input shaft (47) supported on a wooden block, use a light hammer and a pin punch of 0.120 inch (3.0 mm) diameter to carefully drive out pin. Remove torsion bar (44) and spacer (45). Cross pin remains in torsion bar. Remove drive spool ring (46) by placing spool end down on a flat surface then rotating input shaft full right and left until drive ring slips free. Holding spool and input shaft in this same position, rotate input shaft in a clockwise direction until 5/16-inch (7.94 mm) actuator ball is disengaged from groove in input shaft. Lift out shaft, taking care not to lose ball. Valve spool spring (40), should not be removed except for renewal. If required, remove with great care so as not to damage spool finish.

Back out plug (43) from housing assembly (35) and shake out ¼-inch (6.35 mm) ball (41). Renew "O" ring (42).

All removed parts should be carefully cleaned in solvent and inspected for undue wear or damage. Any parts which are scored or galled must be renewed. If needle bearing (56) in cover assembly (58) requires renewal, set cover up in a press, flanged end down, and use a properly fitted mandrel to press old bearing out toward flange. Invert cover and press new bearing into place applying

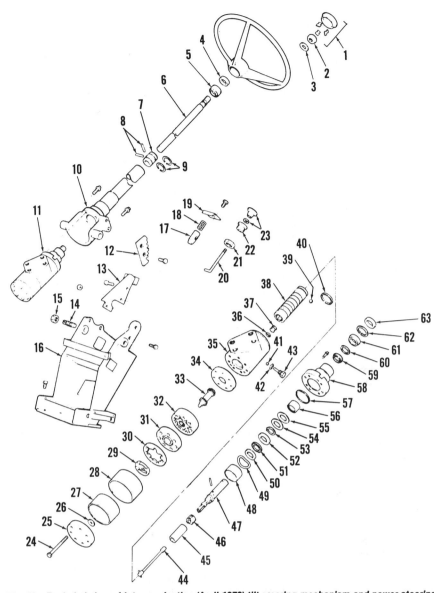

Fig. 53 – Exploded view of later production (April 1973) tilt steering mechanism and power steering motor used on later 5200 and 7200 models and on Models 6600C and 7600C. Steering motor is used on Models 6700, 6710, 7700 and 7710 but bearing (56) is not used on later units.

1. Cap & retainers	18. Latch spring	35. Housing assy.	52. Thrust washer
2. Steering wheel nut	19. Spring retainer	36. "O" ring seal	53. Spool retainer ring
3. Flat washer	20. Tilt latch rod	37. Plug	54. Shim, 0.0025 & 0.005 in.
4. Dust seal	21. Guide nut	38. Valve spool	(0.064 & 0.127 mm)**
5. Bearing & retainer	22. Rod guide	39. Steel ball, 5/16-in.	55. Shim, 0.010 & 0.030 in.
6. Steering shaft	23. Control knob & nut	40. Valve spool spring	(0.254 & 0.762 mm)**
7. Motor coupling	24. Bolt, special, 5/16-18	41. Steel ball, ¼-in.	56. Needle bearing
8. Shaft dowel pin (2)	25. Housing end cover	42. "O" ring, 0.301 ID	57. End cover seal
9. Dowel retainer (2)	26. Rotor pin washer	43. Plug	58. Cover assy.
10. Column assy.	27. Rotor seal	44. Torsion bar & pin	59. Packing
11. Steering motor assy.	28. Shaft bearing retainer	45. Housing spacer	60. Back-up washer
12. Tilt control quadrant	29. Commutator*	46. Spool/drive ring	61. Hub seal ring
13. Tilt control bracket	30. Commutator ring*	47. Input shaft & pin	62. Snap ring
14. Tilt pivot pin (2)	31. Motor manifold	48. Housing spacer	63. Motor shaft dust seal
15. Locknut (2)	32. Rotor/stator assy.	49. Spring washer	
16. Steering support	33. Valve actuating link	50. Thrust washer	*Matched set
17. Tilt control latch	34. Rotor spacer	51. Control valve bearing	**As required

force on mandrel only against lettered end of bearing. See Fig. 54 for bearing installation depth and tool dimensions.

Use a micrometer to measure thickness of commutator ring (30 – Fig. 53) and commutator (29). If commutator ring is thicker by 0.0015 inch (0.038 mm) than commutator (29), renew this matched set.

Place rotor/stator (32) on lapped inner face of housing end cover (25). Be certain that vanes and vane springs are correctly installed in rotor slots. Arched back of each spring must be in contact with its vane. Position one lobe of rotor in a valley of stator as shown at (V – Fig. 55). Center opposite lobe on crown of stator as shown, then use two narrow

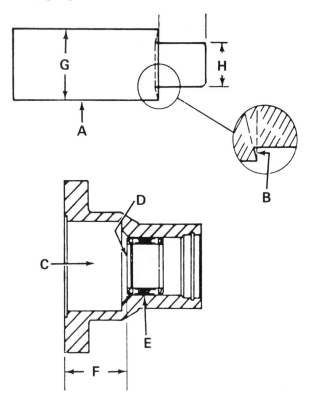

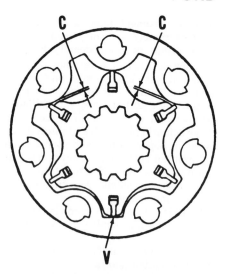

Fig. 54—Sectional view of housing upper cover to show needle bearing installation dimensions. Refer to text for procedure. Note dimensions of installation tool (A).

A. Installation tool (mandrel)
B. Undercut
C. Installation bearing this direction
D. Bearing number this side
E. Needle bearing
F. 1.280-1.300 in. (32.51-33.02 mm)
G. 1.062-1.120 in. (26.97-28.45 mm)
H. 0.870 in. (22.10 mm)

Fig. 55—With rotor positioned in stator as shown, clearance (C) measured at two points as shown must not exceed 0.006 inch (0.152 mm). Refer to text for procedure.

feeler gages to check clearance (C) between rotor lobes and stator. If clearance exceeds 0.006 inch (0.15 mm), renew complete rotor/stator assembly. Use a micrometer to measure thickness of both rotor and stator. If thickness of stator exceeds that of rotor by 0.002 inch (0.05 mm), renew assembly. Stator, rotor, vanes and springs are available only as an assembly.

Prior to reassembly, all parts should be washed in clean solvent and air dried. DO NOT WIPE WITH CLOTH OR RAGS. Except when otherwise indicated, all parts are left dry for reassembly. Insert ball (41 – Fig. 53) into cavity in housing (35) and reinstall pin and plug (43) with a new "O" ring (42). Tighten plug to 10-14 ft.-lbs. (14-19 N·m). Reclamp housing in a vise by use of a threaded fitting as before with top (input shaft side) upward. Assemble thrust washer (50), control valve bearing (51), thrust washer (52) and snap ring (53) on input shaft (47). If valve spool spring (40) was removed for renewal, carefully install new spring on spool. Insert steel ball (39) into ball seat within spool (38). Assemble spring washer (49) over thrust washer (50) on shaft, then insert input shaft (47) into valve spool (38) and engage ball and helix with a counterclockwise twist. Hold assembly horizontal for this operation. Use small mid-section of torsion bar (44) for a gage by placing between end of spool and thrust washer (50). Holding this arrangement, set input shaft/spool assembly upright on

shaft end on a firm, flat surface. Insert spool drive ring (46) into valve spool (38) to engage input shaft spline, then remove torsion bar. Place torsion bar (44) inside spacer (45) and insert assembly into valve spool, aligning cross holes. Use a 0.120 inch (3.0 mm) pin punch to hold alignment, then press cross pin through until flush with surface of shaft. Complete pin installation by driving pin about 1/32-inch (0.794 mm) below flush. Place spacer (48) over spool and install assembly in housing (35). Reinstall original shims (54 and 55) on thrust washer (52), then lubricate new cover seal (57) and set into flange of cover (58). Position cover assembly by aligning match marks to housing, then install cap screws just finger tight. For a cover alignment tool, use a worm drive type hose clamp over joint between cover flange and housing, then tighten cap screws in a cross pattern to a torque of 18-22 ft.-lbs. (24-30 N·m).

IMPORTANT NOTE: If input shaft (47) or end cover (58) have been renewed, this procedure for selection of new shims (54 and 55) must be followed:

Reassemble with new parts installed, using original shims, as outlined in preceding paragraph. Invert unit (input shaft end down) in a vise. Pull end of input shaft down by hand and prevent rotation, then engage valve actuating link (33) by its splines in valve spool (38) and rotate link until spool end is flush with

housing. Lift link away to align its drive slot to drive pin of torsion bar (44), then re-engage link. Now, use a straightedge and a feeler gage to determine if spool end extends from flush to 0.0025 inch (0.0635 mm) above surface of housing. If not, select a new shim pack for placement under cover assembly (58). Shims (54 and 55) are offered in thicknesses of 0.0025, 0.005, 0.010 and 0.030 inch (0.0635, 0.127, 0.254 and 0.762 mm). Recheck with new shim pack in place to be sure that this specification is met before completing reassembly.

To continue reassembly of steering control, with valve actuator link (33) installed, place rotor spacer plate (34) on bottom of housing (35) with its plain side up. Install rotor/stator assembly over link splines with assembly screw holes aligned. Be sure that vanes and vane springs are properly seated. Install motor manifold (31) with circular slotted side up and its assembly holes also aligned. Install commutator ring (30), slotted side up, then install commutator (29), properly engaged, over tip of link (33). Install rotor seal (27) and retainer (28). Now, apply a few drops of hydraulic fluid to commutator (29). Stick washer (26) over pin using a little clean grease, then install housing end cover (25) with bolt holes aligned and start cap screws (24). Tighten all seven cap screws alternately in even steps, while rotating input shaft, to a final torque value of 15-19 ft.-lbs. (20-26 N·m). Check rolling torque needed to turn input shaft by attaching coupling, steering shaft and steering wheel retaining nut. Torque necessary to turn input shaft should not exceed 100 in.-lbs. (11.3 N·m).

Set unit up in vise again with input shaft at top. Lubricate and install a new packing seal (59) and a new brass back-up washer (60). Install hub seal ring (61) flat side up. Install snap ring (62) and dust seal (63). Remove assembly from vise, remove fitting from port and lay unit on its side, ports upward. With valve assembly so positioned, pour clean fluid into inlet port while turning input shaft until fluid runs from outlet port. Plug or cover ports to prevent dirt entry until fluid lines are reconnected during reinstallation of steering system in tractor. Refer to paragraph 40 for final bleeding procedure.

FRONT SYSTEM AND STEERING
(Models 6700-6710-7700-7710)

45. These models are equipped with a front axle which may be adjusted to track widths from 56 to 88 inches. A short (85.5 inches) or long (101.5 inches) tractor wheelbase may be obtained by reversing front axle and replacing spindle steering arms, tie rod sleeve and anchor for power steering cylinder.

Models 7700 and 7710 are equipped with a tilting steering column while Models 6700 and 6710 are available with tilting or non-tilting steering columns. Both models are equipped with hydrostatic power steering which consists of an engine-driven pump with fluid reservoir, a Ross-type steering motor and a steering cylinder.

FRONT AXLE

All Models

46. **REMOVE AND REINSTALL.** To remove front axle, remove front weights and support front of tractor. Disconnect and cap steering cylinder hoses. Unbolt and remove front support bracket (21 – Fig. 56) then roll front axle assembly away from tractor.

To install front axle, reverse removal procedure. Tighten front support bracket (21) screws to 250-270 ft.-lbs. (340-367 N·m). Cycle steering wheel several times in both directions to bleed steering cylinder and check fluid level in reservoir.

NOTE: Tractor wheelbase may be changed from long to short or vice versa by reversing direction of front axle center member (23 – Fig. 56) and replacing steering arms (3 and 7), tie rod sleeve (5) and steering cylinder anchor (11). Steering cylinder hoses must be rerouted and lengthened or shortened.

47. **SPINDLE BUSHINGS.** Spindle bushings (13 and 16 – Fig. 56) are pre-sized and can be renewed without removing axle extension. Pull old bushings and install new ones using a piloted driver to prevent damage to bushing and axle bore. Tighten spindle retaining nuts to 200-250 ft.-lbs. (272-340 N·m).

48. **PIVOT PINS AND BUSHINGS.** Remove front axle as outlined in paragraph 46 for access to axle pivot pins and bushings in front support (1 – Fig. 56) and bracket (21). Renew front axle center member (23) if pivot pins are excessively worn or damaged. Use a suitable driving tool to remove bushings. New bushings are pre-sized.

49. **TIE ROD AND TOE-IN.** Recommended toe-in is 0-½ inch (0-13 mm). To adjust toe-in, remove bolt securing sleeve (5) to tie rod (6) and loosen clamp on sleeve. Turn sleeve (5) until desired toe-in is obtained. Tighten clamp, install bolt securing sleeve to tie rod and recheck toe-in measurement.

FRONT SUPPORT

All Models

50. **REMOVE AND REINSTALL.** To remove front support casting (1 – Fig. 56), remove radiator as outlined in paragraph 149. Remove front axle assembly as outlined in paragraph 46. Attach a hoist or other device to front support casting and unscrew retaining bolts and nuts.

To install front support casting, attach front support casting to engine using four upper bolts and tighten bolts to 250-270 ft.-lbs. (340-367 N·m). Be sure oil pan cap screws are tight and measure gap between front support and oil pan at attaching bolt holes. Loosen the four installed cap screws, then install the two lower screws using shims equal to the measured clearance. Shims are available in thicknesses of 0.014, 0.017, 0.021, 0.024 and 0.027 inch (0.356, 0.432, 0.533, 0.610 and 0.686 mm). Install front axle and radiator by reversing removal procedure.

POWER STEERING SYSTEM

All Models

These models are equipped with a hydrostatic power steering system consisting of a steering cylinder attached to front axle, a Ross-type steering motor and a tilting steering gear mechanism.

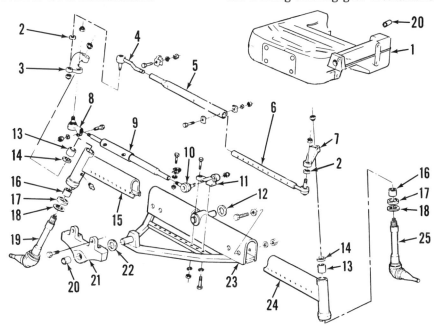

Fig. 56 – Exploded view of front axle assembly used on Models 6700, 6710, 7700 and 7710. Short wheelbase configuration is shown.

1. Front support	8. Rod end	14. Seal	20. Bushing
2. Seal	9. Steering cylinder	15. Axle extension (R.H.)	21. Front support bracket
3. Steering arm	10. Rod end	16. Bushing	22. Thrust washer
4. Tie rod end	11. Steering cylinder anchor	17. Thrust bearing	23. Axle center member
5. Sleeve	12. Thrust washer	18. Thrust washer	24. Axle extension (L.H.)
6. Tie rod	13. Bushing	19. Spindle (R.H.)	25. Spindle (L.H.)
7. Steering arm			

Models 6700 and 6710 may be equipped with a non-tilting steering gear.

51. FLUID AND BLEEDING. Use Ford M-2C41A fluid in this system. Fluid level should be kept at bottom of reservoir filler neck. System capacity is 5.0 U.S. pints (2.4 liters). Reservoir (30 – Fig. 20) should be disassembled for cleaning and new filter (29) and fluid installed after each 600 hours of operation.

System is self-bleeding. Whenever power steering system has been disassembled for any reason, after reassembly, refill fluid reservoir. Start engine and run at low idle rpm. Operate steering from full right to full left through at least five complete cycles. Replenish reservoir fluid as needed to maintain level. System is free of trapped air when no bubbles appear, steering is firm and reservoir level remains steady at full.

52. PRESSURE AND FLOW. System pressure is regulated at 1550-1650 psi (10.70-11.38 MPa) by the pressure relief valve in pump. Rate of flow should be 2.7 gpm (10.2 liters/min.) at 1000 engine rpm on Models 6700 and 7700. Rate of flow should be 3.6 gpm (13.6 liters/min.) at 1000 engine rpm on Models 6710 and 7710.

53. R&R AND OVERHAUL PUMP. Refer to paragraph 15 and Fig. 20 for procedure and exploded view of pump.

54. R&R STEERING MOTOR. Disconnect battery ground cable. Clean joints at junctions between flexible hoses and rigid tubing leading to steering motor. Mark all hoses and tubing to aid when reconnecting. Disconnect hoses from tubing and seal all openings to prevent contamination. Unscrew bolts securing rigid power steering tubing to side of engine. Remove left radiator panel, left engine side panel and left hood-to-cab side panel. Unscrew bolt securing wire loom bracket to steering motor. Detach Dual Power control rod from control pedal and remove pedal by withdrawing pivot pin. Detach main wiring harness connector and cover from front cab panel. Detach throttle control rod from pedal. Remove five screws holding lower steering column shroud and push shroud up for access to steering column coupler. Remove four screws retaining steering motor support bracket. Loosen pinch bolt (4 – Fig. 57) in coupler and remove steering motor and bracket out left side of tractor.

Reverse removal procedure to install steering motor. Refer to paragraph 51 to fill and bleed power steering system.

55. OVERHAUL STEERING MOTOR. Refer to paragraph 44 and Fig. 53 for steering motor overhaul.

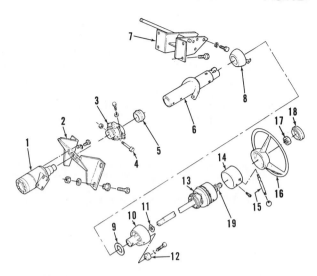

Fig. 57 — Exploded view of tilting steering column used on some 6700 and 6710 models and all 7700 and 7710 models. Refer to Fig. 53 for exploded view of steering motor.

1. Steering motor
2. Column support
3. Coupler
4. Bolt
5. Seal
6. Steering column
7. Column support
8. Column flange
9. Washer
10. Extension
11. Spring washer
12. Column retainer
13. Tilt assembly
14. Cover
15. Steering column lock lever
16. Steering wheel
17. Nut
18. Cap
19. Spring

FRONT-WHEEL DRIVE SYSTEM AND STEERING

(Models 5600-6600-6700-7600-7700)

Front-wheel drive is offered as an option. Major components are the transfer box, front axle, differential, wheel hubs, final drive planetary units and front axle assembly. The front-wheel drive transfer box and transmission handbrake are contained in a common housing that is bolted directly onto the left side of the rear axle center housing. A limited slip differential is available as an option on all models.

Type APL 1351 axle is used on Models

5600, 6600 and 6700 and APL 1551 axle is used on Models 7600 and 7700.

Power steering system is the same as is used on two-wheel drive models.

FRONT AXLE

All Models

55A. R&R FRONT AXLE AND CENTER SUPPORT. Remove eight

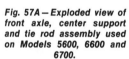

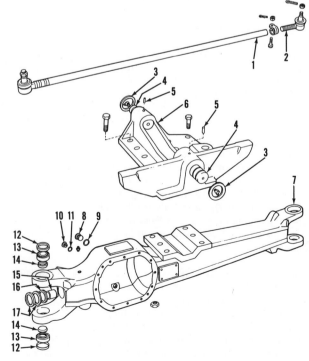

Fig. 57A — Exploded view of front axle, center support and tie rod assembly used on Models 5600, 6600 and 6700.

1. Tie rod
2. Tie rod end
3. Shim
4. Pivot pin
5. Pin
6. Center support
7. Axle housing
8. Plug
9. Gasket
10. Plug
11. Gasket
12. Dust cap
13. Bearing
14. Inner cover
15. Bushing
16. Shaft seal
17. Seal assy.

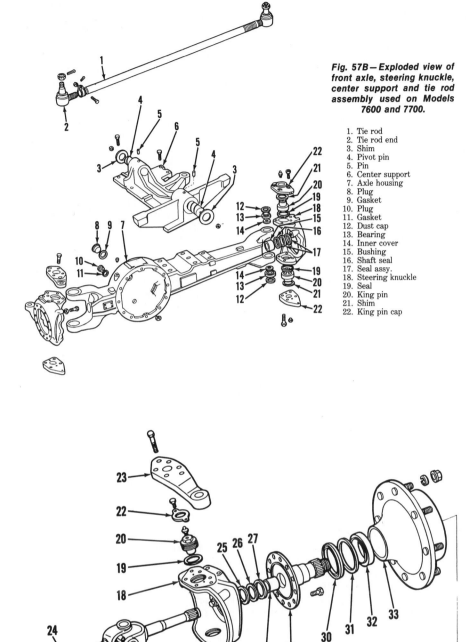

Fig. 57B — Exploded view of front axle, steering knuckle, center support and tie rod assembly used on Models 7600 and 7700.

1. Tie rod
2. Tie rod end
3. Shim
4. Pivot pin
5. Pin
6. Center support
7. Axle housing
8. Plug
9. Gasket
10. Plug
11. Gasket
12. Dust cap
13. Bearing
14. Inner cover
15. Bushing
16. Shaft seal
17. Seal assy.
18. Steering knuckle
19. Seal
20. King pin
21. Shim
22. King pin cap

Fig. 57C — Exploded view of steering knuckle, hub carrier, hub and planetary drive used on Models 5600, 6600 and 6700.

locknuts at differential drive flange, then withdraw drive shaft flange from mounting bolts and lower shaft to the ground. Remove front weights and weight bracket.

On Models 6700 and 7700, remove steering cylinder hose support from center support. Disconnect steering cylinder from left steering arm and center support (6 – Fig. 57B). Temporarily secure steering cylinder to underneath side of tractor. On Models 5600, 6600 and 7600, disconnect drag link from steering arm.

Support front of tractor, then remove fasteners securing front support bracket. Remove front support bracket and roll axle assembly and center support away from tractor as a complete unit.

Inspect components for excessive wear or damage and renew as needed.

To install front axle (7 – Fig. 57A or 57B) and center support (6) assembly, reverse removal procedure. Install shims (3) to reduce center support (6) end play to 0.008-0.016 inch (0.2-0.4 mm). Shims are available in thicknesses of 0.173 to 0.220 inch (4.4-5.6 mm) in increments of 0.012 inch (0.3 mm). Tighten fasteners to the following torques: center support to axle housing cap screws – 302 ft.-lbs. (410 N·m); front support bracket cap screws – 175-225 ft.-lbs. (237-305 N·m); steering cylinder joint end to steering arm or center support nut and drag link end to steering arm nut – 221 ft.-lbs. (300 N·m); drive shaft flange nuts – 36 ft.-lbs. (49 N·m) torque.

55B. **TOE-IN.** Toe-in is adjusted by detaching tie rod end and rotating end. Correct toe-in is 0-1/8 inch (0-3 mm). Check for bent or excessively worn parts if toe-in is not correct.

PLANETARY CARRIER, WHEEL HUB AND HUB CARRIER

All Models

55C. **REMOVE AND REINSTALL.** To remove wheel hub (33 – Fig. 57C or

18. Steering knuckle	39. Lock plate
19. Seal	40. Slotted nut
20. King pin	41. Sun gear
22. Lock plate	42. Snap ring
23. Steering arm	43. Detent plug
24. Axle shaft	44. Spacing washer
25. Retaining ring	45. Snap ring
26. Ring	46. Planetary carrier
27. Seal	47. Roll pin
28. Bushing	48. Roll pin
29. Hub carrier	49. Gasket
30. Seal	50. Plug
31. Washer	51. Snap ring
32. Bearing	52. Pin
33. Hub	53. Plug
34. Bearing	54. Gasket
35. Circlip	55. Thrust washer
36. Ring gear carrier	56. Spacer
37. Ring gear	57. Needle bearings
38. Slotted nut	58. Planetary pinion

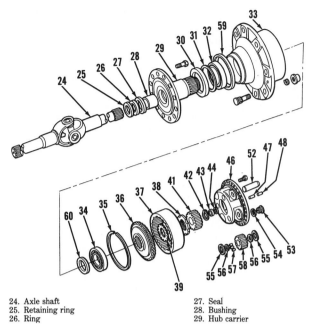

Fig. 57D—Exploded view of hub carrier, hub and planetary drive used on Models 7600 and 7700.

30. Seal
31. Washer
32. Bearing
33. Hub
34. Bearing
35. Circlip
36. Ring gear carrier
37. Ring gear
38. Slotted nut
39. Lock plate
41. Sun gear
42. Snap ring
43. Detent plug
44. Spacing washer
46. Planetary carrier
47. Roll pin
48. Roll pin
52. Pin
53. Plug
54. Gasket
55. Thrust washer
56. Spacer
57. Needle bearings
58. Planetary pinion
59. Dust shield
60. Shim

24. Axle shaft
25. Retaining ring
26. Ring
27. Seal
28. Bushing
29. Hub carrier

57D) and planetary carrier assembly (46), raise tractor and remove wheel on side requiring service. Remove eight locknuts at differential drive flange, then withdraw drive shaft flange from mounting bolts and lower shaft to the ground. Drain oil from hub (33) and scribe alignment marks on planetary carrier (46) and hub (33). Remove carrier (46) to hub (33) securing cap screws. Thread two suitable jackscrews in the appropriate carrier bores to extract carrier (46). Detach snap ring (42) and remove sun gear (41).

On Models 5600, 6600 and 6700, bend back tabs on lock plate (39–Fig. 57C) and remove outer slotted nut (40). Withdraw lock plate (39), then remove inner slotted nut (38). On Models 7600 and 7700, bend back tabs on lock plate (39–Fig. 57D) and remove slotted nut (38).

Remove ring gear (37–Fig. 57C or 57D) and ring gear carrier (36) from hub (33). Pull hub and bearing assembly carefully off hub carrier (29). On Models 7600 and 7700, extract shim (60–Fig. 57D) from hub carrier (29).

Remove fasteners securing hub carrier (29–Fig. 57C or 57D) to steering knuckle and remove carrier.

Reassembly is reverse order of disassembly. Install hub carrier (29) and secure to steering knuckle with hex head cap screws. Tighten cap screws to a torque of 155 ft.-lbs. (210 N·m). On Models 7600 and 7700, install shim (60–Fig. 57D) on hub carrier (29). Install assembled hub (33–Fig. 57C or 57D) on hub carrier (29). Position assembled ring gear carrier (36) and ring gear (37) over splines of hub carrier (29) and push in. On Models 7600 and 7700, install lock plate (39–Fig. 57D) over hub carrier

(29). On all models lightly grease threads of slotted nut (38–Fig. 57C or 57D), then install finger tight onto carrier shaft.

Hub roller bearings (32 and 34) are preloaded by turning slotted nut (38). Refer to Fig. 57E and measure rolling resistance of hub roller bearings. Wrap a strong cord around wheel stud and hub, then attach a pull scale (S–Fig. 57E) to cord. Pull scale and note reading as hub rotates. Nut (38) is properly tightened when rolling resistance is between 8-16 lbs. (3-7 kg) on Models 5600, 6600 and 6700 and 9-14 lbs. (4-6 kg) on Models 7600 and 7700 with new bearings, and 3-8 lbs. (1.4-3.7 kg) on Models 5600, 6600 and 6700 and 4.5-7 lbs. (2-3 kg) on Models 7600 and 7700 with original bearings.

If slotted nut (38–Fig. 57D) locks before proper rolling resistance is obtained on Models 7600 and 7700, it will be necessary to adjust thickness of shim (60). Shim is available in metric sizes from 5.25 to 5.4 mm in increments of 0.05 mm.

After attaining correct rolling resistance, on Models 5600, 6600 and 6700 install lock plate (39–Fig. 57C) and slotted nut (40) on carrier shaft and securely tighten. On all models, bend lock plate (39–Fig. 57C or 57D) to engage nut(s). Install sun gear (41) on axle shaft (24) and secure with snap ring (42).

To check axle shaft (24) end play, push axle shaft inward. While shaft is being turned by an assistant, turn hub (33) to left and right steering lock positions. Turn hub to straight ahead position. With reference to Fig. 57F, measure depth of shaft end (24) from joint face of hub (33). Refer to Fig. 57G and using a similar set-up, determine height of detent plug (43–Fig. 57C or 57D) above joint face of carrier (46–Fig. 57G). Sub-

Fig. 57E—Measure hub bearing preload by measuring rolling resistance using a spring scale (S). Turn nut (38) to adjust preload. Refer to text.

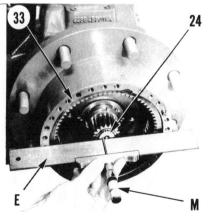

Fig. 57F—View showing procedure for measuring depth of shaft end (24) from joint face of hub (33) using straightedge (E) and depth micrometer (M). Refer to text.

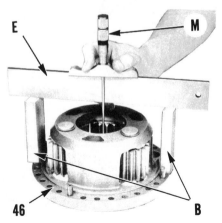

Fig. 57G—Measure height of detent plug as outlined in text.

B. Gage blocks
E. Straightedge
M. Depth micrometer
46. Planetary carrier

tract height of detent plug (43 – Fig. 57C or 57D) from previously measured depth of shaft end to obtain shaft end play. Drive out detent plug (43) and install spacing washer(s) (44) equal to measured shaft end play so end play will be zero. Spacing washer (44) is available in metric thicknesses of 0.3, 0.5 and 1.0 mm. Reinstall detent plug (43).

To install carrier assembly (46), make sure alignment marks on carrier (46) and hub (33) are aligned, then secure carrier assembly to hub (33). Cap screws should be tightened to a torque of 18 ft.-lbs. (25 N·m). Refill hub assembly with 1.6 U.S. pints (0.75 liter) on Models 5600, 6600 and 6700 and 2.1 U.S. pints (1.0 liter) on Models 7600 and 7700 of Ford 134 lubricant or a suitable equivalent. Install drive shaft flange to differential drive flange and tighten to 36 ft.-lbs. (49 N·m) torque.

55D. OVERHAUL. Remove snap ring (35 – Fig. 57C or 57D) to separate ring gear carrier (36) from ring gear (37). Disassemble planetary carrier assembly (46) with reference to Fig. 57C or 57D. Inspect components for damage and excessive wear. Note that tapered roller bearing and cup must be renewed as a complete unit.

AXLE SHAFTS

All Models

55E. REMOVE AND REINSTALL. To remove axle shafts (24 – Fig. 57C or 57D), refer to paragraph 55C and remove planetary carrier, wheel hub and hub carrier as outlined. Axle shaft can now be extracted.

Inspect bushings (15 – Fig. 57A or 57B) and (28 – Fig. 57C or 57D) and seals (16 – Fig. 57A or 57B) and (27 – Fig. 57C or 57D) for excessive wear or any other damage and renew as needed. Inspect axle shaft and universal joint for excessive wear or any other damage and renew as needed.

To install axle shaft, reverse removal procedure. Adjust hub bearing preload and axle shaft end play as outlined in paragraph 55C.

STEERING KNUCKLE AND KING PINS

Models 5600-6600-6700

55F. R&R AND OVERHAUL. Steering knuckle (18 – Fig. 57C) may be removed without the removal of the final planetary drive components. If steering knuckle is removed with final planetary drive components, a suitable hoist or jack must be used to support the

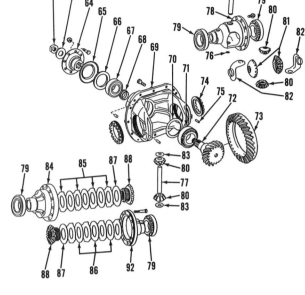

Fig. 57H – Exploded view of differential housing assembly used on Models 5600, 6600 and 6700. Standard differential is shown at upper right while limited slip differential is shown at lower left.

62. Hex nut
63. Washer
64. Drive flange
65. Dust shield
66. Shaft seal
67. Bearing
68. Sleeve
69. Housing
70. Shim
71. Bearing
72. Bevel pinion
73. Bevel ring gear
74. Adjusting nut
75. Pin
76. Snap ring
77. Cross-shaft
78. Differential case
79. Bearing
80. Spider gear
81. Side gear
82. Thrust washer
83. Thrust washer
84. Differential case
85. Drive plate
86. Driven plate
87. Pressure plate
88. Side gear
92. Bearing carrier

weight of the hub so to allow removal of the assembly.

To service steering knuckle (18), remove axle shaft as outlined in paragraph 55E and refer to the following.

Remove upper and lower king pin (20) grease fittings. Remove tie rod end from steering knuckle. If needed, remove steering arm (23). Remove two hex head cap screws securing upper and lower lock plates, then withdraw plates. Unscrew and remove upper and lower king pins (20).

NOTE: King pins (20) may be locked in position by a thread sealant. To ease removal, apply sufficient heat around king pin to break seal.

Lift steering knuckle (18) from axle housing. Withdraw tapered bearing (13 – Fig. 57A), dust cap (12) and seal ring (19 – Fig. 57C) from king pin. Inspect all components for excessive wear or any other damage and renew as needed. If service to bearing (13 – Fig. 57A) outer race is required, then drive out inner cover (14). Use a suitable punch and hammer to drive race from end of axle housing. Use suitable tools to install new outer race and inner cover.

Reassemble steering knuckle to axle housing in reverse order of removal. Prior to installation of king pins (20), coat threads with Ford EM4G-52A thread sealant or a suitable equivalent. Tighten upper and lower king pins equally to ensure correct alignment of axle shaft. Install axle shaft and hub carrier and tighten hub carrier securing cap screws to a torque of 155 ft.-lbs. (210 N·m). Check steering torque by attaching Ford tool T3125 steering

knuckle adapter and a torque wrench to top of steering knuckle just above king pin. Measure torque required to turn steering knuckle. If steering torque is not 8-11 ft.-lbs. (11-15 N·m), adjust upper and lower king pins equally until 8-11 ft.-lbs. (11-15 N·m) steering torque is attained. After adjustment, rotate steering knuckle assembly from full left steering lock to full right steering lock and check for binding. If binding occurs, equally adjust upper and lower king pins (20) until smooth operation is noted and steering torque is to specification. Install and tighten tie rod end retaining nut to 221 ft.-lbs. (300 N·m) torque. Tighten nut securing steering cylinder end to steering arm (23) to 221 ft.-lbs. (300 N·m) torque. Complete reassembly as outlined in paragraph 55E.

Models 7600-7700

55G. R&R AND OVERHAUL. Steering knuckle (18 – Fig. 57B) may be removed without the removal of the final planetary drive components. If steering knuckle is removed with final planetary drive components, a suitable hoist or jack must be used to support the weight of the hub to allow removal of the assembly.

To service steering knuckle (18), remove axle shaft as outlined in paragraph 55E and refer to the following.

Remove upper and lower king pin caps (22) and tie rod (1). If left knuckle assembly is to be serviced, remove steering cylinder end from steering arm. Remove shims (21) and set aside for installation in original location. Remove upper and lower king pins (20), bearing cones (13) and dust cap (12). Mark components for return to original location. Lift steering knuckle (18) from axle housing.

Inspect all components for excessive wear or any other damage and renew as needed. If service to bearing (13) outer race is required, use Ford puller tool 954C or a suitable equivalent to extract race. If damage to inner cover (14) is noted, a suitable punch and hammer must be used to drive inner cover from end of axle housing. Use suitable tools to install new outer race and cover.

To reassemble, install seal ring (19), dust cap (12) and bearing (13) cone on each king pin (20). Mount steering knuckle on axle housing yoke and install upper king pin (20). Coat originally installed shim (21) with grease and position it over king pin (20). Install king pin cap (22) on steering knuckle (18). Tighten socket head screws to 100 ft.-lbs. (135 N·m) and hex head screws to 140 ft.-lbs. (190 N·m). Repeat procedure on lower king pin assembly. Install axle shaft and hub carrier and tighten hub carrier securing cap screws to a torque of 155 ft.-lbs. (210 N·m). Check steering torque by attaching Ford tool T3119 steering knuckle adapter and a torque wrench to upper cap (22) just above king pin. Measure torque required to turn steering knuckle. If steering torque is not 8-11 ft.-lbs. (11-15 N·m), install or remove shims (21) as required to obtain desired torque reading. Shims are available in metric thicknesses from 2.6 to 3.3 mm. Thickness of upper and lower shim packs should be equal; however, it may be necessary to transfer shims between shim packs so drive shaft is centered in axle housing. With steering knuckle in straight ahead position, rotate axle shaft and check for binding. Transfer shims between shim packs (21) to remove any binding.

Install and tighten tie rod end retaining nut to 221 ft.-lbs. (300 N·m) torque. If removed, tighten nut securing steering cylinder end to steering arm to 221 ft.-lbs. (300 N·m) torque. Complete reassembly as outlined in paragraph 55E.

DIFFERENTIAL AND BEVEL GEARS

All Models

55H. REMOVE AND REINSTALL. Support front of tractor. Remove front wheels, disconnect drive shaft from differential input flange and drain axle housing oil into a suitable container. Detach tie rod ends from steering knuckles and, depending on model, steering cylinder end or drag link end from steering arm. Remove cap screws securing hub carriers (29–Fig. 57C or

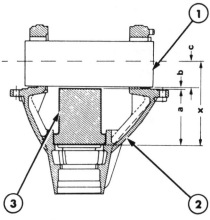

Fig. 57I — Diagram of pinion shimming tools and measuring points. Refer to text.

1. Pinion setting mandrel	2. Differential housing
	3. Dummy pinion

57D) to steering knuckles. Support hubs, then pull hub assemblies and axle shafts out far enough to disengage shaft inner ends from differential side gears. Remove cap screws securing differential housing to axle housing.

NOTE: Remove two differential housing cap screws and replace with guide studs to ease differential housing removal and installation.

Carefully remove differential housing assembly using a suitable floor jack or hoist.

Installation is reverse order of removal. Differential housing to axle housing fasteners should be tightened to a torque of 61 ft.-lbs. (83 N·m). Tighten tie rod ends to steering knuckles and if so equipped, steering cylinder end or drag link end to steering arm to 221 ft.-lbs. (300 N·m) torque. Fill axle housing with 11.0 U.S. pints (5.2 liters) of lubricant on Models 5600, 6600 and 6700 and 13.7 U.S. pints (6.5 liters) of lubricant on Models 7600 and 7700. Recommended lubricant is API GL-5 SAE 90 or a suitable equivalent.

Models 5600-6600-6700

55I. OVERHAUL. Mount differential housing in a vise and loosen ten cap screws that retain bevel ring gear (73–Fig. 57H) to differential case half (84). Drive out roll pins (75) that lock adjusting nuts (74), then unscrew and remove adjusting nuts (74). Remove left and right bearing (79) cups. Using a suitable puller assembly, extract left and right bearing (79) inner races from left and right case halves (84). Extract differential assembly from differential housing and secure in a soft-jawed vise with bevel ring gear retaining screw heads up. Remove ten bevel ring gear retaining screws.

On models without limited slip differential, drive bevel gear from differential case (78) using a suitable brass punch and hammer. Remove snap ring (76) from either end of cross-shaft (77) and remove shaft from differential case (78). Complete disassembly with reference to Fig. 57H.

On models with limited slip differential, lift off bearing carrier (92) and withdraw side gear (88) and clutch pack assembly from bevel ring gear side. Drive or press bevel ring gear (73) from differential case (84). Using a suitable punch and hammer, drive cross-shaft (77) from differential case (84). Complete disassembly with reference to Fig. 57H.

Remove hex nut (62), washer (63) and drive flange (64) from bevel pinion gear (72).

NOTE: Nut (62) may be secured with thread sealant; apply heat to nut (62) to break seal.

Drive or press bevel pinion gear from housing. Remove and discard collapsible sleeve (68). Remove oil seal (66) and bearing (67) from housing. Use suitable tools and extract bearing (71) cone from bevel pinion shaft. Drive bearing (71) cup from housing using a suitable punch and hammer.

Clean and inspect all components and renew as necessary. Bevel ring gear and bevel pinion gear are available only as a matched set. Identical serial numbers will be stamped on outer edge of bevel ring gear (73) and on end of bevel pinion gear (72). If housing, bearings, differential case or bevel ring and pinion gears are renewed, then the following shimming procedure must be followed to determine pinion bearing shim (70) thickness. All measurements should be metric or conversion will be necessary. Refer to Fig. 57I and install dummy pinion tool T3123-1 into bore of differential housing, then install pinion setting mandrel tool T3123 into differential bearing bores.

Apply pressure to mandrel as following measurements shown in Fig. 57I are taken: Height of dummy pinion (a) plus gap between dummy pinion and mandrel (b) plus half the diameter of differential bores (c) equals dimension (x).

Note that dummy pinion height (a) is marked on side of tool. Dimension (c) is 1.67 inches (42.5 mm).

Measure overall width of assembled pinion bearing (71–Fig. 57H) cone and cup, then add bearing width to etched pinion setting number (in millimeters) marked next to serial number on end of bevel pinion (72). Subtract result from dimension (x-Fig. 57I) to obtain required thickness of pinion bearing shim (70–Fig. 57H). Shims are available in thicknesses of 0.1, 0.2, 0.3 and 0.6 mm.

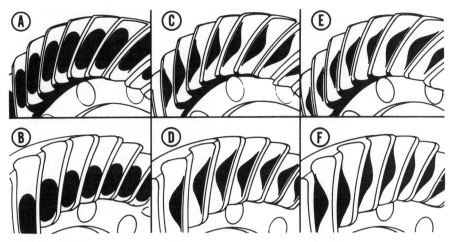

Fig. 57J—Views of bevel ring gear tooth contact patterns. Refer to text to determine thickness of shim (70—Fig. 57H or Fig. 57K) if appropriate tools are available.

A. Proper tooth
 contact-coast side
 pattern
B. Proper tooth
 contact-drive side
 pattern

C. Pinion gear requires
 thicker shim-coast
 side pattern
D. Pinion gear requires
 thicker shim-drive
 side pattern

E. Pinion gear requires
 thinner shim-coast
 side pattern
F. Pinion gear requires
 thinner shim-drive
 side pattern

Install shim (70) and bearing (71) cup in housing and press bearing (71) cone in pinion gear shaft until seated against shoulder.

To determine pinion bearing preload, install bearing (67) cup into small bore at rear of housing (69) and install bevel pinion gear (72). Install a new collapsible sleeve (68) on bevel pinion shaft, then install bearing (67) cone and oil seal (66). Install drive flange (64) and washer (63). Coat threads of hex nut (62) with Ford thread sealant EM4G-52A or a suitable equivalent, then install nut (62) finger tight on pinion shaft (72). While holding drive flange (64) with a suitable tool, tighten hex nut (62) in small increments until torque required to turn assembly is between 9-18 in.-lbs. If desired torque reading is exceeded, then bearing preload is excessive and a new sleeve (68) must be installed and bearing preload readjusted.

On models with limited slip, install pressure plate (87) on side gear (88) opposite bevel ring gear so polished side is towards gear. Assemble clutch disc plates (85 and 86) on side gear (88) in an alternating sequence as shown in Fig. 57H. Place assembled side gear and clutch pack assembly on a suitable piece of tube with clutch plates at top and tabs aligned. Carefully lower differential case (84) onto side gear and clutch pack assembly so clutch plate tabs engage slots in case. Invert differential case and install cross-shaft (77), spider gears (80) and thrust washers (83) in reverse order of disassembly. Position bevel ring gear (73) on differential case (84) using two retaining cap screws to ensure alignment of screw holes. Press bevel ring gear (73) onto differential case (84) using suitable tools. Install remaining side gear and clutch pack assembly. Locate bearing carrier (92) on differential case (84) and install the ten retaining cap screws. Tighten cap screws to a torque of 155 ft.-lbs. (210 N·m). Check clutch pack free play as follows: Mount a dial indicator gage on differential case so gage plunger contacts outer clutch plate. Use two screwdrivers to move clutch plate assembly up and down and note movement of gage needle. Allowable free play is 0.004-0.008 inch (0.1-0.2 mm). Adjust free play by changing thickness of pressure plate (87). Pressure plates (87) are available in thicknesses of 2.8, 2.9 and 3.0 mm. If after installation of thickest pressure plate (87) free play remains excessive, a new set of clutch plates (85 and 86) must be installed. Repeat instructions for opposite differential case half. Disassemble differential case as previously outlined to renew clutch plates or adjust pressure plate.

On models without limited slip differential, reassembly is reverse order of disassembly. Tighten bevel ring gear retaining screws to 155 ft.-lbs. (210 N·m) torque.

Position assembled differential case in housing. Press left and right bearing cones (79) on differential case end using suitable tools. Install left and right bearing (79) cups and adjusting nuts (74) in differential housing.

Backlash between bevel ring gear and bevel pinion gear should be 0.005-0.007 inch (0.13-0.18 mm). Backlash is adjusted by rotating adjusting nuts (74). Backlash is reduced by moving bevel ring gear towards bevel pinion gear.

To check preload of left and right differential bearings (79), mount a dial indicator so tip of gage contacts back (flat) side of bevel ring gear (73). While prying differential side to side, turn adjusting nut (74) opposite bevel ring gear just until no movement of differential assembly is noted on dial gage. Preload bearings by tightening adjusting nut (74) opposite bevel ring gear 1½-2½ slots, as necessary to align locking roll pin (75) with a slot in nut (74). Recheck bevel ring gear to bevel pinion gear backlash and repeat previous procedure if needed. With dial indicator gage tip contacting back of bevel ring gear, check for bevel ring gear runout. If runout exceeds 0.003 inch (0.08 mm), then gear is seated improperly on differential case and should be removed and inspected.

Refer to Fig. 57J for proper bevel ring and pinion gear tooth contact pattern. If pinion shimming procedure wasn't performed properly, ideal tooth pattern will not be obtained and shimming procedure will have to be repeated.

To install differential assembly, refer to paragraph 55H.

Models 7600-7700

55J. **OVERHAUL.** Mount differential housing in a vise and loosen twelve cap screws that retain bevel ring gear (73—Fig. 57K) to differential case half (84). Drive out roll pins (75) that lock adjusting nuts (74), then unscrew and remove adjusting nuts (74). Remove left and right bearing (79) cups. Using a suitable puller assembly, extract left and right bearing (79) inner races from left and right case halves (84). Note that differential components must be disassembled in differential housing and cannot be removed as a complete unit. Remove twelve bevel ring gear retaining screws, then separate left and right case halves and remove differential components with reference to Fig. 57K.

Remove lock plate (89), hex nut (62), washer (63) and drive flange (64) from bevel pinion gear (72). Drive or press bevel pinion gear from housing. Remove and discard collapsible sleeve (68). Remove oil seal (66) and bearing (67) from housing. Use suitable tools and extract bearing (71) cone from bevel pinion shaft. Drive bearing (71) cup from housing using a suitable punch and hammer.

Clean and inspect all components and renew as necessary. Bevel ring gear and bevel pinion gear are available only as a matched set. Identical serial numbers will be stamped on outer edge of bevel ring gear (73) and on end of bevel pinion gear (72). If housing, bearings, differential case or bevel ring and pinion gears are renewed, then the following shimm-

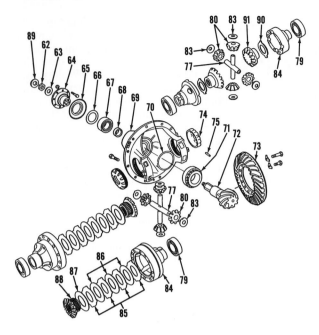

Fig. 57K — Exploded view of front-wheel differential housing assembly used on Models 7600 and 7700. Standard differential is shown at upper right while limited slip differential is shown at lower left.

62. Hex nut
63. Washer
64. Drive flange
65. Dust shield
66. Shaft seal
67. Bearing
68. Sleeve
69. Housing
70. Shim
71. Bearing
72. Bevel pinion
73. Bevel ring gear
74. Adjusting nut
75. Pin
77. Cross-shaft
79. Bearing
80. Spider gear
83. Thrust washer
84. Case half
85. Drive plate
86. Driven plate
87. Pressure plate
88. Side gear
89. Lock plate
90. Thrust washer
91. Side gear

ing procedure must be followed to determine pinion bearing shim (70) thickness. All measurements should be metric or conversion will be necessary. Refer to Fig. 57I and install dummy pinion tool T3123-2 into bore of differential housing, then install pinion setting mandrel, tool T3123 into differential bearing bores.

Apply pressure to mandrel as following measurements shown in Fig. 57I are taken: Height of dummy pinion (a) plus gap between dummy pinion and mandrel (b) plus half the diameter of differential bores (c) equals dimension (x).

Note that dummy pinion height (a) is marked on side of tool. Dimension (c) is 1.77 inches (45 mm).

Measure overall width of assembled pinion bearing (71 – Fig. 57K) cone and cup, then add bearing width to etched pinion setting number (in millimeters) marked next to serial number on end of bevel pinion (72). Subtract result from dimension (x – Fig. 57I) to obtain required thickness of pinion bearing shim (70 – Fig. 57K). Shims are available in thicknesses of 0.1, 0.2, 0.3 and 0.5 mm. Install shim (70) and bearing (71) cup in housing and press bearing (71) cone on pinion gear shaft until seated against shoulder.

To determine pinion bearing preload, install bearing (67) cup into small bore at rear of housing (69) and install bevel pinion gear (72). Install a new collapsible sleeve (68) on bevel pinion shaft, then install bearing (67) cone and oil seal (66). Install drive flange (64), washer (63), then install hex nut (62) finger tight. While holding drive flange (64) with a

suitable tool, tighten hex nut (62) in small increments until torque required to turn assembly is between 9-18 in.-lbs. Secure with a new lock plate (89). If desired torque reading is exceeded, then bearing preload is excessive and a new sleeve (68) must be installed and bearing preload readjusted.

On models with limited slip, install bevel ring gear (73) onto differential case half (84). Install pressure plates (87) on side gears (88) so polished side is towards gear and assemble clutch disc plates (85 and 86) on side gears (88) in an alternating sequence as shown in Fig. 57K. Complete reassembly of differential components, then mate differential halves together. Install bevel ring gear retaining screws and tighten to a torque of 155 ft.-lbs. (210 N·m). Check clutch pack free play as follows: Mount a dial indicator gage on differential case so gage plunger contacts outer clutch plate. Use two screwdrivers to move clutch plate assembly up and down and note movement of gage needle. Allowable free play is 0.004-0.008 inch (0.1-0.2 mm). Adjust free play by changing thickness of pressure plate (87). Pressure plates (87) are available in thicknesses of 2.8, 2.9 and 3.0 mm sizes. If after installation of thickest pressure plate (87) free play remains excessive, a new set of clutch plates (85 and 86) must be installed. Repeat instructions for opposite differential case half. Separate case halves for installation in differential housing.

Reassembly of components in differential housing is reverse order of removal. Press left and right bearing

cones (79) on their respective case halves (84) using suitable tools. Install left and right bearing (79) cups and adjusting nuts (74) in differential housing. Tighten bevel ring gear retaining screws to 155 ft.-lbs. (210 N·m) torque.

Backlash between bevel ring gear and bevel pinion gear should be 0.006-0.008 inch (0.15-0.20 mm). Backlash is adjusted by rotating adjustment nuts (74). Backlash is reduced by moving bevel ring gear towards bevel pinion gear.

To check left and right differential bearings (79) preload, mount a dial indicator so tip of gage contacts back (flat) side of bevel ring gear (73). While prying differential side to side, turn adjusting nut (74) opposite bevel ring gear just until no movement of differential assembly is noted on dial gage. Preload bearings by tightening adjusting nut (74) opposite bevel ring gear 1½-2½ slots, as necessary to align locking roll pin (75) with a slot in nut (74). Recheck bevel ring gear to bevel pinion gear backlash and repeat previous procedure as needed. With dial indicator gage tip still contacting back of bevel ring gear, check for bevel ring gear runout. If runout exceeds 0.003 inch (0.08 mm), then gear is seated improperly on differential case and should be removed and inspected.

Refer to Fig. 57J for proper bevel ring and pinion gear tooth contact pattern. If pinion shimming procedure wasn't performed properly, ideal tooth pattern will not be obtained and shimming procedure will have to be repeated.

To install differential assembly, refer to paragraph 55H.

POWER STEERING SYSTEM

All Models

55K. The power steering system is the same as used on two-wheel drive models. Refer to paragraph 10 or 51 for service.

DRIVE SHAFT

All Models

55L. **R&R AND OVERHAUL.** Remove eight locknuts at transfer box output flange, then withdraw drive shaft flange from mounting bolts and lower rear of shaft to the ground. Remove eight bolts and nuts at differential input flange and lower shaft to the ground.

Inspect all components for damage or excessive wear and renew as necessary. If either universal joint is renewed, it is recommended to balance drive shaft.

To install drive shaft, reverse removal procedure. Tighten all securing nuts to 36 ft.-lbs. (49 N·m) torque.

TRANSFER BOX

All Models

55M. OPERATION. The transfer box assembly is actuated by a control rod that passes through the cab floor and connects to a control valve designed to divert oil from the tractor hydraulic system to clutch piston (31 – Fig. 57L). When the rod is pushed down, the control valve allows hydraulic oil to pressurize clutch piston (31) and disengage the clutch. When the rod is pulled up, the oil will return to the sump through an oil passage leading from the control valve to the front of the transfer box case. In this position, clutch plates (22 and 23) are engaged as pressure of Belleville washers (37) against pressure piece (24) forces clutch plates together. With clutch plates engaged, plate carrier (20) and output shaft (39) rotate and transfer power to front-wheel drive assembly.

55N. R&R TRANSMISSION HANDBRAKE AND TRANSFER BOX. Jack up left rear wheel and support axle, then remove wheel. Remove transfer box drain plug (19 – Fig. 57L) and allow case lubricant to drain into a suitable container. Remove eight locknuts at transfer box output flange, then withdraw drive shaft flange from mounting bolts and lower rear of shaft to the ground. Remove cotter pin, washer and clevis pin from operating rod. Disconnect main oil feed pipe at the side of control valve. Disconnect lower end of handbrake cable from control rod. Remove the two hex head cap screws securing transfer box assembly to support bracket. While supporting handbrake/transfer box assembly, remove the four hex head cap screws securing assembly to rear axle center housing. Remove complete unit.

Installation is reverse order of removal. Be sure to maintain correct alignment of spur gear to rear axle pinion gear. Tighten transfer box-to-center housing cap screws to 70 ft.-lbs. (95 N·m) torque and transfer box-to-support bracket cap screws to 63 ft.-lbs. (86 N·m) torque. Tighten screws securing drive shaft flange to transfer box output flange to 36 ft.-lbs. (49 N·m) torque. Refill rear axle center housing to "FULL" mark on dipstick with M2C134-A lubricant or a suitable equivalent.

NOTE: Check for proper alignment of drive shaft splines if drive shaft has been separated.

55O. OVERHAUL. Remove two cap screws securing control valve assembly to transfer box case, then extract control valve assembly. Remove four cap screws retaining rear cover (52 – Fig.

Fig. 57L – Exploded view of transfer box used on Models 5600, 6600, 6700, 7600 and 7700.

1. Case
2. Thrust washer
3. Needle bearing
4. Spacer
5. Gear
6. Pin
7. Shaft
8. Pin
9. Seal
10. Tapered roller bearing
11. Drum & shaft assy.
12. Needle bearing
13. Needle bearing
14. Thrust washers
15. Needle bearing
16. Retainer ring
17. Split rings
18. Gasket
19. Plug
20. Plate carrier
21. Pressure plate
22. Drive plate
23. Driven plate
24. Pressure piece
25. Seal ring
26. Thrust ring
27. Seal ring
28. Dowel pin
29. Detent spring
30. Seal ring
31. Piston
32. Shim
33. Bearing
34. Thrust ring
35. Snap ring
36. Snap ring
37. Belleville washers
38. Shim
39. Output shaft
40. Oil deflector
41. Seal ring
42. Tapered roller bearing
43. Shim
44. Oil seal
45. Cover

46. Output flange
47. Disc
48. Pin

49. Lock plate
50. Cap screw
51. Drive shaft assy.

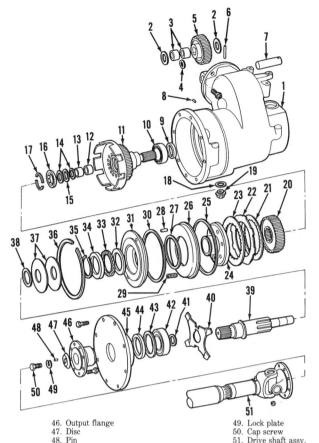

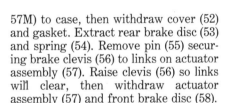

57M) to case, then withdraw cover (52) and gasket. Extract rear brake disc (53) and spring (54). Remove pin (55) securing brake clevis (56) to links on actuator assembly (57). Raise clevis (56) so links will clear, then withdraw actuator assembly (57) and front brake disc (58).

Using special tool T3122 or a suitable equivalent to prevent rotation of output flange (46 – Fig. 57L). Remove hex head screw (50), lock plate (49), disc (47) and pin (48) from end of output shaft, then remove flange (46). Mark case (1) and cover (45) for proper alignment during reassembly. Remove cap screws retaining cover (45), then using a soft-faced mallet, remove cover (45). Remove clutch and output shaft assembly snap ring (36). Drive roll pin (8) into transfer box case from outside. Drive or press clutch and output shaft assembly and drum and shaft assembly (11) out of case from the rear. Separate drum and shaft assembly (11) from output shaft assembly.

Use special puller tool 943 and slide hammer tool 943S or suitable or equivalents to extract bearing race from cover (45). Note shim(s) (43).

Using suitable tools, remove bearing inner race and oil deflector (40) from output shaft. Remove retainer ring (16)

---

from rear of output shaft. Place output shaft assembly under a press and use clutch compressor tool N.775 and slave ring tool CT9056 or suitable equivalents to compress Belleville washers (37). Support output shaft and remove split rings (17). Note that output shaft is free to fall when split rings are removed. Remove output shaft components. Remove snap ring (35) and disassemble piston (31), springs (29), dowel pins (28), seals (25, 27 and 30) and thrust ring (26). If necessary, remove oil seal (9) at rear face of transfer box case (1).

If necessary, remove needle bearings (12 and 13) from drum and shaft assembly (11). If necessary for removal, split outer bearing (13) with a suitable cold-chisel while using caution not to damage inner surface. Use bearing remover tool T3127 to remove inner bearing. Remove all other components using suitable tools.

Inspect all components for excessive wear or any other damage and renew as needed. Control valve must be renewed as a complete unit if found defective.

Reassembly is reverse order of disassembly. If oil seal (9) is renewed, coat periphery of new seal with Ford Loctite or a suitable equivalent and drive seal in case bore until top surface of seal is 7/16-15/32 inch (11-12 mm) below surface of case. Be sure lip of seal faces toward the front.

Install new seal ring (30) on piston (31) and new seal rings (25 and 27) on thrust ring (26). Coat seal rings with petroleum jelly prior to installation. Install two dowel pins (28) and four springs (29) in piston (31) and mate thrust ring (26) with piston. Install piston and thrust ring assembly on pressure piece (24) and assemble shim (32), thrust bearing (33) and small thrust ring (34) on pressure piece. Compress unit and retain with snap ring (35).

Install clutch plate carrier (20) in clutch drum (11). Coat drive plates (22) with M2C134-A oil or a suitable equivalent, then insert components in the following order: pressure plate (21) with flat side up, drive plate (22) and driven plate (23), then continue alter-

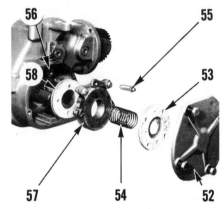

Fig. 57M – View showing handbrake components contained in transfer box housing used on Models 5600, 6600, 6700, 7600 and 7700 equipped with front-wheel drive.

52. Cover
53. Rear brake disc
54. Spring
55. Pin
56. Clevis
57. Actuator assy.
58. Front brake disc

nating installation of plates (22 and 23) until remaining plates are installed. If old Belleville washers are to be installed, disregard following paragraph. However, if new Belleville washers are necessary use procedure in following paragraph to determine shim thickness.

To determine shim (38) thickness, place clutch assembly on output shaft with clutch drum up. Remove drum without moving plates. Coat split rings (17) with petroleum jelly and install rings to secure clutch components. Carefully turn assembly over so shaft collar (C – Fig. 57N) is facing up and measure gap between pressure piece (24) and underside of output shaft collar (C) using a gage block (G) and feeler gage (F). Record this measurement. Note dimension of Belleville washer pack as indicated on packing slip, then subtract dimension of washer pack from previously measured gap between shaft collar and pressure piece. Result is thickness of shim (38 – Fig. 57L). Shim is available in metric thicknesses from 2.0 mm through 5.0 mm in 0.5 mm increments. Separate output shaft from clutch and piston assemblies. Install oil deflector (40) with lip engaging flat on collar, then press roller bearing (42) cone on front of shaft until oil deflector (40) and bearing (42) are seated against shaft collar. Install original or new shim (38), then install Belleville washers (37) with concave sides facing each other. Install clutch and piston assemblies. Remove drum and shaft assembly (11) without disturbing components. Using suitable tools, compress clutch and install split rings (17) on output shaft. Cover retainer ring (16) with petroleum jelly and install over split rings (17). Reinstall drum and shaft assembly (11).

Coat transfer box housing bores

(where piston and thrust ring are located) with petroleum jelly and lower case (1) over clutch and output shaft components. Align one recess in outer edge of thrust ring (26) with clutch piston oil pressure feed hole, then drive roll pin (8) into feed hole 0.006 inch (15 mm) from outer edge of hole. Roll pin must engage thrust ring recess to prevent turning. Secure clutch piston with snap ring (36).

Coat periphery of new oil seal (44) with Ford Loctite or a suitable equivalent and press into transfer box cover (45) with sealing lip towards tapered roller bearing.

To establish recommended output shaft end play of 0-0.004 inch (0-0.10 mm), it is necessary to follow the shimming procedure outlined in the following paragraphs.

Measure height of bearing recess above flange surface. Assemble gage blocks (B – Fig. 57O), straightedge (S) and depth gage (D) as shown in Fig. 57O. Bearing recess height is obtained by taking height of gage block (B) plus height of straightedge (S) minus depth gage (D) measured distance from top of straightedge to base of bearing recess. Record difference.

Measure the projection of case surface (1 – Fig. 57P) above bearing assembly (R). Position straightedge (S) and depth gage (D) as shown in Fig. 57P. Measure distance from top of straightedge to bearing race surface and subtract height of straightedge to obtain the distance.

Subtract recorded height of bearing recess from the distance case surface projects above bearing assembly to obtain thickness of shim (43 – Fig. 57L) to give a 0.004 inch (0-0.10 mm) end play. Shims are available in the following sizes: 0.020 inch (0.50 mm) and 0.098 inch

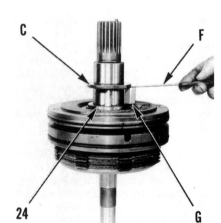

Fig. 57N – View showing transfer box clutch shimming procedure.

C. Output shaft collar
F. Feeler gage
G. Gage block
24. Pressure piece

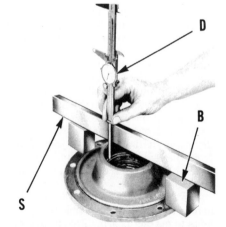

Fig. 57O – View showing shimming procedure for checking height of bearing recess. Refer to text.

B. Gage block
D. Depth gage
S. Straightedge

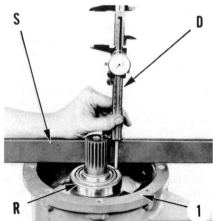

Fig. 57P – View showing shimming procedure for checking projection of case surface above bearing assembly. Refer to text.

D. Depth gage
R. Bearing assy.
S. Straightedge
1. Case surface

(2.50 mm) through 0.114 inch (2.90 mm) in increments of 0.004 inch (0.10 mm).

Install selected shim (43) and bearing outer race in front cover (45). Install new seal ring (41) on output shaft. Coat mounting surface of front cover (45) with a suitable sealer, then install front cover on case using alignment marks made during disassembly. Tighten securing cap screws to 32 ft.-lbs. (44 N·m) torque.

Install flange (46), disc (47), pin (48), lock plate (49) and securing cap screw (50). Tighten cap screw (50) to 88 ft.-lbs. (120 N·m) torque. Bend lock plate (49) over against cap screw head to prevent loosening.

Reassemble handbrake components in reverse order of removal. Install valve assembly with new "O" rings and drain plug (19) with a new gasket (18).

FRONT-WHEEL DRIVE SYSTEM AND STEERING
(Models 5610-6610-6710-7610-7710)

56. Front-wheel drive is offered as an option on 5610, 6610, 6710, 7610 and 7710 models. Major components are the transfer box, drive shaft, front axle, differential, axle shafts and axle hub assemblies. The transfer box bolts to underneath side of rear axle center housing. Transmission/rear axle oil lubricates transfer box assembly.

The differential assembly is center mounted in axle housing. A limited slip differential is available as a factory option.

At each axle end is a hub that contains a planetary reduction gear set. Each hub is mounted to a steering knuckle. The steering knuckle is mounted to axle housing by use of top and bottom king pins. On Models 5610, 6610 and 7610 an integral power steering system is used. On Models 6710 and 7710 an axle-mounted, power steering cylinder with hydrostatic steering is used.

FRONT AXLE

Models 5610-6610-7610

57. REMOVE AND REINSTALL. Position tractor on a suitable level surface and apply parking brake. Remove cap screws securing drive shaft flange to differential pinion flange. Withdraw drive shaft and swing shaft to the side. Remove front weights. Remove nuts securing left and right tie rod assemblies to steering spindle arms, then detach rod ends. Support front of tractor. Remove cap screws securing pivot pin brackets and withdraw pivot pins and brackets. Raise front of tractor enough to allow axle assembly to be rolled away from tractor.

Inspect components for excessive wear or damage and renew as needed. Refer to specific service sections for detailed information. To install front axle, reverse removal procedure. Tighten fasteners to following torques: pivot pin bracket securing cap screws—40 ft.-lbs. (54 N·m); tie rod end retaining nut—90 ft.-lbs. (122 N·m); drive shaft flange cap screws—55 ft.-lbs. (75 N·m). Check differential and axle hub fluid levels and refill with Ford EOAZ-19580 lubricant or a suitable equivalent.

Models 6710-7710

58. REMOVE AND REINSTALL. Position tractor on a suitable level surface and apply parking brake. Remove cap screws securing drive shaft flange to differential pinion flange. Withdraw drive shaft and swing shaft to the side. Remove front weights. Remove power steering hoses from cylinder assembly and plug openings. Temporarily tie hoses clear of axle assembly. Remove cap screws securing front support bracket, then withdraw bracket. Slide rear pivot pin from axle support housing and roll axle assembly away from tractor.

Inspect components for excessive wear or damage and renew as needed. Refer to specific service sections for detailed information. To install front axle, reverse removal procedure. Tighten fasteners to following torques: front support bracket retaining cap screws—200 ft.-lbs. (272 N·m); drive shaft flange cap screws—55 ft.-lbs. (75 N·m). Check differential and axle hub fluid levels and refill with Ford EOAZ-19580 lubricant or a suitable equivalent. Refill power steering system with Ford M-2C41A fluid or a suitable equivalent, then cycle system until completely bled of air. Check and refill reservoir to specific level.

All Models

59. TOE-IN. Toe-in is adjusted by detaching tie rod end and rotating end. Correct toe-in is 0-0.40 inch (10.16 mm). Check for bent or excessively worn parts if toe-in is not to specification.

CARRIER HOUSING, PLANETARY GEARS, RING GEAR AND WHEEL HUB

All Models

60. R&R AND OVERHAUL. Raise front of tractor and remove wheel on side requiring service. Turn carrier housing (3—Fig. 58) so drain plug (1) is at bottom. Remove drain plug (1) and seal (2) and allow oil to drain into a suitable container. Remove screws (4) securing carrier housing (3) and insert pry levers in slots provided on side of wheel hub (21). Pry housing from hub. Remove snap ring (11), thrust washer (12) and cap screws (14) retaining ring gear (5). Using a suitable puller, withdraw ring gear (15) from roll pins (16).

NOTE: A step plate must be used for puller center screw to force against. Damage to axle drive components could result if step plate is not used.

Pull hub assembly (21) with outer bearing (18) from spindle (24) using a suitable puller and step plate.

Remove snap ring (10) retaining planetary gear (9) and bearing (8) on mounting shaft, then withdraw assembly while noting correct mounting position. Bearing is retained in gear by circlip (7); remove circlip and slide bearing from gear. Bearing (18) located on spindle (24) should not be removed unless damage is noted. Bearing must be heated to aid in removal.

Clean all parts with a suitable solvent and blow dry with clean compressed air. Inspect all bearings for corrosion, cracks, binding, excessive wear or any other damage. Inspect gears for cracks, chipped teeth, excessive wear or any other damage. Examine wheel hub and carrier housing for cracks or any other damage. Renew parts as needed. If a roller bearing cone is renewed, mating cup must also be renewed. Renew all seals for reassembly.

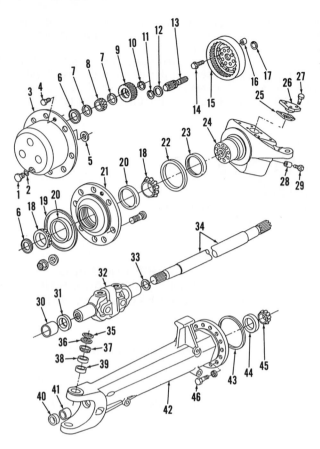

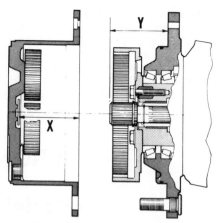

Fig. 58—Exploded view of front axle, wheel hub, spindle and planetary carrier assembly used on Models 5610, 6610, 6710, 7610 and 7710.

1. Drain plug
2. Seal
3. Carrier housing
4. Cap screw
5. Shim
6. Spacer ring
7. Circlip
8. Bearing
9. Planetary gear
10. Snap ring
11. Snap ring
12. Thrust washer
13. Sun gear
14. Cap screw
15. Ring gear
16. Roll pin
17. Circlip
18. Bearing cone
19. Seal
20. Bearing cup
21. Hub
22. Seal
23. Shield
24. Spindle & knuckle assy.
25. Shim (upper kingpin only)
26. Kingpin
27. Cap screw
28. Bushing
29. Stop bolt
30. Bushing
31. Seal
32. Joint assy.
33. Circlip
34. Axle shaft
35. "O" ring
36. Retainer
37. Bearing cone
38. Bearing cup
39. Plug
40. Seal
41. Bushing
42. Axle housing
43. Seal
44. Bearing cup
45. Bearing cone
46. Cap screw

Fig. 60—Measure dimensions X and Y to establish axle shaft end play. Refer to text.

Reassemble in reverse order of removal. Tighten cap screws retaining ring gear to 65 ft.-lbs. (88 N·m). Using a spring scale (S—Fig. 59) and a length of cord (C) wrapped around wheel hub studs (B), check wheel hub bearing rolling resistance. Measured reading should be 6.6-10.6 pounds (3.0-4.75 kg). If reading is incorrect, renew hub bearings and cups and recheck rolling resistance. Thrust washer (5—Fig. 58) is located in center of carrier housing (3) and is used to control axle shaft end play. With a thrust washer located in housing, measure distance (X—Fig. 60) using a straightedge and depth gage. Push in sun gear shaft completely and measure distance from end of shaft to carrier housing mounting surface (Y). Subtract distance (Y) from distance (X) to establish shaft end play. Correct axle shaft end play is 0.008-0.020 inch (0.2-0.5 mm). Shims are available in thicknesses of 0.060, 0.079 and 0.098 inch (1.5, 2.0 and 2.5 mm). Adjust shim thickness until correct end play is attained. Install carrier housing and tighten securing screws. Fill hub with 1.3 U.S. pints of Ford EOAZ-19580 oil or a suitable equivalent. Install plug with seal and tighten.

KING PINS AND BEARINGS

All Models

61. **R&R AND OVERHAUL.** Position tractor on a suitable level surface and apply parking brake. Raise front of tractor and remove wheel on side requiring service. Remove tie rod end securing

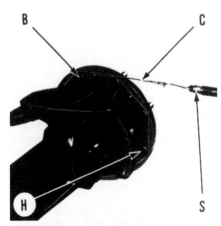

Fig. 59—Wheel hub (H) bearing rolling resistance is checked by wrapping cord (C) around hub studs (B) and measuring rolling resistance with spring scale (S).

nut, then detach end from knuckle arm. Securely support weight of hub assembly using a sling or other suitable holding fixture. Remove cap screws (27—Fig. 58) retaining upper and lower king pins (26). Withdraw upper and lower king pins (26), shim (25) (upper king pin only) and "O" rings (35) from knuckle bores. Separate hub/knuckle assembly with axle shaft from axle housing. Remove upper and lower retainers (36). Lift upper and lower cone and roller bearing (37) from cup (38) in axle housing.

Inspect king pins for excessive wear or any other damage. Inspect bearings for corrosion, roughness, cracks, excessive wear or any other damage. If roller bearing cones are renewed, cups must be renewed as follows:

Using a suitable punch and hammer, knock grease retaining plugs (39) inward from cup bores. Bearing cups (38) may now be driven outward from bores. Renew bearing cups and install grease retaining plugs. Fill bearing cups with grease (specification ITMIC 137A or a good quality, multi-purpose lithium base grease). Pack bearings with grease and position in cups, then install new retainers (36).

Reassemble in reverse order of removal. Ensure axle shaft splines engage side gear splines. Align steering knuckle bores with bores in axle housing. If original roller bearing cone and king pins are being used, original shim may be reused. Complete reassembly by installing new "O" rings on king pins and position shim on upper king pin, then using a suitable soft-faced mallet, drive king pins into position. Tighten securing cap screws to 90 ft.-lbs. (122 N·m) torque. If either roller bearing cone or king pin is renewed, then the following procedure must be used to determine shim thickness.

Remove axle shaft from hub assembly as outlined in paragraph 62. Align steer-

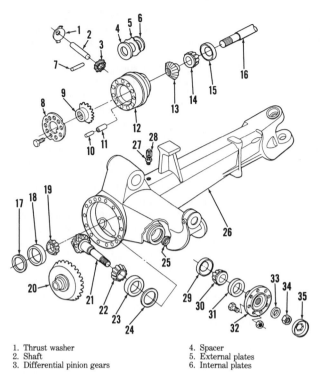

Fig. 61 – Exploded view of front-wheel drive differential assembly used on Models 5610, 6610, 6710, 7610 and 7710. Housing assembly for Models 5610, 6610 and 7610 is shown. Components 4, 5 and 6 are used on models with limited slip differential.

7. Pin	
8. Locking plate	
9. Side gear	
10. Pin	
11. Pin	
12. Differential case	
13. Side gear	
14. Bearing cone	
15. Bearing cup	
16. Axle shaft	
17. Shim	
18. Bearing cup	
19. Bearing cone	
20. Ring gear	
21. Pinion gear	
22. Bearing cone	
23. Bearing cup	
24. Shim	
25. Sleeve	
26. Axle housing	
27. Plug	
28. Breather	
29. Bearing cup	
30. Bearing cone	
31. Seal	
32. Flange	
33. Washer	
34. Nut	
35. Retainer	

1. Thrust washer
2. Shaft
3. Differential pinion gears
4. Spacer
5. External plates
6. Internal plates

ing knuckle bores with bores in axle housing. Install lower king pin and tighten retaining cap screws to 90 ft.-lbs. (122 N·m) torque. Install upper king pin with a test shim of 0.055 inch (1.4 mm) thickness. Tighten retaining cap screws to 90 ft.-lbs. (122 N·m) torque. Install special tool (Churchill FT.3149 or Nuday 4802) over heads of cap screws securing king pin. Using a suitable torque wrench, check amount of torque required to turn king pin bearings. Correct reading is 6-8 ft.-lbs. (8-11 N·m). Adjust shim thickness until correct reading is attained. Shims are available in thicknesses of 0.024-0.079 inch (0.6-2.0 mm) in 0.008 inch (0.2 mm) increments. Noting correct shim thickness, remove king pins. Reinstall axle shaft and reassemble king pins and hub assembly following recommended procedure.

SUN GEAR SHAFT, UNIVERSAL JOINT AND AXLE SHAFT

All Models

62. **R&R AND OVERHAUL.** Remove carrier housing (3 – Fig. 58) and snap ring (11) and thrust washer (12) as outlined in paragraph 60. Remove king pins and tie rod end as outlined in paragraph 61. Withdraw hub assembly, then separate axle shaft from hub or axle housing.

If universal joints are to be renewed, follow standard shop procedures. Separate axle shaft (34) from joint assembly

(32) by squeezing circlip (33) and separating universal joint from axle shaft. Separate sun gear shaft (13) from universal joint by pulling shaft free from joint.

Examine oil seal (31) and bushing (30) located in wheel hub (21) and oil seal (40) and bushing (41) located in axle housing for excessive wear or any other damage. Inspect universal joint yokes for excessive wear, roughness or any other damage. Renew all parts as needed.

Reassemble in reverse order of removal.

DIFFERENTIAL AND BEVEL GEAR ASSEMBLY

All Models

63. **REMOVAL.** Position tractor on a suitable level surface and apply parking brake. Raise and support front of tractor, then remove left front wheel. Remove differential housing drain plug and allow oil to drain into a suitable container. Remove nut retaining left tie rod end, then detach end from steering knuckle. Support left axle assembly using a suitable holding fixture. Remove cap screws (46 – Fig. 58) securing axle housing assembly to differential housing. Separate housings, then withdraw and place axle housing assembly to the side. Extract differential assembly from housing. Remove bolts and nuts securing drive shaft flange to pinion shaft flange. Separate components and swing drive shaft to the side. Remove retainer (35 – Fig. 61) and nut (34) retaining pin-

ion shaft flange. Using a suitable soft-faced hammer drive pinion shaft into differential housing. Lift flange (32) from housing. Remove pinion shaft (21), roller bearing cone (22) and sleeve (25) from differential housing.

64. **OVERHAUL.** Using puller (Churchill tool 1003 or Nuday 9516), bearing knife (Churchill tool 951 or Nuday 9190) and shaft end protector (Churchill tool 625A or Nuday 9212) or suitable equivalents, withdraw differential carrier bearings (14 and 19 – Fig. 61). Pry locking plate (8) from cap screws securing ring gear (20), then remove cap screws. Press ring gear from differential case with differential and ring gear pressing set (Nuday tool 4853) or by using four long bolts installed into differential case and pressing differential assembly from ring gear as shown in Fig. 62.

On models without limited slip, drive out roll pin (7 – Fig. 61) and push spider shaft (2) from differential case. Rotate side gears (9 and 13) and withdraw spider gears (3), side gears (9 and 13) and thrust washers (1) from opening in case.

On models with limited slip, lift out side gear and clutch plates as an assembly. Drive out roll pin (7) and push spider shaft (2) from differential case. Lift out spider gears, thrust washers and other side gear with clutch plates.

Inspect all bearings, cups, gears, splines, thrust washers and limited slip clutch plates (models so equipped) for excessive wear, scoring, corrosion, roughness or any other damage. Inspect differential casing and housing for excessive wear, cracks or any other damage. Renew all parts as needed. Beveled ring gear (20) and pinion gear (21) are renewable only as a matched set. Sleeve (25) must be renewed to ensure correct pinion bearing preload is obtained. If inner pinion bearing (22) is to be renewed, use puller (Churchill tool 1003 or Nuday tool 9516), bearing knife

Fig. 62 – If special tools are not available, use four bolts and plate assembly to press ring gear from differential case.

(Churchill tool 951 or Nuday tool 9190) and shaft end protector (Churchill tool 625A or Nuday tool 9212) or suitable equivalents as shown in Fig. 63.

Reassembly is reverse of disassembly procedure. Coat all components with recommended oil during reassembly.

On models equipped with limited slip, the following should be noted when assembling clutch packs. Spacers (4 – Fig. 61) must be installed with machined surface side towards side gears (13). Clutch pack clearance must be measured prior to installation. On side with ring gear, measure clearance using a depth gage as shown in Fig. 64. On opposite side, measure clearance through access holes in differential case using a feeler gage as shown in Fig. 65. Correct clearance is 0.004-0.008 inch (0.1-0.2 mm) and is adjusted by varying thickness of spacers (4 – Fig. 61). Both spacers must be of equal thickness. Spacers are available in thicknesses of 0.110-0.130 inch (2.8-3.3 mm) in 0.004 inch (0.1 mm) increments. Side gear should turn slowly, but lock up with clutch pack when turned fast.

On all models, complete differential reassembly and tighten cap screws securing ring gear to 50 ft.-lbs. (68 N·m) torque. Install new locking plate over cap screw heads.

If differential case, bearings or either axle housing is renewed, the following procedure must be followed to determine thickness of shims (17 – Fig. 61) to be located behind left and right differential bearing cups (18).

NOTE: Shims are used to provide correct bearing preload and location of shims determines backlash between ring gear and pinion gear.

Withdraw axle shafts from housings. Install original shims behind differential bearing cups (18) and press into position. Wrap a suitable cord of single thickness around differential case and install differential assembly into housing. Route cord end out pinion bore and secure end on outside of housing. Install left axle housing and secure with at least four retaining cap screws located diagonally opposite each other. Tighten securing cap screws to 140 ft.-lbs. (190 N·m) torque. Connect a spring scale to cord end as shown in Fig. 66 and measure amount of pull required to turn differential assembly. Correct range is 4-12 pounds (2-6 kg.). If measured reading is below correct range, then increase shim thickness. If measured reading is above correct range, then decrease shim thickness. Shims are available in thicknesses of 0.024-0.071 inch (0.6-1.8 mm) in 0.004 inch (0.1 mm) increments. Divide shims equally and adjust shim thickness until correct reading is attained, then remove differential assembly, bearing cups and shims and identify for reassembly.

Gear mesh position is adjusted using shim (24 – Fig. 61). Shim adjustment is required if ring and pinion gears, pinion bearings, or housing is renewed; shim adjustment is not required if original and new pinion gears are marked the same and bearings and housing are not renewed.

If only ring and pinion gears are to be renewed, shim thickness may be determined by noting marked number on pinion gear. If number marked on new pinion is larger than number on original pinion, subtract the difference from original shim pack to determine needed shim pack thickness. If number marked on new pinion is smaller than number on original pinion, add the difference to original shim pack to determine needed shim pack thickness.

If bearings or housing are renewed, or if original shim pack (24) is not available, shim (24) thickness may be determined using the following procedure (measurements should be in millimeters): Install dummy pinion (Churchill tool FT.3148-1 or Nuday tool 4799) into housing bore.

NOTE: Be sure dummy pinion is securely seated against shoulder in housing bore.

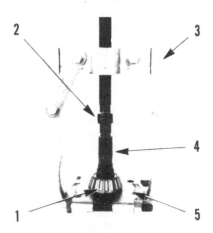

Fig. 63 – View showing procedure for pulling inner bearing (1) from pinion shaft (4) by use of puller (3), bearing knife (5) and shaft end protector (2).

Fig. 65 – On models with limited slip differential, clutch pack clearance on side opposite ring gear is measured by using a feeler gage (1) inserted in access holes (3) of differential case (2).

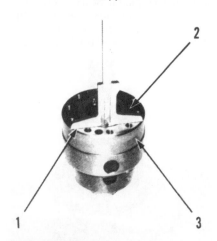

Fig. 64 – On models with limited slip differential, clutch pack clearance on ring gear side is measured between top of clutch pack (2) and top of differential case (3) by use of depth gage (1).

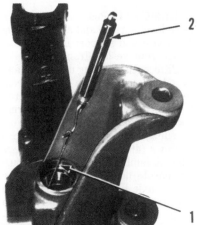

Fig. 66 – Hook spring scale (2) in cord end (1) to measure amount of pull required to turn differential assembly. Refer to text.

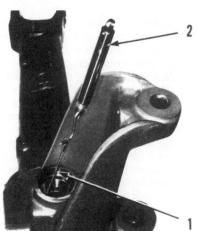

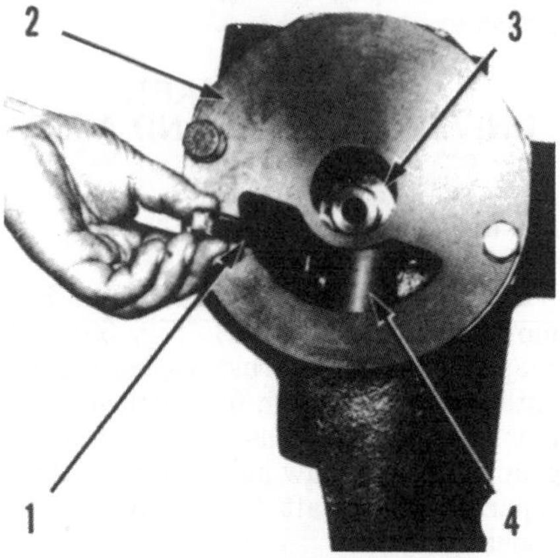

Fig. 67 – Use special plate (2), arbor (3), dummy pinion (4) and feeler gage (1) to measure distance (D – Fig. 68).

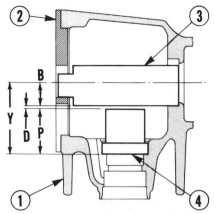

Fig. 68 — Dimensions B, D, P and Y are used in set of formulas to determine thickness of shim (24 — Fig. 61).

1. Housing
2. Plate
3. Arbor
4. Dummy pinion

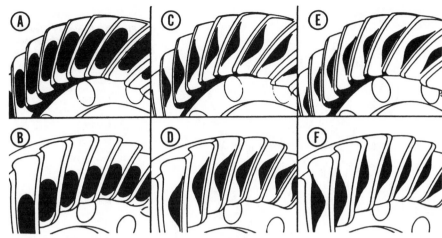

Fig. 70 — Views of ring gear tooth contact patterns. Adjust shims as needed until proper tooth pattern is attained.

A. Proper tooth contact-coast side pattern
B. Proper tooth contact-drive side pattern
C. Pinion gear requires thicker shim-coast side pattern
D. Pinion gear requires thicker shim-drive side pattern
E. Pinion gear requires thinner shim-coast side pattern
F. Pinion gear requires thinner shim-drive side pattern

Install special plate (2 – Fig. 67) (Churchill tool FT.3148 or Nuday tool 4798) onto differential housing. Insert test arbor (3) (Churchill tool FT.3148 or Nuday tool 4801) through plate (2) opening. Position arbor in bearing bore of housing. Figure 68 shows correct positioning of arbor and dummy pinion.

NOTE: Arbor end will fit loosely in bearing bore.

Using a feeler gage, measure distance between top of dummy pinion and adjacent side of arbor. With measured distance recorded, use the following set of formulas to determine thickness of shim (24 – Fig. 61).

Height (P – Fig. 68) of dummy pinion (Dimension is marked on side of tool) + distance (D) measured with feeler gage + one-half diameter (B) of bearing bore (Dimension is 1.47 inches or 36.7 mm) = Dimension Y.

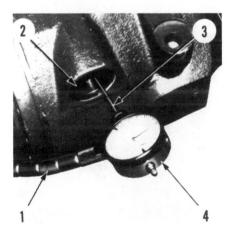

Fig. 69 — To check backlash between ring gear tooth and pinion gear tooth, mount dial indicator stand (1) on housing and position dial indicator (4) so plunger (3) protrudes through drain plug hole (2) and contacts ring gear tooth in a perpendicular position.

Place inner pinion bearing cup (23 – Fig. 61) on a solid flat surface, install bearing cone (22) and rotate cone to be sure all bearing rollers are seated in cup. Measure distance from surface to top of bearing. The measured height will be called dimension "X". Subtract dimension "X" from dimension "Y" and record as dimension "Z". Subtract millimeter dimension located at end of pinion serial number from dimension "Z" to obtain desired thickness of shim (24).

Install correct pinion shim(s) (24 – Fig. 61) in housing, then using suitable tools drive bearing cups (23 and 29) into housing bores until completely seated. Press inner pinion bearing (22) onto pinion shaft until bearing seats against pinion shoulder. Install a new bearing preload adjustment sleeve on pinion shaft. Lubricate pinion assembly with suitable oil and position in housing. Lubricate outer pinion bearing (30) with suitable oil and place onto pinion shaft. Using a suitable tool, drive oil seal (31) into housing with oil seal lip facing in. Position flange (32) onto pinion shaft splines and install washer (33) and nut (34).

NOTE: Coat threads of nut (34) with Ford EM46-52A thread sealant or a suitable equivalent before installation.

To prevent flange (32) rotation while tightening securing nut (34), use flange wrench (Churchill tool T.3122 or Nuday tool 8170). Tighten nut in small increments so bearing preload is 12-18 in.-lbs. (1-2 N·m).

NOTE: Wrench holding flange must be removed during bearing preload check.

After adjusting pinion shaft bearing preload, install nut retainer (35), reconnect drive shaft and tighten securing bolts and nuts to 40 ft.-lbs. (54 N·m).

Install correct differential bearing shims (17) behind left and right differential bearing cups (18) and press cups into position. Install right axle shaft. Lubricate differential components with a suitable oil and install differential assembly into housing while engaging side gear and axle shaft splines and ring gear and pinion teeth. Install left axle housing and tighten securing cap screws to 140 ft.-lbs. (190 N·m) torque. Install left axle shaft, then remove differential housing drain plug. Securely locate dial indicator stand (1 – Fig. 69) on housing and position dial indicator (4) so plunger (3) protrudes through drain plug hole (2) and contacts ring gear tooth in a perpendicular position. Hold pinion flange, then rotate one axle shaft. Ring gear will rotate from axle shaft movement. While noting dial indicator needle, check backlash between ring gear teeth and pinion gear tooth. Correct backlash is 0.006-0.010 inch (0.15-0.25 mm). If backlash is not within limits, shims (17 – Fig. 61) behind differential bearing cups must be adjusted. Transfer differential bearing adjusting shims (17) from side to side to obtain correct gear backlash; DO NOT change total shim pack thickness as previously adjusted bearing preload will be affected.

Refer to Fig. 70 for proper ring gear tooth contact pattern. If ring gear or pinion gear shimming procedure wasn't performed properly, ideal tooth pattern will not be obtained and shimming procedure will have to be repeated.

Install drain plug and fill differential housing through fill plug hole with 6½ U.S. quarts of Ford EOAZ-19580 oil or a suitable equivalent, then install fill plug. Check axle hub fluid levels and refill as needed.

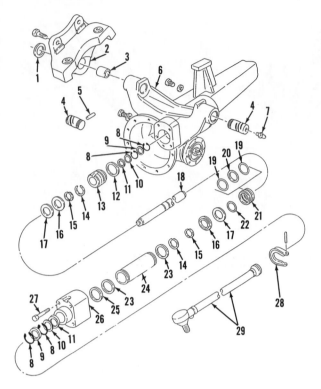

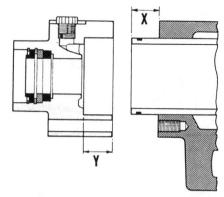

Fig. 71 — Exploded view of power steering cylinder and front support used on Models 6710 and 7710.

1. Shim
2. Front support
3. Bushing
4. Pivot pin
5. Pin
6. Housing
7. Grease fitting
8. Snap ring
9. Dust seal
10. Back-up washer
11. Seal
12. Seal
13. Gland
14. Snap ring
15. Split ring
16. Washer
17. Spacer
18. Cylinder rod
19. Sealing rings
20. Seal
21. Piston
22. Seal
23. Seal
24. Tube
25. Shim
26. End cap
27. Cap screw
28. Clamp assy.
29. Tie rod assy.

Fig. 72 — Subtract distance X from depth Y to determine shim (25 — Fig. 71) thickness.

POWER STEERING SYSTEM

Models 5610-6610-7610

65. The power steering system is the same as used on two-wheel drive models. Refer to appropriate service section for repair procedures.

Models 6710-7710

66. **OVERHAUL.** Position tractor on a suitable level surface and apply parking brake. Place a suitable fluid holding container under cylinder housing. Disconnect power steering hose from fitting on top of end cap (26 – Fig. 71). Detach tie rod ends from steering knuckles. Release clamps (28) on each side and unscrew tie rod assemblies (29) from cylinder rod (18). Remove cap screws (27) securing end cap (26), then withdraw end cap from cylinder rod. Note shims (25) located in end cap. Extract cylinder rod (18) and piston assembly (21) as a unit from housing. Using a suitable tool, push gland (13) and tube (24) from housing.

NOTE: Piston (21 – Fig. 71) and cylinder rod (18) must be separated to renew seal (22).

Clean all components with a suitable solvent and blow dry with clean compressed air. Inspect cylinder tube (24), piston (21) and cylinder rod (18) for excessive wear, scoring or any other damage and renew as needed. If tube (24), gland (13) or end cap (26) is renewed, the following procedure must be used to establish thickness of shim (25).

Push gland (13) and tube (24) completely into housing. Measure protrusion (X – Fig. 72) of tube past housing surface. Use a depth gage or other suitable measuring device and measure depth of end cap bore – distance Y. Subtract measurement X from distance Y; difference is shim (25) thickness needed to properly locate tube in end cap. Shims are available in thicknesses of 0.040-0.072 inch (1.0-1.8 mm) in increments of 0.004 inch (0.1 mm).

Reassembly is reverse of disassembly. Install new seals and sealing rings. Coat all components with Ford M-2C41A power steering fluid or a suitable equivalent during reassembly. Tighten end cap screws to a torque of 50 ft.-lbs. (68 N·m).

67. **FLUID AND BLEEDING.** Use Ford M-2C41A fluid in power steering system. Fluid level should be maintained at bottom of reservoir filler neck. System capacity is 5.0 U.S. pints (2.4 liters). System is self-bleeding; start engine and cycle steering until all air is bled. Check reservoir fluid level and refill as needed.

TRANSFER BOX

All Models

68. The transfer box is attached to underneath side of rear axle center housing and is lubricated by transmission/rear axle oil. Transfer box may be equipped with an optional transmission hand brake. A gear on rear axle pinion is used to drive transfer box idler gear. An

electric solenoid controlled by a switch opens or closes control valve oil passage. When solenoid is de-energized, control valve oil passage is closed and pressurized oil acting on clutch piston is dumped, allowing Belleville washers to force clutch piston forward thereby compressing clutch plates and engaging transfer box drive.

69. **REMOVE AND REINSTALL.** Position tractor on a suitable level surface and securely block wheels to prevent rolling. Remove drain plug (4 – Fig. 73) and drain oil. Detach solenoid cover (1), then unscrew solenoid wire protective tube (3) fitting from solenoid housing and disconnect wire from solenoid. Disconnect oil supply line (5) from housing fitting. On models equipped with a transmission hand brake, disconnect control cable from actuating lever. Place a suitable jack under transfer box housing and remove cap screws (2) retaining transfer box assembly to center housing. Lower assembly and slide housing rear-

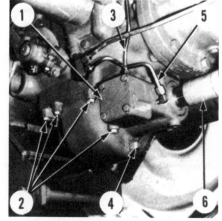

Fig. 73 — View showing transfer box assembly without transmission hand brake. Models with transmission hand brake are similar.

1. Solenoid cover
2. Cap screws
3. Tube
4. Drain plug
5. Oil supply line
6. Yoke

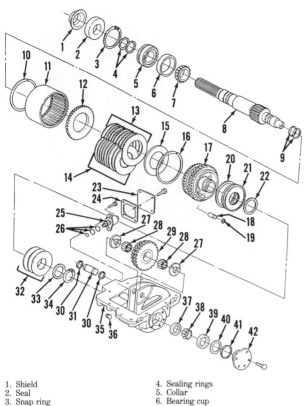

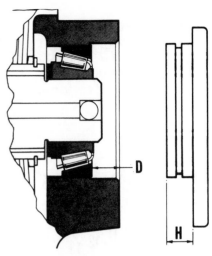

Fig. 74 — Exploded view of transfer box used on models without transmission hand brake.

7. Bearing cone
8. Shaft
9. Sealing rings
10. Snap ring
11. Clutch housing
12. Spacer plate
13. Drive plates
14. Driven plates
15. Shim
16. Snap ring
17. Actuating cylinder/ drive gear housing
18. Plunger
19. "O" ring
20. Outer seal
21. Piston
22. Inner seal
23. Plate
24. Gasket
25. Solenoid & control valve assy.
26. "O" rings
27. Thrust washer
28. Needle bearings
29. Idler gear
30. Snap ring
31. Shaft
32. Belleville washers
33. Spacer ring
34. Snap ring
35. Housing
36. Plug
37. Thrust washer
38. Bearing cone
39. Bearing cup
40. Shim
41. "O" ring
42. End plate

1. Shield
2. Seal
3. Snap ring
4. Sealing rings
5. Collar
6. Bearing cup

Fig. 77 — Measure distance "D" and height "H" to determine shim (40 — Fig. 74) thickness for correct output shaft end play. Refer to text.

ward to disconnect output shaft splines from drive shaft yoke (6).

Installation is reverse of removal. Tighten cap screws securing housing to 35 ft.-lbs. (48 N·m) torque. Refill transmission/rear axle unit with 57¾ U.S. quarts on models with single-speed pto and 61¾ U.S. quarts on models with dual-speed pto. Recommended oil is Ford 134.

70. OVERHAUL (Models Without Transmission Hand Brake). Refer to Fig. 74 for an exploded view of transfer box. Remove snap ring (30) at one end of

shaft (31) and slide shaft from housing. Withdraw idler gear (29) with needle bearings (28) and thrust washers (27). Remove shield (1) from output shaft, then remove oil seal (2) and snap ring (3) from housing bore. Remove cap screws securing end plate (42), then lift plate and "O" ring seal (41) from housing noting bearing preload shim (40). Using a press and suitable tools, press output shaft (8) out front of transfer box assembly. Withdraw clutch assembly and actuating cylinder/drive gear assembly from housing as a unit. Compress

Belleville washers (32) and remove snap ring (34). Compress snap ring (16) through access holes (H – Fig. 75) in clutch housing (11), then separate clutch housing from actuating cylinder/drive gear assembly. Withdraw piston (21 – Fig. 74) and plungers (18) from actuating cylinder (17).

Clean all gears, shafts, bearings and housings with a suitable solvent and blow dry with clean compressed air. Inspect all gears and shaft splines for excessive wear, chipped teeth, burrs or any other damage. Inspect housings for excessive wear, cracks or any other damage. Inspect bearings for excessive wear, corrosion, roughness or any other damage. Inspect shaft sealing rings for excessive wear or any other damage. Inspect clutch plates for wear, cracks, distortion, damaged splines or any other damage. Inspect bushing in actuating cylinder/drive gear housing for excessive wear. Examine actuating cylinder piston and plungers for excessive wear or any other damage. Check filter at end of control valve for blockage. Be sure all oil passages are clean and free of restrictions. Renew all parts as needed. Renew all seals, "O" rings and gaskets. Control valve and solenoid assembly is renewable only as a complete unit. Clutch plates must be renewed as a set.

Reassemble in reverse order of disassembly. Check clutch plate clearance by positioning clutch assembly (A – Fig. 76) in a press and compressing clutch plates with a suitable tool. Measure gap between front snap ring (10) and spacer plate (12) using feeler gage (G). Correct clearance is 0.027-0.098 inch (0.7-2.7 mm). If measured clearance is not within limits, adjust shim (15 – Fig. 74) thickness until correct clearance is attained. Shim (15) is available in thicknesses of 0.18-0.32 inch (4.7-8.1 mm) in increments of 0.008 inch (0.2 mm).

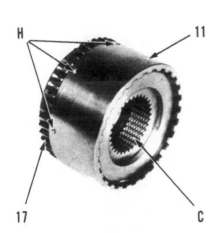

Fig. 75 — View showing access holes (H) used in compressing snap ring to separate clutch housing (11) from actuating cylinder/drive gear assembly (17). Clutch plates (C) are shown in housing.

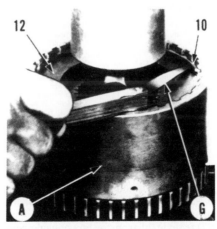

Fig. 76 — View showing procedure for checking clutch plate clearance on models without transmission hand brake.

A. Clutch assy.
G. Feeler gage
10. Snap ring
12. Spacer plate

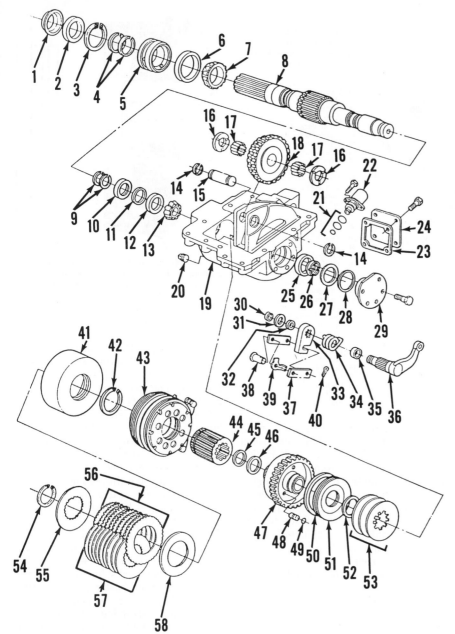

Fig. 79 — When pressing output shaft from transfer box assembly on models equipped with transmission hand brake, snap ring (54) must be compressed as outlined in text.

B. Support block
S. Snap ring pliers
T. Pressing tool
54. Snap ring

a gap of 0.001-0.003 inch (0.025-0.075 mm). Shims are available in thicknesses of 0.040-0.088 inch (1.00-2.25 mm) in increments of 0.002 inch (0.05 mm). Install end plate and tighten cap screws to 35 ft.-lbs. (48 N·m) torque.

NOTE: Be sure output shaft and clutch assembly can be rotated by hand. If not, recheck and adjust shims (40).

Complete reassembly. Tighten solenoid cover plate cap screws to 20 ft.-lbs. (27 N·m) torque.

71. OVERHAUL (Models With Transmission Hand Brake). Refer to Fig. 78 for an exploded view of transfer box. Remove snap ring (14) at one end of shaft (15) and slide shaft from housing. Withdraw idler gear (18) with needle bearings (17) and thrust washers (16). Remove shield (1) from output shaft, then remove oil seal (2) and snap ring (3) from housing bore. Remove cap screws securing end plate (29), then lift plate and "O" ring seal (28) from housing noting bearing preload shim (27). Use special tool (Churchill tool FT.3145 or Nuday 4645) and remove slotted nut (26) from shaft end. Position a support block (B – Fig. 79) on each side of clutch assembly. While holding clutch retaining snap ring (54) open with pliers (S), press output shaft out front of transfer box using suitable tools. Slide clutch assembly off splined sleeve and lift from housing.

Remove snap ring (42 – Fig. 78) from splined sleeve (44). Remove outer brake discs, detach brake actuating linkage, then remove brake actuator assembly and remaining brake discs. Lift out splined sleeve (44), needle bearing (45) and thrust washer (46). Remove actuating cylinder/drive gear housing (47)

Fig. 78 — Exploded view of transfer box used on models with transmission hand brake.

1. Shield
2. Seal
3. Snap ring
4. Sealing rings
5. Collar
6. Bearing cup
7. Bearing cone
8. Shaft
9. Sealing rings
10. Step ring
11. Needle bearing
12. Shim
13. Bearing cone
14. Snap ring
15. Shaft
16. Thrust washer
17. Needle bearing
18. Idler gear
19. Housing
20. Drain plug
21. "O" rings
22. Solenoid & control valve assy.
23. Gasket
24. Plate
25. Bearing cup
26. Slotted nut
27. Shim
28. "O" ring
29. End plate
30. Ring
31. Shim
32. Seal
33. Lever
34. Spring
35. Ring
36. Actuating lever
37. Link
38. Pin
39. Stop
40. Cotter pin
41. Clutch housing
42. Snap ring
43. Brake assy.
44. Splined sleeve
45. Needle bearing
46. Thrust washer
47. Actuating cylinder/drive gear housing
48. Plunger
49. "O" ring
50. Outer seal
51. Piston
52. Inner seal
53. Belleville washers
54. Snap ring
55. Spacer plate
56. Drive plates
57. Driven plates
58. Shim

Check output shaft end play as follows: Install clutch assembly and actuating cylinder assembly in transfer box housing. Install bearing (7) and sealing rings (4 and 9) on output shaft (8), then install shaft while aligning splines. Install bearing cup (6), oil collar (5) and snap ring (3). Position thrust washer (37) on rear of output shaft, then press rear

bearing (38) onto shaft and install bearing cup (39) in housing rear bore. Using a suitable depth gage or other measuring device, measure distance (D – Fig. 77) from lip of transfer box housing rear bore to surface of bearing cup. Measure height of end plate protrusion (H) and subtract from distance (D). Install shim(s) (40 – Fig. 74) which will provide

with plungers (48), piston assembly (51) and Belleville washers (53).

Clean all gears, shafts, bearings and housings with a suitable solvent and blow dry with clean compressed air. Inspect all gears and shaft splines for excessive wear, chipped teeth, burrs or any other damage. Inspect housings for excessive wear, cracks or any other damage. Inspect bearings for excessive wear, corrosion, roughness or any other damage. Inspect shaft sealing rings for excessive wear or any other damage. Inspect clutch plates or brake discs for wear, cracks, distortions, damaged splines or any other damage. Separate brake actuator by removing return springs, then split halves and inspect operating balls and ramp tracks. Inspect

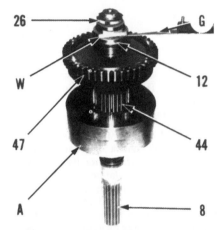

Fig. 80—On models with transmission hand brake, assemble components designated in text and measure clutch plate clearance using feeler gage (G).

A. Clutch assy.	
G. Feeler gage	26. Slotted nut
W. Washer	44. Splined sleeve
8. Output shaft	47. Actuating cylinder &
12. Shim	drive gear housing

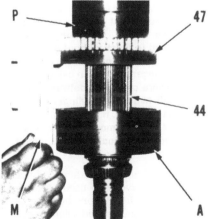

Fig. 81—Assemble components designated in text and measure transmission hand brake clearance using measuring tool (M).

A. Clutch assy.	44. Splined sleeve
M. Measuring tool	47. Actuating cylinder/
P. Pressing tool	drive gear housing

bushing in actuating cylinder/drive gear housing for excessive wear. Examine actuating cylinder piston and plungers for excessive wear or any other damage. Check filter at end of control valve for blockage. Be sure all oil passages are clean and free of restrictions. Renew all parts as needed. Renew all seals, "O" rings and gaskets for reassembly. Control valve and solenoid assembly is renewable only as a complete unit. Clutch plates and brake discs must be renewed as a set.

Before reassembly, correct shim thicknesses must be established. Check clutch plates and brake discs clearance as follows: Place assembled clutch unit on output shaft (8) against snap ring (54), then stand shaft vertically as shown in Fig. 80. Position splined sleeve (44—Fig. 78) with snap ring (42) on shaft. Install needle bearing (45) and thrust washer (46) onto shaft, then install actuating cylinder/drive gear housing (47), plungers (48) and piston (51). DO NOT install Belleville washers. Position step ring (10), needle bearing (11) and shim (12) onto shaft. In place of bearing (13), install a suitable spacer (W—Fig. 80) that will butt against shaft step. Retain components with slotted nut (26). With reference to Fig. 80 measure clearance between shim (12) and spacer (W). Correct clearance is 0.027-0.098 inch (0.7-2.7 mm). Shim (12) is available in thicknesses of 0.287-0.309 inch (7.30-7.85 mm) in increments of 0.002 inch (0.05 mm). Adjust shim as needed until correct measured gap is attained.

Position shaft assembly in a press and using a suitable tool located against clutch piston (51—Fig. 78), compress clutch plates. With reference to Fig. 81 measure distance between top surface of actuating cylinder/drive gear housing (47) and clutch housing (A). Correct distance is 2.374-2.390 inches (60.3-60.7 mm). If distance is incorrect, shim (58—Fig. 78) must be adjusted. Shim (58) is available in thicknesses of 0.18-0.32 inch (4.7-8.1 mm) in increments of 0.008 inch (0.2 mm). Adjust shim as

needed until correct distance is attained. Disassemble components and note correct shims for reassembly.

To reassemble, stand transfer box housing (19) on rear surface. Position an alignment sleeve 2⅛ inches (53 mm) long in rear bearing bore. Install plungers (48), "O" rings (49), piston (51) and seals (50 and 52) in actuating cylinder/drive gear housing (47), then install housing and Belleville washers (53) over sleeve. Install in correct sequence two internally splined brake discs (59—Fig. 82) and one stationary plate (62) on face of housing (47—Fig. 78). Hook and secure linkage to brake actuator (60—Fig. 82) and install actuator. Install splined sleeve (44) through actuator and engage splines of brake discs. Install thrust washer (46—Fig. 78) and needle bearing (45) into sleeve (44). Install remaining internally splined discs (63—Fig. 82) and stationary plates (61) over sleeve and retain with snap ring (42—Fig. 78). Position assembled clutch assembly in housing on splined sleeve, then place snap ring (54) on spacer plate (55) of clutch assembly (due to limited space, shaft is inserted through snap ring). Install bearing cone (7) and sealing rings (4 and 9) on output shaft (8). Hold snap ring (54) open and insert shaft into housing while aligning splines.

NOTE: Ensure snap ring correctly engages groove in shaft. Alignment sleeve will be forced from bearing bore as shaft is installed.

Install step ring (10), needle bearing (11), shim (12) and bearing (13) onto shaft. Secure components with slotted nut (26) and tighten to a torque of 120 ft.-lbs. (163 N·m). Install front bearing cup (6), oil collar (5) and snap ring (3).

Check output shaft end play as follows: Install bearing cup (25) until completely positioned against bearing (13). Using a suitable depth gage or other measuring device, measure distance (D—Fig. 77) from lip of transfer box housing rear bore to surface of bearing cup. Measure height of end plate

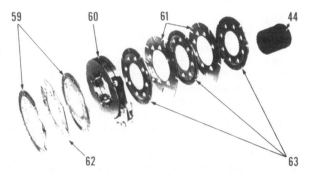

Fig. 82—Exploded view of transmission hand brake assembly.

44. Splined sleeve
59. Internally splined discs
60. Brake actuator
61. Stationary plates
62. Stationary plate
63. Internally splined discs

protrusion (H) and subtract from distance (D). Select shim(s) (27–Fig. 78) which will provide a gap of 0.001-0.003 inch (0.025-0.075 mm). Shims are available in thicknesses of 0.040-0.088 inch (1.00-2.25 mm) in increments of 0.002 inch (0.05 mm). Install selected shim(s), then place "O" ring (28–Fig. 78) on end plate (29). Install end plate and tighten

cap screws to 35 ft.-lbs. (48 N·m) torque.

NOTE: Be sure output shaft and clutch assembly can be rotated by hand. If not, recheck and adjust selected shims.

Complete reassembly as needed, tighten solenoid cover plate cap screws to 20 ft.-lbs. (27 N·m) torque.

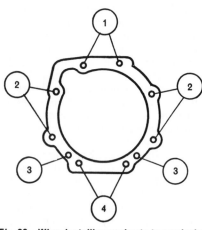

Fig. 83—When installing engine to transmission housing, tighten bolts shown at locations in drawing to the following torque specifications:

1. 55-75 ft.-lbs. (75-102 N·m)	3. 110-135 ft.-lbs. (150-184 N·m)
2. 220-300 ft.-lbs. (299-408 N·m)	4. 125-140 ft.-lbs. (170-190 N·m)

ENGINES AND COMPONENTS

R&R ENGINE WITH CLUTCH

All Models

72. To remove engine and clutch assembly, first remove front axle, front support or pedestal, radiator and radiator shell assembly as outlined in paragraph 83 or 84. Then, proceed as follows:

Disconnect battery cables and on all models except 6700, 6710, 7700 and 7710, remove battery, battery tray and support brackets. If engine is to be disassembled, drain oil pan. Disconnect all interfering wiring and sheet metal. Disconnect power steering hoses on models so equipped. Disconnect heater and air conditioning hoses on models so equipped. Shut off fuel supply valve, remove fuel supply line and diesel excess fuel return line. Disconnect oil cooler lines. Disconnect throttle linkage, choke wire or diesel stop wire and proofmeter drive cable. Remove flywheel timing hole cover screw. Remove any engine accessories as required for indicated engine overhaul, attach hoist to engine, then unbolt and remove engine and clutch assembly from transmission.

To reinstall engine, reverse removal procedure. Refer to Fig. 83 for engine to transmission bolt tightening torques. Bleed diesel fuel system as outlined in paragraph 123, fill and bleed power steering system as outlined in paragraph 13, 31, 40, 51 or 67.

ENGINE COMPRESSION PRESSURES

All Models

73. The Ford Motor Company recommends that compression pressures be checked only at cranking speeds. Non-diesel engine compression pressures should be checked with all spark plugs removed, throttle in wide open position and choke knob pushed in. Insufficient cranking speed can cause low compression pressure readings. For this reason, battery should be well charged and starting system functioning properly to

assure a cranking speed of 150 to 200 rpm.

The following table gives compression pressure ranges and maximum allowable cylinder-to-cylinder variation for a cranking speed of approximately 200 rpm:

Non-Diesel Engines:
Compression Pressure
  Range . . . . 110-210 psi
      (759-1449 kPa)
Max. Allowable Variation . . . . 25 psi
      (173 kPa)

Diesel Engines:
Compression Pressure
  Range . . . . 300-400 psi
      (2070-2760 kPa)
Max. Allowable Variation . . . . 50 psi
      (345 kPa)

R&R CYLINDER HEAD

All Models

74. To remove cylinder head, proceed as follows: Remove vertical exhaust muffler, if so equipped, and remove engine hood. On models with horizontal exhaust system, disconnect exhaust pipe from exhaust manifold. Remove turbocharger (diesel models so equipped) and exhaust manifold. Remove battery, battery tray and battery support brackets on all models except 6700, 6710, 7700 and 7710. Drain cooling system and remove upper radiator hose and heater hoses on models so equipped. Disconnect interfering wiring and detach any braces bolted to head.

On diesel models, disconnect air intake hose from intake manifold, shut off fuel supply valve and remove fuel filter assembly from intake manifold. Remove fuel injectors as outlined in paragraph 130. Remove intake manifold.

On gasoline models, shut off fuel supply valve, disconnect air inlet hose, fuel line and throttle rod from carburetor, disconnect vacuum line from intake manifold and distributor and remove intake manifold and carburetor as a unit. Remove spark plugs.

Disconnect ventilation tube from rocker arm cover and remove rocker arm cover, rocker arms assembly and push rods. Remove remaining cylinder head bolts and remove cylinder head from engine block.

NOTE: If cylinder head gasket failure has occurred, check cylinder head and block mating surfaces for flatness. Maximum allowable deviation from flatness is 0.006 inch (0.15 mm) overall or 0.003 inch (0.076 mm) in any six inches. If cylinder head is not within flatness specified or is rough, surface may be lightly machined. If cylinder block is not within flatness specified, it may be machined providing distance between top of pistons at top dead center and top surface of cylinder block is not less than 0.002 inch (0.05 mm) after machining. If cylinder head is also skimmed to restore flatness, check clearance of lower edge of valve seat insert to face of cylinder head. This clearance must be at least 0.117 inch (2.97 mm). To determine if cylinder head bolts will bottom after machining, install cylinder head to block without gasket, install rocker arm supports, washers and all head bolts finger tight. Then using a feeler gage, measure clearance between underside of bolt heads and cylinder head or rocker arm supports. If clearance is 0.010 inch (0.254 mm) or more, cut threads of bolt hole deeper in block with a ½-inch, 13 UNC-2A tap.

75. Two different thickness cylinder head gaskets have been used on Model 5000 engines. On tractors built before mid-1969, head gasket had a compressed thickness of 0.029 inch (0.737 mm). Later gaskets have a compressed thickness of 0.037 inch (0.940 mm). Because of construction differences, earlier gasket should not be used on later engines.

The cylinder head gasket for Model 7000 engines has a steel plate sandwiched between the two outer layers. Thickness is 0.068 inch (1.727 mm). See Fig. 84 for comparative view of new head gasket. Note increase in number and size of coolant passage holes to accommodate 50 per cent increase in flow volume from water pump. Fire ring of new gasket is made of stainless steel.

When reassembling, do not use gasket sealer or compound and be sure that gasket is properly positioned on the two dowel pins. Using the sequence shown in Fig. 85 for 5000 and 7000 models and Fig. 86 for all other models, tighten cylinder head bolts in three steps as follows:

First Step . . . . 90 ft.-lbs.
(122 N·m)
Second Step . . . . 100 ft.-lbs.
(136 N·m)
Final Step . . . . 110 ft.-lbs.
(150 N·m)

NOTE: Torque values given are for lubricated threads; tighten cylinder head bolts only when engine is cold.

Adjust valve gap as indicated in paragraph 78. Complete reassembly of engine by reversing disassembly procedure. Tighten intake manifold bolts to a torque of 23-28 ft.-lbs. (31-38 N·m), exhaust manifold bolts to a torque of 25-30 ft.-lbs. (34-41 N·m). With reassembly complete, bleed diesel fuel system as outlined in paragraph 123, start engine and bring to normal operating temperature. Shut off engine and readjust valve gap to 0.015 inch (0.38 mm) on intake valves and 0.018 inch (0.46 mm) on exhaust valves. The cylinder head bolts should be retorqued and valve gap readjusted after 50 hours of operation.

VALVES, STEM SEALS AND SEATS

All Models

76. Exhaust valves are equipped with positive type rotators and an "O" ring type seal is used between valve stem and rotator body. Intake valve stems of most models are fitted with umbrella type oil seals. Both intake and exhaust valves seat on renewable type valve seat inserts that are a shrink fit in cylinder head. Inserts are available in oversizes of 0.010, 0.020 and 0.030 inch (0.254, 0.508 and 0.762 mm) as well as standard size.

Intake and exhaust valve face angle for early 5000 models is 44 degrees and valve seat angle is 45 degrees resulting in a one degree interference angle. Intake and exhaust valve face angle for 5600, 5610, 6600, 6610, 6700 and 6710 models is 44.5 degrees and valve seat angle is 45 degrees resulting in a 0.5 degree interference angle. Renew valve if margin is less than 1/32-inch (0.794 mm) after valve is refaced. Desired valve seat width is 0.080-0.102 inch (2.032-2.590 mm) for intake valves and 0.084-0.106 inch (2.133-2.692 mm) for exhaust valves.

Seats can be narrowed and centered by using 30 and 60 degree stones. Total seat runout should not exceed 0.0015 inch (0.038 mm).

For late 5000 models and 7000 models the intake valve face angle is 29 degrees and valve seat angle is 30 degrees. Exhaust valve face angle is 44 degrees and valve seat angle is 45 degrees. Be sure to maintain one degree interference angle. Intake valve face angle for 7600, 7610, 7700 and 7710 models is 29.5 degrees and valve seat angle is 30 degrees. Exhaust valve face angle is 44.5 degrees and valve seat angle is 45 degrees

resulting in a 0.5 degree interference angle. Renew valve if margin is less than 1/16-inch (1.587 mm) for intake, or less than 1/32-inch (0.784 mm) for exhaust after valve is refaced. Desired valve seat width is 0.080-0.102 inch (2.032-2.590 mm) for intake valves and 0.084-0.106 inch (2.133-2.696 mm) for exhaust valves. Seats for intake valves can be narrowed and centered by using 15 and 45 degree stones. Seats for exhaust valves can be narrowed and centered by using 30 and 60 degree stones. Total seat runout should not exceed 0.0015 inch (0.038 mm).

Desired stem to guide clearance is 0.001-0.0027 inch (0.025-0.069 mm) for intake valves and 0.002-0.0037 inch (0.051-0.094 mm) for exhaust valves. New (standard) stem diameter is 0.3711-0.3718 inch (9.426-9.444 mm) for intake valves and 0.3701-0.3708 inch (9.401-

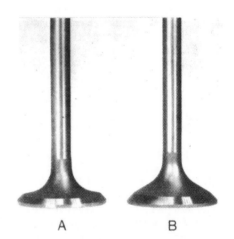

Fig. 87—Comparison view to show difference in design between intake valve (B) used on later models and older type (A). Changed profile improves heat tolerance and flow of air into combustion chamber.

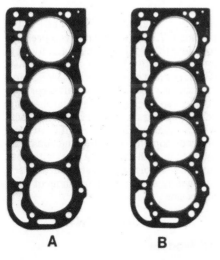

Fig. 84—Comparative view of new (B) and older style cylinder head gasket (A) used on engines with 4.4 inch (111.8 mm) bore. See text.

Fig. 85—Drawing showing cylinder head bolt tightening sequence for 5000 and 7000 models.

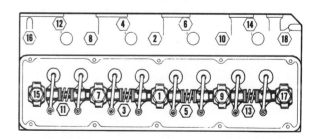

Fig. 86—Drawing showing cylinder head bolt tightening sequence for all models except 5000 and 7000 models.

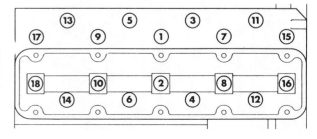

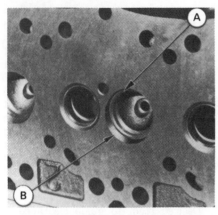

Fig. 88—View showing underside of new style cylinder head. Note increased relief area for valve seats. New style cylinder head is not interchangeable with earlier design. Turbocharged models have additional relief cut into injector port area.

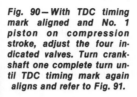

Fig. 90—With TDC timing mark aligned and No. 1 piston on compression stroke, adjust the four indicated valves. Turn crankshaft one complete turn until TDC timing mark again aligns and refer to Fig. 91.

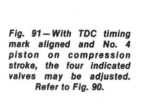

Fig. 91—With TDC timing mark aligned and No. 4 piston on compression stroke, the four indicated valves may be adjusted. Refer to Fig. 90.

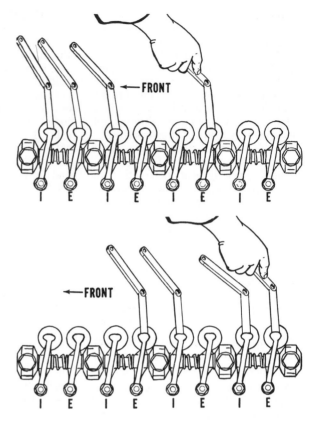

9.418 mm) for exhaust valves. Valves with 0.003, 0.015 and 0.030 inch (0.076, 0.381 and 0.762 mm) oversize stems are available as well as reamers (Nuday tool SW502) for enlarging valve guide bore to these oversizes.

Refer to Fig. 87 for comparative view of new type intake valve with increased thickness at valve head for better heat dissipation. If new valve is installed in older engines, seats must be ground to fit.

See Fig. 88 for view of new cylinder head to show increased relief area at valve seats. Turbocharged Models 7600, 7610, 7700 and 7710 have an additional relief as a fuel mixture recess cut into injector port area between valve seats. These heads also have coring changes for increased coolant flow, and are not interchangeable with earlier production cylinder heads.

NOTE: Although valves for gasoline and diesel engines are dimensionally the same, exhaust valve material is different.

Fig. 89—Valve guides are integral with cylinder head. Reamers in 0.003, 0.015 and 0.030 inch (0.076, 0.381 and 0.762 mm) oversize are available (Nuday tool SW-502) for repairing worn guides to fit oversize valve stems. Refer to text.

For this reason, gasoline and diesel exhaust valves should not be interchanged. Exhaust valves for gasoline engines may be identified by an 0.18 inch (4.572 mm) diameter depression in center of top face of valve head whereas exhaust valves for diesel engines have a flat top face on valve head.

VALVE GUIDES AND SPRINGS

All Models

77. Intake and exhaust valve guides are an integral part of the cylinder head and are non-renewable. Standard valve guide diameter is 0.3728-0.3735 inch (9.469-9.487 mm). Valve guides may be reamed to 0.003, 0.015 or 0.030 inch (0.076, 0.381 or 0.762 mm) oversize and valves with oversize stems installed if stem-to-guide clearance is excessive. Desired stem-to-guide clearance is 0.001-0.0027 inch (0.025-0.069 mm) for intake valves and 0.002-0.0037 inch (0.051-0.094 mm) for exhaust valves. A valve stem guide reamer kit (Nuday tool SW502 as shown in Fig. 89) is available.

Intake and exhaust valve springs are interchangeable. Valve spring free length should be 2.15 inches (54.61 mm). Springs should exert a force of 61 to 69 pounds (27.7-31.3 kg) when compressed to a length of 1.74 inches (44.196 mm), and a force of 125-139 pounds (57.0-63.0 kg) when compressed to a length of 1.32 inches (33.53 mm). Valve springs should also be checked for squareness by setting spring on flat surface and check with a square; renew spring if clearance between top end of spring and square is more than 1/16-inch (1.59 mm) with bottom end of spring against square. Also, renew any spring showing signs of rust or erosion.

ADJUSTMENT

All Models

78. **TAPPET GAP ADJUSTMENT.** Recommended initial (cold) tappet gap

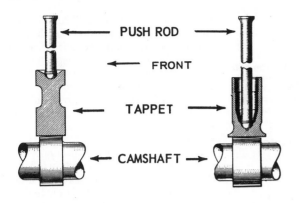

PUSH ROD

FRONT

TAPPET

CAMSHAFT

PREVIOUS TYPE   **CURRENT TYPE**

Fig. 92 — Views showing previous and current types of push rods, tappets and camshafts for diesel engines.

adjustment is 0.017 inch (0.43 mm) for intake valves and 0.021 inch (0.53 mm) for exhaust. Recommended setting at operating temperature is 0.015 inch (0.38 mm) for intake valves and 0.018 inch (0.46 mm) for exhaust valves.

Valves can be statically adjusted using the two-position method and Figs. 90 and 91 as a guide; proceed as follows: Turn crankshaft until "TDC" flywheel timing mark is aligned with timing pointer, then check the two front (No. 1 Cylinder) rocker arms. If rocker arms are loose, No. 1 cylinder is on compression stroke; adjust the valves shown in Fig. 90. If front rocker arms are tight, No. 1 cylinder is on exhaust stroke; adjust valves shown in Fig. 91. In either event, complete the adjustment by turning the crankshaft one complete revolution and using the appropriate alternate diagram.

EXHAUST VALVE ROTATORS

All Models

79. The positive type valve rotators require no maintenance, but each ex-

haust valve should be observed while engine is running to be sure valve rotates slightly. Renew rotator on any exhaust valve that fails to turn.

VALVE TAPPETS (CAM FOLLOWERS)

Gasoline Engines

80. The tappets used in gasoline engines are fabricated cylindrical or cast webbed cylindrical type of furnace hardened iron. Desired tappet-to-bore clearance for fabricated tappets is 0.0006-0.0021 inch (0.015-0.053 mm). Desired clearance for cast tappets is 0.0005-0.0023 inch (0.013-0.058 mm). Tappet bore diameter is 0.990-0.991 inch (25.146-25.171 mm). Tappets may be removed from above with magnet after removing cylinder head as outlined in paragraph 74.

Tappet height is 2.25 inches (5.715 cm) and length of push rod is 10.675 inches (27.14 cm) for all non-diesel engines.

Diesel Engines

81. On early production Model 5000 diesel engines, barrel type tappets were used; late production Model 5000 and all

other models, engines are equipped with semi-mushroom type tappets. Refer to Fig. 92. Tappet bore diameter in cylinder block is 0.990-0.991 inch (25.146-25.171 mm). Tappet diameter is 0.9889-0.9894 inch (25.118-25.130 mm); desired tappet-to-bore clearance is 0.0006-0.0021 inch (0.015-0.053 mm).

The early barrel type tappets can be removed from above with a magnet after removing cylinder head as outlined in paragraph 74. To remove late semi-mushroom type tappets, first remove camshaft as outlined in paragraph 90.

Lobe width on camshaft used with early barrel type tappets is 0.545-0.575 inch (13.84-14.61 mm); if renewing early camshaft with later type having lobe width of 0.825-0.855 inch (20.96-21.72 mm), late semi-mushroom type tappets and longer push rods should also be installed. Early type barrel tappets require push rods of 10.63-10.67 inches (27.0-27.1 cm) in length; push rods used with late semi-mushroom type tappets are 12.15-12.19 inches (30.86-30.96 cm) long.

ROCKER ARMS

All Models

82. To remove rocker arms, lift hood and swing battery tray out, if so equipped. Remove rocker arm cover and unscrew, but do not remove, the five cylinder head bolts that retain rocker arm assembly to cylinder head. Lift out rocker arm assembly and head bolts as a unit.

To disassemble, withdraw cylinder head bolts. Rocker arm-to-shaft clearance should be 0.002-0.004 inch (0.05-0.10 mm). Shaft diameter is 1.000-1.001 inch (25.40-25.43 mm); rocker arm inside diameter is 1.003-1.004 inch (25.48-25.50 mm). Renew rocker arm if clearance is excessive or if valve contact pad is worn more than 0.002 inch (0.05 mm).

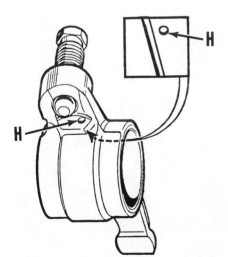

Fig. 93 — Drawing of exhaust rocker arm used on turbocharged engines. Note location of oil holes (H).

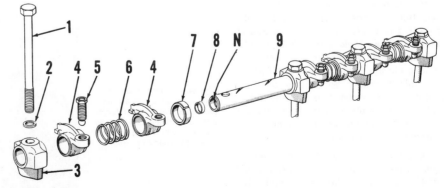

Fig. 94 — Exploded view of rocker arm assembly. Cylinder head bolts (1) are used to retain rocker arm supports. Notch (N) in end of shaft must be installed upward and to front of engine. Refer to note in text concerning washers (2).

N. Notch
1. Cylinder head bolts
2. Flat washers
3. Rocker arm supports
4. Rocker arms
5. Adjusting screws
6. Springs
7. Spacers
8. Shaft end plugs
9. Rocker arm shaft

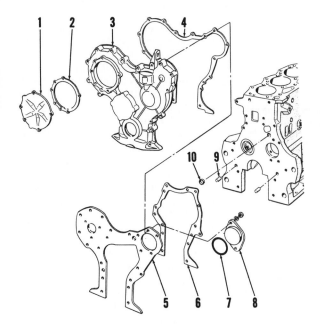

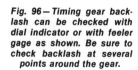

Fig. 95 — View of timing gear cover and engine front plate installation. On non-diesel engines, governor assembly is used instead of plate (1). On models without power steering, plate (8) covers hole in engine front plate (5).

1. Cover plate
2. Gasket
3. Timing gear cover
4. Gasket
5. Engine front plate
6. Gasket
7. "O" ring
8. Cover plate
9. Dowel pins
10. Oil gallery plug

Torque required to turn valve adjustment screw in rocker arm should be 9 to 26 ft.-lbs. (12-35 N·m); renew rocker arm and/or screw if torque required to turn screw is less than 9 ft.-lbs. (12 N·m).

NOTE: Rocker arms are not interchangeable on turbocharged engines. Oil flow is restricted on exhaust rocker arms. Note location of oil holes in rocker arm shown in Fig. 93.

When reassembling, be sure notch (N – Fig. 94) in one end of rocker arm shaft is up and towards front end of engine; this will correctly position rocker arm oiling holes. Back each rocker arm adjusting screw out two turns, then tighten all retaining bolts evenly until valve springs are compressed and rocker arm supports are snug against cylinder head; then, tighten all bolts to a torque of 110 ft.-lbs. (150 N·m).

NOTE: A few engines were built with two thin hardened washers under valve rocker arm shaft support bolt heads instead of the single 0.16-0.17 inch (4.06-4.31 mm) thick washer (2 – Fig. 94) which is listed in the Ford Tractor Parts Catalog for this location. When one of these engines is serviced, it is very important when reassembling, that either the correct washer is installed or both the thinner washers are reinstalled.

After rocker arm shaft is installed, adjust valve clearance, cold, to 0.017 inch (0.43 mm) on intake valves and 0.021 inch (0.53 mm) on exhaust valves. After engine is assembled, start and bring to normal operating temperature and adjust valve clearance as outlined in paragraph 78.

Reinstall rocker arm cover with new gasket and tighten retaining bolts to a torque of 13 ft.-lbs. (18 N·m).

R&R TIMING GEAR COVER

All Models Except 6700-6710-7700-7710

83. To remove timing gear cover, remove hood and grille, drain cooling system and disconnect radiator and air cleaner hoses. Disconnect battery ground cable and wiring to headlights. On Models 5100 and 7100 without power steering, disconnect drag link from spindle arm. On Models 5100, 5600, 5610, 6600, 6610, 7100, 7600 and 7610 with power steering, disconnect drag link from steering gear arm and the power steering tubes to steering cylinder. On

Fig. 96 — Timing gear backlash can be checked with dial indicator or with feeler gage as shown. Be sure to check backlash at several points around the gear.

Models 5200, 6600C, 7200 and 7600C, disconnect hydraulic tubes leading to steering motor. Support tractor under front end of transmission. Depending on type of assembly, detach front axle and support as a unit or as separate components from tractor. Remove radiator assembly. If used, note location of spacer shims between support and engine oil pan.

With front axle and front support removed from tractor proceed as follows: Remove fan belt and generator or alternator front mounting bolt. Drain and remove oil pan. Remove cap screw and washers from front end of crankshaft, then using a suitable puller and shaft protector, remove crankshaft pulley. Unbolt and remove timing gear cover.

With timing gear cover removed, crankshaft front oil seal and dust seal can be renewed as outlined in paragraph 100.

To reinstall timing gear cover, reverse removal procedure. Tighten timing gear cover retaining cap screws to a torque of 13-18 ft.-lbs. (18-24 N·m). Tighten crankshaft pulley retaining cap screw to a torque of 165 ft.-lbs. (224 N·m). Install oil pan as outlined in paragraph 105.

With timing gear cover reinstalled, reinstall front support and front axle assemblies by reversing removal procedure and observing the following: Reinstall front support to oil pan bolts with same spacer shims as were removed. However, if shims were lost or if engine block and/or oil pan were renewed, select new shims as follows: Tighten front support to engine bolts to a torque of 280 ft.-lbs. (381 N·m) and measure clearance between front support and oil pan at bolting points. Select shim thickness equal to clearance, loosen support to engine bolts and install support to oil pan bolts with spacer

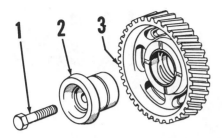

Fig. 97—Camshaft drive gear and adapter shaft as removed from front end of cylinder block. The bushing for gear (3) is not available separately.

1. Retainer bolt
2. Adapter shaft
3. Camshaft drive gear

shims. Then, tighten all front support retaining bolts to a torque of 280 ft.-lbs. (381 N·m). Refer to paragraph 13, 31 or 40 on models where power steering lines were disconnected.

Models 6700-6710-7700-7710

84. To remove timing gear cover, disconnect battery and drain cooling system. If so equipped, disconnect air conditioning hoses. Remove muffler, hood and grille panels. Disconnect interfering wiring. Disconnect oil cooler, radiator and air cleaner hoses. Remove fuel tank and frame rails for hood. Disconnect power steering tubes from power steering cylinder manifold. Support tractor under front end of transmission. Depending on type of assembly, detach front axle and support as a unit or as separate components from tractor. Remove radiator assembly. If used, note location of spacer shims between support and engine oil pan.

With front axle assembly separated from tractor, proceed as follows: Remove fan, shroud and fan belt. Drain and remove oil pan as outlined in paragraph 105. Remove cap screw and washers from crankshaft, then using a suitable puller and shaft protector, remove crankshaft pulley. Unbolt and remove timing gear cover.

With timing gear cover removed, crankshaft front oil seal and dust seal can be renewed as outlined in paragraph 100.

To reinstall timing gear cover, reverse removal procedure. Tighten timing gear cover retaining cap screws to 13-18 ft.-lbs. (18-24 N·m). Tighten crankshaft pulley retaining cap screw to 165 ft.-lbs. (224 N·m). Install oil pan as outlined in paragraph 105.

Install front axle assembly by reversing separation procedure. If shims between front support and oil pan were lost or if engine block or oil pan was renewed, refer to paragraph 50 to determine required shim sizes. Refer to paragraph 51 or 67 to fill and bleed power steering system.

TIMING GEARS

All Models

85. Before removing any gears in timing gear train, first remove rocker arm assembly as covered in paragraph 82, to avoid the possibility of damage to piston or valve train if either camshaft or crankshaft should be turned independently of the other.

The timing gear train consists of the crankshaft gear, camshaft gear, injection pump drive gear or distributor drive gear and a camshaft drive gear (idler gear) connecting the other three gears of the train. Refer to Fig. 98.

Timing gear backlash between crankshaft gear and camshaft drive gear, or between camshaft drive gear and cam-

shaft gear should be 0.001-0.006 inch (0.025-0.150 mm) on gasoline engines and 0.001-0.009 inch (0.025-0.230 mm) on diesel engines. Backlash between camshaft drive gear and fuel injection pump drive gear or distributor drive gear should be 0.001-0.012 inch (0.025-0.300 mm). If backlash is not within recommended limits, renew camshaft drive gear, camshaft drive gear shaft and/or any other gears concerned.

86. CAMSHAFT DRIVE (IDLER) GEAR AND SHAFT. To remove, unscrew self-locking cap screw and remove camshaft drive gear and shaft (adapter) from front face of cylinder block. Renew shaft and/or gear if bushing to shaft clearance is excessive, or if bearing sur-

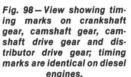

Fig. 98—View showing timing marks on crankshaft gear, camshaft gear, camshaft drive gear and distributor drive gear; timing marks are identical on diesel engines.

Fig. 99—Model 5600, 5610, 6600 and 6700 diesel engines, equipped with distributor-type injection pump are fitted with the same pump drive gear as used on three-cylinder models. Align numeral 4 on gear with camshaft drive gear timing mark when timing injection pumps on these four-cylinder engines.

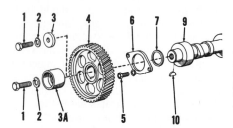

Fig. 100—Drawing showing assembly of cam-shaft gear to camshaft. Flat washer (3) is used on diesel engines; fuel pump eccentric (3A) is used on gasoline engines.

1. Cap screws	5. Cap screws (2)
2. Lockwasher	6. Thrust plate
3. Flat washer (diesel)	7. Spacer
3A. Eccentric shaft	9. Camshaft
4. Camshaft gear	10. Woodruff key

faces are scored. Inspect gear teeth for wear or score marks; small burrs can be removed with fine carborundum stone.

To reinstall camshaft drive gear, turn crankshaft so number 1 piston is at top dead center on compression stroke and turn camshaft and fuel injection pump drive gear or distributor drive gear so timing marks point to center of cam-shaft drive gear location. Place cam-shaft drive gear within the other three gears so all timing marks are aligned as shown in Fig. 98; also see Fig. 99. Install adapter (shaft) and tighten self-locking cap screw to a torque of 100-105 ft.-lbs. (136-143 N·m).

87. CAMSHAFT GEAR. To remove camshaft gear, remove cap screw (1–Fig. 100), lockwasher (2) and washer (3) on diesel models or fuel pump eccentric (3A) on gasoline models; then, pull gear from shaft. With gear removed, inspect drive key (10), thrust plate (6) and spacer (7) and renew if damaged in any way.

To reinstall gear, first install spacer, thrust plate and drive key, then install gear, washer or gasoline fuel pump eccentric, lockwasher and gear retaining cap screw. Tighten cap screw to a torque of 40-45 ft.-lbs. (54-61 N·m).

88. CRANKSHAFT GEAR. If not removed with timing gear cover and seal assembly, slide spacer (5–Fig. 101) from crankshaft; then using remover-replacer (Fig. 102) (Nuday tool SW-501) and insert (Nuday tool SW501-1) or equivalent tool, pull gear from crank-shaft. Inspect gear and crankshaft pulley drive key (3–Fig. 101) and renew if damaged in any way.

To reinstall gear, first drive key (3) into crankshaft keyway until fully seated, then install gear with timing mark outward using remover-replacer tools as used in removal procedure, or push gear onto shaft with bolt threaded into front end of crankshaft and using nut, large washer and a sleeve.

Fig. 101—Crankshaft gear installation is shown for all models. Dust seal (6) and oil seal (7) are pressed into timing gear cover from inside and ride on spacer (5) on late production tractors; on early units, dust seal and oil seal ride on crankshaft pulley hub.

1. Cap screw	5. Spacer
2. Flat washer	6. Dust seal
3. Drive key	7. Oil seal
4. Crankshaft pulley	8. Oil slinger
	9. Crankshaft gear
	12. Crankshaft

89. FUEL INJECTION PUMP OR DISTRIBUTOR DRIVE GEAR. For diesel models, refer to paragraph 140. On gasoline models, refer to paragraph 146 for information concerning distributor drive unit and governor.

CAMSHAFT AND BEARINGS

All Models

90. To remove camshaft, first remove engine as outlined in paragraph 72 and timing gear cover as outlined in paragraph 83 or 84. On gasoline engines and Model 5000 diesel engines with early barrel type tappets, remove cylinder head as in paragraph 74, then lift tappets (cam followers) from openings in top of cylinder block with a magnet. On Model 5000 diesel engines with late semi-mushroom type tappets and on all other models, remove rocker arm assembly and push rods, then invert engine assembly to allow cam followers to fall away from camshaft.

NOTE: On Model 5000 diesel engines, measure push rods before removing cylinder head; if push rods are approximately 10½ inches (26.7 cm) long, engine is equipped with early barrel type tappets. If diesel engine push rods are approximately 12-5/32 inches (30.8 cm) long, engine is equipped with semi-mushroom type tappets.

Remove clutch, flywheel and engine rear plate. Remove hydraulic pump drive gear cover (models so equipped) from rear end of cylinder block and push the camshaft rear cover plate from cylinder block with punch as shown in Fig. 104. On all other models, pry cover from rear of cylinder block. Remove the oil filter and oil pump drive gear (Fig. 122).

NOTE: If engine is inverted, floating shaft will be removed along with drive gear.

Using a feeler gage or dial indicator, measure camshaft end play. Desired end play is 0.001-0.007 inch (0.025-0.178 mm). If end play is excessive, renew thrust plate during reassembly. On models equipped with an engine mounted hydraulic pump, remove cap screw (1–Fig. 103), lockwasher (2) and flat washer (3) (diesel models) or fuel pump eccentric (3A) (gasoline models) and pull camshaft gear from shaft. Remove Woodruff key (10), thrust plate (6) and spacer (7). Withdraw camshaft and hydraulic pump drive gear from rear of cylinder block, then remove pump drive gear if necessary.

On all other models, work through holes in camshaft gear, remove thrust plate cap screws and pull camshaft and gear assembly from front of cylinder block.

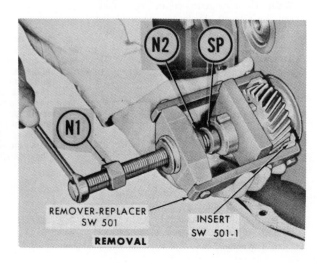

Fig. 102—Using special remover-replacer (Nuday Tool SW-501) to remove crank-shaft gear. To install gear, nut (N2) and shaft protector (SP) are removed, threaded shaft is screwed into end of crankshaft and nut (N1) is turned to push gear onto shaft.

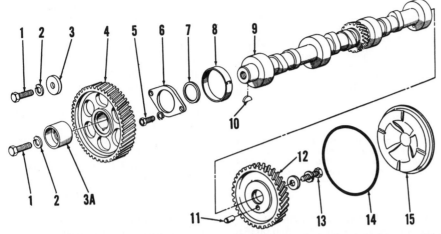

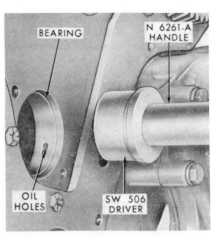

Fig. 105 — View showing special tool for removing and installing camshaft bearings. Be sure that bearings are installed with oil holes aligned with oil holes in cylinder block.

Fig. 103 — Exploded view drawing of camshaft, camshaft gear and hydraulic pump drive gear. End play is controlled by thrust plate (6). Renewable camshaft bearings (8) are used in cylinder block for all camshaft journals. Washer (3) is used on diesel models; fuel pump eccentric (3A) is used on gasoline models. Items 11 through 13 are only used on models equipped with an engine mounted hydraulic pump. Item 14 is used on Model 5000 only.

1. Cap screw	5. Cap screws (2)	9. Camshaft
2. Lockwasher	6. Thrust plate	10. Woodruff key
3. Flat washer	7. Spacer	11. Dowel pin
3A. Fuel pump eccentric	8. Camshaft bearing	12. Hydraulic lift pump
4. Camshaft gear	(4 or 5)	drive gear
		13. Cap screw
		14. "O" ring
		15. Hydraulic pump drive
		gear cover

The camshaft is supported in five bearings. Check camshaft and bearings against the following values:

Camshaft journal
diameter . . . . 2.3895-2.3905 inch
(60.69-60.72 mm)
Desired journal to bearing
clearance . . . . 0.001-0.003 inch
(0.025-0.076 mm)
Camshaft end play . . . . 0.001-0.007 inch
(0.025-0.178 mm)

If excessive bearing wear is indicated, bearings can be removed and new bearings installed with bearing driver (Nuday tool SW-506) and handle (Nuday tool N6261-A) or equivalent tools. It will be necessary to invert engine and remove oil pan to remove and install bearings. Pay particular attention that oil holes in bearings are aligned with oil passages in cylinder block. New bearings are pre-sized and should not require reaming if carefully installed.

Reinstall camshaft by reversing removal procedure. Tighten hydraulic pump drive gear retaining cap screw to a torque of 40-45 ft.-lbs. (54-61 N·m), (5000 models), tighten thrust plate cap screws to a torque of 12-15 ft.-lbs. (16-20 N·m) and tighten camshaft gear retaining cap screw to a torque of 40-45 ft.-lbs. (54-61 N·m). Place hydraulic pump drive gear cover "O" ring (models prior to 8-67) in groove in cylinder block, lubricate "O" ring and push cover into place as shown in Fig. 106. On later model engines, use a suitable sealant around cover plate.

NOTE: Be sure camshaft gear is timed as shown in Fig. 98 and Fig. 99.

CONNECTING ROD AND PISTON UNITS

All Models

91. Connecting rod and piston units are removed from above after removing

cylinder head and oil pan. Be sure to remove top ridge from cylinder bores before attempting to withdraw the assemblies.

Connecting rod and bearing cap are numbered to correspond to their respective cylinder bores. When renewing connecting rod, be sure to stamp cylinder number on new rod and cap.

When reassembling, it is important that identification number or notch in top face of piston is towards front end of engine. It is standard practice to assemble connecting rod to piston with cylinder numbers to right side of engine (away from camshaft); however, rod is symmetrical and rod can be installed with numbers in either direction (away from or towards camshaft) without affecting performance or durability of engine.

When installing connecting rod cap, be sure that bearing liner tangs, and

Fig. 104 — Camshaft rear cover plate can be removed with a punch after removing flywheel, rear engine plate and hydraulic pump flange cover.

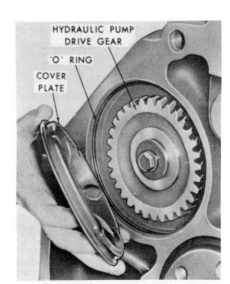

Fig. 106 — Installing hydraulic pump drive gear cover plate (15 — Fig. 103).

cylinder identification number, of rod and cap are towards same side of engine. Tighten connecting rod nuts to a torque of 60-65 ft.-lbs. (82-88 N·m).

PISTON RINGS

Non-Diesel Engines

92. Each cam ground piston is fitted with two compression rings and one oil control ring. Top compression ring and oil control ring are chrome plated. Top compression ring is barrel face type and must be installed with identification mark up. Second compression ring has straight face with inner step and must be installed with step on inside circumference up. The oil control ring and oil ring expander may be installed either side up.

Piston ring sets are available in oversizes of 0.020, 0.030 and 0.040 inch (0.508, 0.762 and 1.016 mm) as well as standard size. The standard size rings are to be used with both grades of standard size pistons and with 0.004-inch (0.102 mm) oversize pistons. Refer to the following specifications for checking piston ring fit:

Ring End Gap:

Top compression ring . . . 0.015-0.030 in.
(0.38-0.76 mm)
Second compression
ring . . . . 0.013-0.028 in.
(0.33-0.71 mm)
Oil control ring . . . . 0.015-0.038 in.
(0.38-0.97 mm)

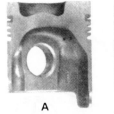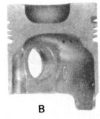

A   B

Fig. 108—Comparative view of later type piston crown (B). Note that combustion chamber has a flat bottom instead of previous toroidal design (A). Later type piston (B) is only piston available for service.

Ring Side Clearance in Groove:

Top compression ring . 0.0029-0.0046 in.
(0.074-0.117 mm)
Second compression
ring . . . . 0.0025-0.0045 in.
(0.064-0.114 mm)
Oil control ring . . . . . . 0.0014-0.0031 in.
(0.036-0.079 mm)

Diesel Engines

93. Pistons are fitted with three compression rings and one oil control ring. The two top compression rings and oil control ring are chrome plated. Top compression ring on 5000 models is barrel face type and must be installed with identification mark up. On all other engines except turbocharged engines, top compression ring may have beveled edge on the inside diameter that must face upward on assembly; if top compression ring is not equipped with a beveled edge, either side may be installed up. Second compression ring has a step on the inside diameter that must face upward on assembly. Third compression ring has either a step or beveled edge on the inside diameter that must face upward on assembly. Inspect surface of second and third compression rings for a punched dot or the word "TOP", marking must face upward on assembly.

The oil control ring and expander may be installed with either side up.

On turbocharged engines, top compression ring is keystone design and may be installed either side up. Second compression ring has a step on the in-

side diameter that must face upward on assembly. Third compression ring has a step on the outside diameter that must face downward on assembly. Inspect surface of second and third compression rings for a punched dot or the word "TOP", marking must face upward on assembly.

The oil control ring and expander may be installed with either side up.

Piston ring sets are available in oversizes of 0.020, 0.030 and 0.040 inch (0.508, 0.762 and 1.016 mm) as well as standard size. The standard size rings are to be used with both grades of standard size pistons and also with 0.004-inch (0.102 mm) oversize pistons. Refer to the following specifications for checking piston ring fit:

Ring End Gap:

Top compression ring . . . 0.015-0.030 in.
(0.38-0.76 mm)
Second & third com-
pression rings . . . . 0.013-0.028 in.
(0.33-0.71 mm)
Oil control ring . . . . 0.015-0.038 in.
(0.38-0.97 mm)

Ring Side Clearance in Groove:

Top compression ring . 0.0044-0.0061 in.
(0.112-0.155 mm)
Second & third com-
pression rings . . . . . . 0.0039-0.0056 in.
(0.099-0.142 mm)
Oil control ring . . . . . . 0.0024-0.0041 in.
(0.061-0.104 mm)

PISTONS AND CYLINDERS

All Models

94. Non-diesel engines have cam ground aluminum alloy pistons with a cast iron insert containing the top ring groove. Diesel engines have trunk type aluminum alloy pistons with a continuous skirt and a Ni-Resist insert containing the top ring groove. Early design pistons were straight in the upper land area; whereas, later production and service pistons are tapered in the upper land areas.

Late model tractors are fitted with a piston having a change in design of its combustion chamber. See Fig. 108.

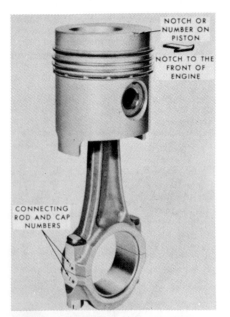

Fig. 107—View showing recommended assembly of piston to connecting rod; however, connecting rod can be installed on crankshaft with numbers facing either way. Piston must be installed with notch or number to the front of engine.

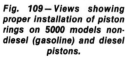

Fig. 109—Views showing proper installation of piston rings on 5000 models non-diesel (gasoline) and diesel pistons.

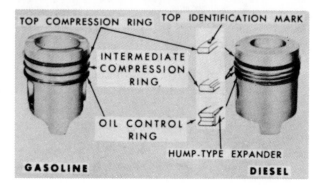

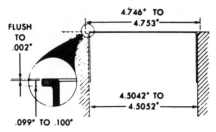

Fig. 110 — Top of installed sleeve flange should be flush to 0.002 inch (0.050 mm) above gasket surface of block as shown.

These new pistons do not interchange with those used in Models 5000 or 7000.

Cylinder bores in engine block are normally unsleeved; however, repair sleeves have been factory installed in engines with 4.2-inch (106.68 mm) diameter cylinders. When factory installed, all cylinders are sleeved, and block is stamped "SB" on left side oil pan flange.

Models with 4.4-inch (111.76 mm) cylinder bore have available for service a thin wall cast sleeve which is not used in production and which cannot be overbored.

Piston, pin and ring sets are available in standard size and 0.004-inch (0.102 mm) oversize for use in standard bores, and oversizes of 0.020, 0.030 and 0.040 inch (0.508, 0.762 and 1.016 mm) for use in rebored blocks.

95. INSTALLING SLEEVES. To install sleeves in unsleeved 4.2-inch (106.68 mm) bore engines, finish-bore block to an inside diameter of 4.358 inches (110.69 mm). Selectively fit sleeves to provide, as nearly as possible, the optimum 0.004-inch (0.102 mm) interference fit for all sleeves.

Thoroughly clean and dry new sleeve and block bore and carefully apply a 3-inch (7.62 cm) band of "Loctite" Sleeve Retainer to top of bore. Install sleeve, outside chamfered edge down and press into position until top of sleeve is flush to 0.001 inch (0.0254 mm) above block surface. Allow block to cure overnight before completing assembly. Reexamine installation and mill block if necessary, to provide correct sleeve height.

NOTE: DO NOT remove more than 0.005 inch (0.127 mm) from top surface.

Finish-bore and hone sleeve after milling, to 4.200-4.2024 inches (106.68-106.74 mm) for standard piston or to appropriate oversize if oversize piston is installed. Refer also to paragraph 96 for additional data on piston and ring installation.

The flanged, thin-wall cast sleeves used in 4.4-inch (111.76 mm) diameter bore engines can be used only with standard size or 0.004-inch (0.102 mm) oversize pistons, and cannot be rebored. Thin-wall sleeves are a light press fit in engine block and should be installed as follows:

First make sure that sleeve and block bore are clean and dry. Chill sleeves for 15 minutes in dry ice and push sleeve as far as possible into block bore. Seat sleeve, if necessary, using a puller and suitable installing plate. Top of installed sleeve flange should be flush to 0.002 inch (0.051 mm) above gasket surface of block as shown in Fig. 110. Machining dimensions for installing sleeves in unbored block are shown in Fig. 110.

96. FITTING PISTONS. Recommended method for fitting pistons is as follows: Before checking piston fit, deglaze cylinder wall using a hone or deglazing tool. Using a micrometer, measure piston diameter to centerline of and at right angle to piston pin bore. Then, using an inside micrometer, measure cylinder bore diameter at a distance of 2¾ inches (6.9 cm) from top of cylinder block crosswise with block. Subtract piston diameter from cylinder bore diameter to obtain piston-to-cylinder bore clearance. Clearance should be within the following specification for proper piston fit:

Non-Diesel Engines:
4.2-inch bore	0.0027-0.0037 in.
(106.68 mm)	(0.069-0.094 mm)
4.4-inch bore	0.0032-0.0042 in.
(111.76 mm)	(0.081-0.107 mm)

Diesel Engines:
4.2-inch bore	0.0075-0.0085 in.
(106.68 mm)	(0.191-0.216 mm)
4.4-inch bore	0.0080-0.0090 in.
(111.76 mm)	(0.203-0.229 mm)

NOTE: Two grades of standard pistons are available. Grade "D" standard size pistons (color coded blue) are larger in diameter than grade "B" standard size piston (color coded red).

If grade "D" piston fits too loosely in standard cylinder bore, try a 0.004-inch (0.102 mm) oversize piston. If not possible to fit the 0.004-inch (0.102 mm) oversize piston by honing, re-bore cylinder to 0.020-inch (0.508 mm).

NOTE: After honing or deglazing cylinder bore, wash bore thoroughly with hot water and detergent until a white cloth can be rubbed against cylinder wall without smudging, then rinse with cold water, dry thoroughly and oil to prevent rusting.

PISTON PINS

All Models

97. Diameter of piston pin is 1.4997-1.5000 inches (38.092-38.100 mm) for 5000, 5600, 5610, 6600, 6610, 6700 and 6710 models and 1.6247-1.6250 inches (41.267-41.275 mm) for the 7000, 7600, 7610, 7700 and 7710 models. The floating type piston pins are retained in piston pin bosses by snap rings and are available in standard size only. Piston pin should have a clearance of 0.0005-0.0007 inch (0.0127-0.0178 mm) in connecting rod bushing and a clearance of 0.0003-0.0005 inch (0.0076-0.0127 mm) in piston bosses. After installing new piston pin bushings in connecting rods, oil hole in bushing must be drilled as shown in Fig. 111 and bushings final sized with a spiral expansion reamer to obtain specified pin to bushing clearance. When assembling, identification notch or number in top face of piston must be to front end of engine and identification number on rod and cap towards right side of engine (away from camshaft); however, rod is symmetrical and can be installed either way without affecting engine performance or durability.

CONNECTING RODS AND BEARINGS

All Models

98. Connecting rod bearings are precision type, renewable from below after removing oil pan and connecting rod bearing caps. When removing bearing

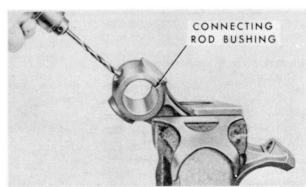

Fig. 111 — Oil hole in connecting rod bushing must be drilled after bushing is installed, but before reaming or honing bushing to fit pin.

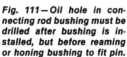

CONNECTING ROD BUSHING

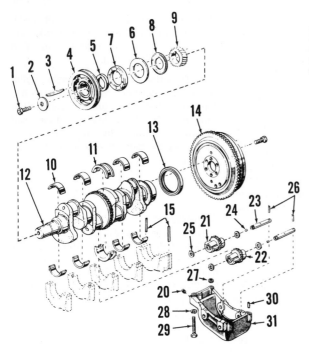

1. Cap screw
2. Flat washer
3. Drive key
4. Crankshaft pulley
5. Spacer
6. Dust seal
7. Oil seal
8. Oil slinger
9. Crankshaft gear
10. Main bearing liners
11. Thrust bearing liners
12. Crankshaft
13. Rear oil seal
14. Flywheel
15. Rear bearing cap seals
20. Pipe plug
21. Drive gear assy.
22. Driven gear assy.
23. Shafts
24. Tapered plugs
25. Thrust washers
26. Spring pins
27. Gasket
28. Lockwashers (4)
29. Cap screws (4)
30. Dowel pins (2)
31. Support assy.

caps, note which way cylinder identification numbers are placed on each rod assembly. It is standard practice that numbers face away from camshaft side of engine; however, connecting rods are symmetrical and may be placed with identification numbers to either side of engine. It is very important that pistons be installed in cylinder bore with identification number or notch in top face of piston towards front end of engine; therefore, do not attempt to realign rod identification numbers if they are to camshaft side of engine.

Crankpin bearing liners may be of two different materials, copper-lead or aluminum-tin alloy. The bearings will have an identification marking as follows:

Copper-lead . . . . .PV or G
Aluminum-tin . . . . .G and AL

Standard size bearing liners of each material are available in two different thicknesses are color-coded to indicate thickness as follows:

Copper-Lead Bearing Thickness:
Red . . . . .0.0943-0.0948 in.
(2.3952-2.4079 mm)
Blue . . . . .0.0947-0.0952 in.
(2.4054-2.4181 mm)

Aluminum-Tin Alloy Bearing Thickness:
Red . . . . .0.0941-0.0946 in.
(2.390-2.403 mm)
Blue . . . . .0.0945-0.0950 in.
(2.400-2.413 mm)

In production, connecting rods and crankshaft crankpin journals are color-coded to indicate bore and journal diameters as follows:

Connecting Rod Bore Diameter:
Red . . . . .2.9412-2.9416 in.
(74.707-74.717 mm)
Blue . . . . .2.9416-2.9420 in.
(74.717-74.727 mm)

Crankpin Journal Diameter:
Red . . . . .2.7500-2.7504 in.
(69.85-69.86 mm)
Blue . . . . .2.7496-2.7500 in.
(69.84-69.85 mm)

When installing a new crankshaft and color-code marks are visible on connecting rods, crankpin bearing liners may be fitted as follows: If color-code markings on both rod and crankshaft crankpin journal are red, install two red bearing liners; if both color-code markings are blue, install two blue coded bearing liners. If color-code marks on rod and crankpin do not match (one is red and the other is blue), install one red and one blue bearing liner.

NOTE: Be sure both bearing liners are of the same material; that is, either both are copper-lead or both are aluminum-tin alloy.

If color code mark is not visible on connecting rod or crankshaft, bearing fit should be checked for proper clearance according to bearing material as follows:

Crankpin Journal To Bearing Liner Clearance:
Copper-lead
bearings . . . . 0.0017-0.0038 in.
(0.0432-0.0965 mm)
Aluminum-tin
bearings . . . . 0.0021-0.0042 in.
(0.053-0.107 mm)

As well as being available in either red-coded or blue-coded standard size, bearing liners are also available in undersizes of 0.002, 0.010, 0.020, 0.030 and 0.040 inch (0.051, 0.254, 0.508, 0.762 and 1.016 mm). When installing undersize crankpin bearing liners, crankpin must be reground to one of the following exact undersizes:

Bearing Undersize	Crankpin Journal Diameter
0.002-in. (0.051 mm) . .	2.7476-2.7480 in. (69.789-69.799 mm)
0.010-in. (0.254 mm) . .	2.7400-2.7404 in. (69.590-69.606 mm)
0.020-in. (0.508 mm) . .	2.7300-2.7304 in. (69.342-69.352 mm)
0.030-in. (0.762 mm) . .	2.7200-2.7204 in. (69.088-69.098 mm)
0.040-in. (1.016 mm) . .	2.7100-2.7104 in. (68.834-68.844 mm)

NOTE: When regrinding crankpin journals, maintain a 0.12-0.14 inch (3.05-3.56 mm) fillet radius and chamfer oil hole after journal is ground to size.

When reassembling, tighten connecting rod nuts to a torque of 60-65 ft.-lbs. (82-88 N·m).

CRANKSHAFT AND MAIN BEARINGS

All Models
99. Crankshaft is supported in five main bearings. Crankshaft end thrust is controlled by flanged main bearing liner which is used on third journal. Before removing main bearing caps, check to see that they have an identification number so they can be installed in same position from which they are removed.

Main bearing liners may be of two different materials, copper-lead or aluminum-tin alloy. The bearings will have an identification marking to indicate bearing material as follows:

Copper-lead . . . . .PV or G
Aluminum-tin alloy . . . . .G and AL

Standard size bearing liners are available in two different thicknesses and are color-coded to indicate thickness as follows:

Red . . . . .0.1245-0.1250 in.
(3.1623-3.1750 mm)
Blue . . . . .0.1249-0.1254 in.
(3.1724-3.1861 mm)

In production, main bearing bores in block and main bearing journals on crankshaft are color-coded to indicate journal diameter as follows:

Main journal diameter:
Red . . . . .3.3718-3.3723 in.
(85.643-85.656 mm)
Blue . . . . .3.3713-3.3718 in.
(85.631-85.643 mm)

When installing a new crankshaft and color-code marks are visible in crankcase at main bearing bores, new main bearing liners may be fit as follows: If color-code marks on bore and journal are both red, install two red coded bearing liners; if both marks are blue, install two blue coded bearing liners. If color-code mark on bore is not the same as color code on journal (one is blue and the other is red), install one red coded bearing liner and one blue coded bearing liner.

NOTE: Be sure both liners used at one journal are of the same material; however, copper-lead bearing liners may be used on one or more journals with aluminum-tin alloy liners on the remaining journals.

If color-code marks are not visible at main bearing bores in block, check bearing fit and install red or blue, or one red and one blue line to obtain a bearing journal to liner clearance of 0.0022-0.0045 inch (0.056-0.114 mm).

NOTE: Recommended clearance is for either copper-lead or aluminum-tin alloy bearing material.

As well as being available in either red coded or blue coded standard size, new main bearing liners are also available in undersizes of 0.002, 0.010, 0.020, 0.030 and 0.040 inch (0.051, 0.254, 0.508, 0.762 and 1.016 mm). When installing undersize main bearing liners, crankshaft journals must be reground to one of the following exact undersizes:

Bearing Undersize	Main Journal Diameter
0.002-in.	3.3693-3.3698 in.
(0.051 mm)	(85.581-85.592 mm)
0.010-in.	3.3618-3.3623 in.
(0.254 mm)	(85.390-85.402 mm)
0.020-in.	3.3518-3.3523 in.
(0.508 mm)	(85.136-85.148 mm)
0.030-in.	3.3418-3.3423 in.
(0.762 mm)	(84.882-84.894 mm)
0.040-in.	3.3318-3.3323 in.
(1.016 mm)	(84.628-84.640 mm)

NOTE: When regrinding crankshaft main bearing journals, maintain a fillet radius of 0.12-0.14 inch (3.05-3.56 mm) and chamfer oil holes after journal is ground to size.

When reinstalling main bearing caps, proceed as follows: Be sure bearing bores and rear main bearing oil seal area are thoroughly clean before installing bearing liners. Be sure tangs on bearing inserts are in slots provided in cylinder block and bearing caps. Refer to paragraph 102 for rear cap side seals. Tighten bearing cap bolts to a torque of 115-125 ft.-lbs. (156-170 N·m).

CRANKSHAFT OIL SEALS

All Models

100. **FRONT OIL SEAL.** Crankshaft front oil seal is mounted in timing gear cover and cover must be removed to renew seal. Timing gear cover removal procedure is outlined in paragraph 83 or 84. To renew seal, drive dust seal and oil seal out towards inside of timing gear cover. Install new dust seal in timing gear cover first, then using a seal driver (OTC No. 630-16 step plate or equivalent), install new oil seal in cover with spring loaded lip towards inside of cover. Refer to Fig. 113.

On early production engines, crankshaft front seal rides on crankshaft pulley hub; later production engines are fitted with a revised crankshaft pulley and pulley spacer so oil seal rides on spacer instead of pulley hub. Carefully inspect pulley hub (early production) or pulley spacer (later production) for wear at seal contact surface and renew pulley or spacer if wear or scoring is evident. Later production crankshaft pulley and spacer can be installed on early production engine if desired.

101. **REAR OIL SEAL.** Crankshaft rear oil seal can be renewed after removing clutch assembly, flywheel and engine rear plate. Pry old seal from bore in cylinder block and rear main bearing cap and thoroughly clean crankshaft seal journal. Apply a light coat of high temperature grease to seal bore, crankshaft seal journal, lip of seal and outer circumference of seal. Install new seal with a 4⅞-inch (12.4 cm) ID sleeve so rear face of seal is 0.060 inch (1.52 mm) below flush with rear face of block. Using a dial indicator, check runout of rear face of seal; runout should not exceed 0.015 inch (0.38 mm). A special seal installation tool (Nuday tool SW 520) is available; using three flywheel bolts, press seal in with tool until flange on tool bottoms and tighten the bolts to 25 ft.-lbs. (34 N·m). See Fig. 114. Then, remove bolts and seal installation tool.

NOTE: Some early models may have a stop installed in front of seal; if so, remove and discard stop before installing new seal.

102. To install new rear main bearing cap side seals, remove cap and proceed as follows: Place side seals in grooves of cap so they extend about ½-inch (1.25 cm) from top surface of bearing cap. Apply sealing compound to top face of cap making sure chamfers on edge of cap

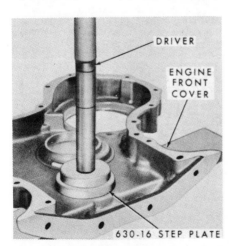

Fig. 113—Installing the crankshaft front oil seal in engine timing gear cover.

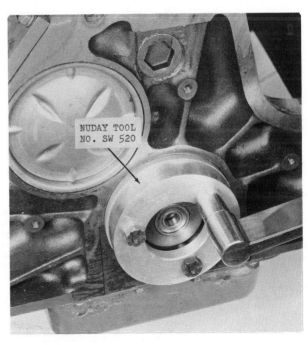

Fig. 114—Installing rear oil seal using special installation tool.

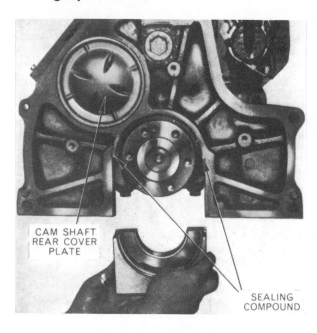

CAM SHAFT REAR COVER PLATE

SEALING COMPOUND

Fig. 115 — Installing rear main bearing cap and side seals. Refer to text for side seal instructions.

balancer gears if excessively worn or scored. Desired shaft to bushing clearance is 0.0002-0.0008 inch (0.005-0.020 mm). Shaft diameter is 0.9995-1.0000 inch (25.387-25.400 mm). When installing new bushings in balancer gears, ream as necessary to obtain specified shaft to bushing clearance. To reassemble place gears in housing with timing marks aligned as shown in Fig. 118 and with marks towards front (roll pin) end of housing. Place a new thrust washer at each side of gear and insert shafts. Check gear backlash which should be 0.0015-0.008 inch (0.038-0.203 mm); if backlash is not within limits, renew balancer gears. If backlash is okay, drive roll pins into place and reinstall balancer unit with new gasket and with crankshaft gear and balancer gear timing marks aligned as shown in Fig. 117.

NOTE: Timing mark on crankshaft gear will be visible when No. 2 piston is at top dead center.

Tighten balancer retaining cap screws to a torque of 60-70 ft.-lbs. (82-95 N·m). Check to be sure timing marks are aligned before reinstalling oil pan.

are covered. See Fig. 115. Insert and lubricate bearing liner, then install cap and tighten retaining bolts to a torque of 115-125 ft.-lbs. (156-170 N·m). Tap side seals into block so they extend about 0.005-0.015 inch (0.127-0.381 mm) from block and cap. With rear cap and side seals installed, soak side seals with penetrating oil and install rear oil seal as in paragraph 101.

ENGINE BALANCER

All Models

103. A Lanchester type engine balancer is used on all 4-cylinder diesel and non-diesel engines. The balancer is driven at twice engine speed by a gear machined on the crankshaft immediately in front of center main bearing.

To remove balancer, first remove oil pan as in paragraph 105, then unbolt and remove balancer unit from lower side of crankcase. Drive out roll pins at front side of balancer casting and remove shafts and gears. See Fig. 116. Renew shafts and/or bushings in

FLYWHEEL

All Models

104. The flywheel can be removed after splitting tractor between engine

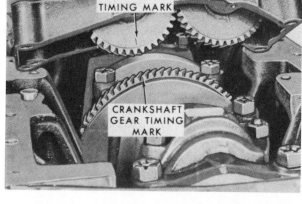

Fig. 117 — View showing timing of balancer drive gear to crankshaft gear. Refer to text for installation procedure.

DRIVE GEAR TIMING MARK

CRANKSHAFT GEAR TIMING MARK

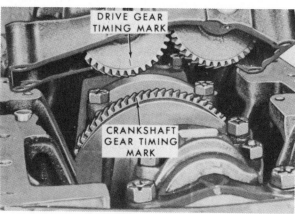

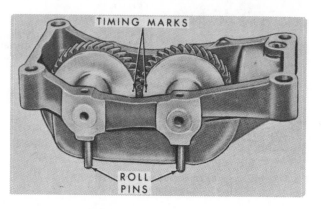

Fig. 116 — Exploded view of dynamic balancer assembly used on four-cylinder engines. Refer also to Figs. 117 and 118.

Fig. 118 — View showing timing of balancer drive gear to balancer driven gear. These timing marks are on opposite side of gear from timing mark on drive gear shown in Fig. 117.

TIMING MARKS

ROLL PINS

20. Pipe plug
21. Drive gear assy.
22. Driven gear assy.
23. Shafts
24. Tapered plugs
25. Thrust washers
26. Spring pins
27. Gasket
28. Lock washers (4)
29. Cap screws (4)
30. Dowel pins (2)

and transmission as outlined in paragraphs 164, 165, 166 and 167 under CLUTCH.

Unbolt and remove clutch from flywheel on all models equipped with 8-speed transmission and on models equipped with Dual Power. On 5000 models equipped with 10-speed Select-O-Speed transmission, remove clutch torque limiter housing from flywheel.

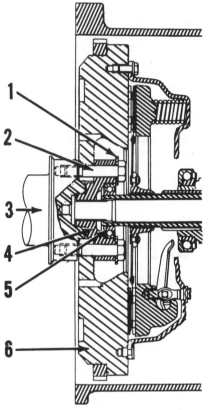

Fig. 119—On Models 5000, 5600, 6600, 6700, 7000, 7600 and 7700 with 8-speed transmission or Dual Power, two flat washers (1) are installed diametrically opposite each other on two of the bolts (2) that hold pto drive plate (4) and flywheel (6) to crankshaft (3).

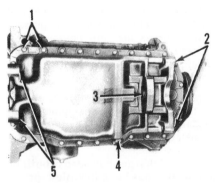

Fig. 120—Bottom view of cast iron oil pan. Engine front support attaching points require a definite shimming procedure which is outlined in appropriate front support removal paragraph.

1. Pan bolts
2. Engine support points
3. Axle support bore
4. Drain plug
5. Engine–transmission support points

Flywheel can be installed in one position only. On models with clutch, flywheel retaining cap screws also retain clutch shaft pilot bearing retainer and on models with independent power take-off, flywheel bolts retain power take-off drive adapter. When reinstalling flywheel, tighten retaining cap screws to a torque of 100-110 ft.-lbs. (136-150 N·m) on Models 5000, 5600, 6600, 6700, 7600 and 7700 and 160 ft.-lbs. (218 N·m) on all other models. On Models 5000, 5600, 6600, 6700, 7000, 7600 and 7700, install the two pilot bearing retainer washers (1–Fig. 119).

Starter ring gear is installed from front face of flywheel; therefore, flywheel must be removed to renew ring gear. Heat gear to be removed with a torch from front side of gear and knock gear off of flywheel. Heat new gear evenly until gear expands enough to slip onto flywheel. Tap gear all the way around to be sure it is properly seated and allow it to cool.

NOTE: Be sure to heat gear evenly; if any portion of gear is heated to a temperature higher than 500°F (260°C), rapid wear will result. Ring gears for diesel engines have 128 teeth; those for gasoline engines have 162 teeth.

OIL PAN (SUMP)

All Models

105. To remove oil pan, drain pan and remove oil level indicator. Disconnect oil

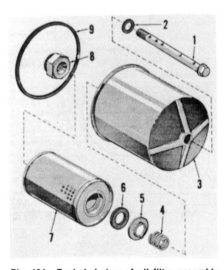

Fig. 121—Exploded view of oil filter assembly used on all models prior to 10-69 production. Note correct placement of filter retaining spring (4), washer (5) and gasket (6). A by-pass valve is incorporated in the cover retaining bolt (1). Models after 10-69 use a 1-quart spin-on type filter.

1. Bolt & valve assy.
2. Gasket
3. Filter cover
4. Spring
5. Washer
6. Gasket
7. Filter element
8. Adapter
9. Gasket

cooler lines on turbocharged models. Detach hood and side panels and on Models 6700, 6710, 7700 and 7710 detach hood frame rails from radiator shell. On Models 6700, 6710, 7700 and 7710, disconnect power steering tubes from front axle manifold. On all models place a support under front of transmission and one at a time replace front support to cylinder block bolts with bolts which are 8 inches longer. Move front axle assembly forward approximately 1½ inches, place a floor jack under oil pan, unbolt and lower pan.

When installing oil pan, tighten screws in center first and work outward. Tighten screws to 30-35 ft.-lbs. (41-48 N·m) on Models 5000 and 7000. On all other models, tighten screws to 22 ft.-lbs. (30 N·m) on stamped oil pans and 28 ft.-lbs. (38 N·m) on cast iron oil pans. Tighten the oil pan to transmission bolts to 275-340 ft.-lbs. (374-462 N·m). Tighten front support bolts to 250-280 ft.-lbs. (340-381 N·m). If a new oil pan is installed refer to paragraphs 3, 30 or 50

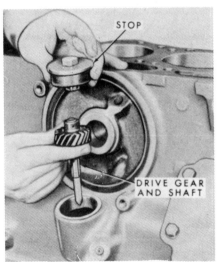

Fig. 122—Removing the oil pump drive gear and shaft assembly. Floating shaft connecting the drive gear and shaft assembly to oil pump rotor can be removed at this time. Refer to Fig. 124.

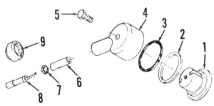

Fig. 123—Exploded view of proofmeter (tachometer) drive assembly used on late model tractors. Assembly is adaptable to all models produced after 4-71 with spin-on oil filter and counterclockwise rotating proofmeter.

1. Drive base
2. Expansion ring
3. "O" ring
4. Adapter body
5. Bolt, 5/16-24 x ¾
6. Bearing
7. Seal
8. Cable/shaft assy.
9. Cover

to determine number and thickness of front support shims.

OIL PUMP AND RELIEF VALVE

All Models

106. To remove oil pump, first remove oil pan as outlined in paragraph 105, then remove the two retaining cap screws and remove pump from cylinder block. Refer to Fig. 124 for exploded view of oil pump and oil pump drive gear

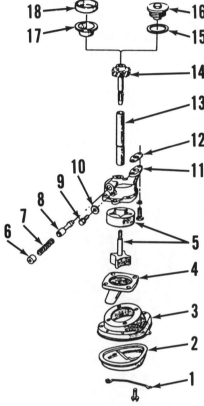

Fig. 124—Exploded view of oil pump, floating shaft and drive gear assembly used on all models. Floating shaft (13) may be removed from either below or above.

1. Retainer clip	10. Retainer washer
2. Screen	11. Pump body
3. Screen cover	12. Gasket
4. Pump cover	13. Floating shaft
5. Rotor set	14. Shaft & gear assy.
6. Plug	15. Gasket
7. Relief valve spring	16. Shaft stop
8. Relief valve	17. Shaft stop
9. Cap screw	18. Plug

assembly. The floating drive shaft (13) will usually be removed with the pump.

To disassemble pump, remove clip (1) and oil screen (2), then remove screws retaining screen cover (3) and pump cover (4) to pump body (11). Remove covers and pump rotor set (5), noting which direction outer rotor was placed in pump body. Remove retainer screw (9) and if necessary, thread a self-tapping screw into plug (6) and pull plug from pump body. Remove spring (7) and oil pressure relief valve (8).

Check pump for wear as shown in Figs. 125 or 126. Pump cover-to-rotor clearance (rotor end play) should be 0.001-0.0035 inch (0.025-0.089 mm); pump body-to-rotor clearance should be 0.006-0.011 inch (0.152-0.279 mm); and rotor clearance should be 0.001-0.006 inch (0.025-0.152 mm) when measured as shown in Fig. 126. Renew rotor set and/or pump body if clearances are excessive. Renew pump cover plate if excessively worn or scored. Relief valve spring should exert a force of 10.7 to 11.9 pounds (4.8-5.4 kg) when compressed to a length of 1.07 inches (25.57 mm). Engine oil pressure should be 60-70 psi (414-483 kPa) at 1000 engine rpm.

NOTE: On turbocharged Models 7000, 7610, 7700 and 7710 oil flow rate is in-

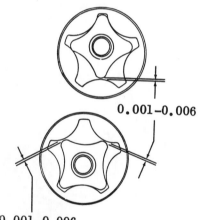

Fig. 126—Checking pump rotor-to-rotor clearance; renew rotor set if clearance exceeds 0.006 inch (0.1524 mm). Refer also to Fig. 125.

 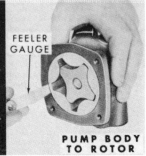

Fig. 125—Checking pump cover-to-rotor clearance and pump body-to-rotor clearance. Refer to text for specifications. Refer also to Fig. 126.

creased from 6 gpm (22.8 lpm) to 10 gpm (38.0 lpm) to insure turbocharger lubrication.

Assemble oil pump and reinstall by reversing removal and disassembly procedure. Tighten oil pump retaining cap screws to a torque of 33-38 ft.-lbs. (45-52 N·m). Prime oil pump by immersing in clean oil and turning rotor shaft prior to installing pump in engine.

SPECIAL NOTE: Tachometer Drive. Because late model tractors are equipped with alternators instead of generators, tachometer (proofmeter) drive is now taken from oil pump drive gear. See Fig. 123 for identification of tachometer drive parts. To remove this drive from engine block, loosen assembly bolt (5) and lightly tap body (4) to release. Lift entire drive free, disengaging drive cable end from square hole in oil pump drive gear. Reverse this procedure to reinstall tachometer drive assembly. Align cable with care after making sure that cable end engages oil pump drive gear properly, then tighten bolt (5) to 10-15 ft.-lbs. (14-20 N·m).

TURBOCHARGER

All 7000, 7600 and 7700 models built prior to February 1978 are equipped with Schwitzer turbochargers. All 7600 and 7700 models built after January 1978 and all 7610 and 7710 models are equipped with AiResearch turbochargers. Turbocharger is mounted directly on engine exhaust manifold. The major components of a turbocharger are: turbine, turbine housing, compressor, compressor housing, bearing and bearing housing. The turbine and compressor wheels are mounted on a common shaft which is supported by a bearing in bearing housing.

Exhaust gases drive turbine and shaft, thereby turning compressor which forces air into engine. Pressurized oil, furnished by engine oil pump, lubricates and cools turbine shaft and bearing.

NOTE: If turbocharger shaft is allowed to turn without oil being furnished by engine oil pump, unit will be damaged due to high rpm at which unit operates.

If turbocharger has been rebuilt or a new unit has been installed, prime unit as outlined in paragraph 108.

NOTE: Schwitzer type turbocharger used on 7600 and 7700 models has been redesigned for improvement in performance. Changes are as follows:

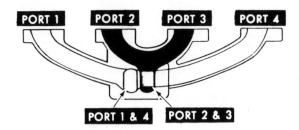

Fig. 127—View of exhaust manifold with dual ports to mate with turbocharger used on late models.

26. Worn or damaged injectors.
27. Valve timing.
28. Burned valves.
29. Worn piston rings.
30. Burned pistons.
31. Leaking oil feed line.
32. Excessive engine pre-oil.
33. Excessive engine idle.
34. Coked or sludged center housing.
35. Oil pump malfunction.
36. Oil filter plugged.

New intake and exhaust manifolds, not interchangeable with those used on Model 7000, have been installed. New exhaust manifold has twin outlet ports mated to twin entry of new turbine housing. Turbocharger mounting on exhaust manifold is changed to correspond. See paragraph 76 for intake valve and cylinder head changes. Small internal parts of turbocharger remain generally unchanged. Service procedures and assembly specifications are not affected; however, new turbocharger and manifolding provide a boost increase to 15 psi (103.5 kPa) over previous 12 psi (82.8 kPa) and peak turbine rotation speed has increased from 90,000 to 105,000 rpm.

TROUBLESHOOTING

107. The following table can be used for locating difficulty with turbocharger. Many of the probable causes are related to the engine air intake and exhaust systems which should be maintained in good condition at all times. The turbocharger should be allowed to cool by idling engine for 2-5 minutes after engine has been operated under load. Never shut down engine at high operating speed, this will cause turbocharger to continue to spin in excess of one hundred thousand rpm without lubrication and damage turbocharger bearings. Use extreme care to avoid damaging any of the turbocharger moving parts when servicing.

SYMPTOMS PROBABLE CAUSES
Engine Lacks Power 1,4,5,6,7,8,
    9,10,11,18,20,21,22,25,26,27,28,29,30
Black Exhaust Smoke 1,4,5,6,7,8,
        9,10,11,18,20,21,22,25,26,27
Blue Exhaust Smoke 1,2,4,8,9,17,
        19,21,22,29,32,33,34
Excessive Oil Consumption 2,8,19
        20,33,34
Excessive Oil in the
    Turbine End 2,7,8,16,17,19,20,
        22,32,34,36
Excessive Oil in the
    Compressor End 1,2,4,5,6,8,9,
        16,19,20,21,33,36
Insufficient Lubrication 12,15,16
        23,24,31,36

Oil In Exhaust Manifold 2,7,19,20,
        22,28,29,30,33,34
Damaged Compressor Wheel 3,6,
        8,20,21,23,24,36
Damaged Turbine Wheel 7,8,18,
        20,21,22,34,36
Drag or Bind in Rotating
    Assembly . . 3,6,7,8,13,14,15,16,20,21,
        22,31,34,36
Worn Bearings, Journals, Bearing
    Bores 6,7,8,12,13,14,15,16,20,
        23,24,31,35,36
Noisy Operation 1,3,4,5,6,7,8,9,10
        11,18,20,21,22
Sludged or Coked Center
    Housing 2,15,17

KEY TO PROBABLE CAUSES
1. Dirty air cleaner element.
2. Plugged crankcase breathers.
3. Air cleaner element missing, leaking, not sealing correctly, loose connections to turbocharger.
4. Collapsed or restricted air tube before turbocharger.
5. Restricted (damaged) crossover pipe from turbocharger to inlet manifold.
6. Foreign object between air cleaner and turbocharger.
7. Foreign object in exhaust system from engine (check engine).
8. Turbocharger flanges, clamps or bolts loose.
9. Inlet manifold cracked, gaskets missing, connections loose.
10. Exhaust manifold cracked, burned, gaskets blown or missing.
11. Restricted exhaust system.
12. Oil lag (oil delay to turbocharger at start up).
13. Insufficient lubrication.
14. Lubricating oil contaminated.
15. Improper type lubricating oil used.
16. Restricted oil feed line.
17. Restricted oil drain line.
18. Turbine housing damaged or restricted.
19. Turbocharger seal leakage.
20. Worn journal bearings.
21. Excessive dirt build-up in compressor housing.
22. Excessive carbon build-up behind turbine wheel.
23. Too fast acceleration at initial start (oil lag).
24. Too little warm-up time.
25. Fuel pump malfunction.

108. **REMOVE AND REINSTALL.** During removal be sure to cap or plug all manifold, turbocharger and oil line openings to prevent damage to turbocharger due to foreign matter.

To remove turbocharger, remove muffler and both hood side panels. Remove exhaust pipe (elbow) and heat shield from exhaust pipe flange, then unbolt flange from manifold and remove flange and sealing rings. Disconnect air cleaner and intake manifold lines from turbocharger. Remove oil supply line, disconnect oil return line from cylinder block, then cap all tube and line openings. Unbolt and remove heat shield from manifold, then unbolt and remove turbocharger with attached oil line. Remove oil return line from turbocharger.

To reinstall, secure oil return line to turbocharger, then while aligning oil return line with adapter to cylinder block, install turbocharger on manifold and tighten nuts to 30-35 ft.-lbs. (41-48 N·m) torque. At this time, fill center housing with oil at oil inlet port and install oil supply line. Use new lock tabs and install heat shield. Position seal ring in turbocharger and install exhaust flange. Connect air cleaner and intake manifold tubes and tighten clamp bolts to 15-20 in.-lbs. (1.69-2.26 N·m). Be sure there is no strain on turbocharger compressor cover from intake manifold tube. If necessary, loosen compressor cover bolts and realign cover with intake manifold tube. Be sure compressor cover is seated correctly and tighten cover bolts to 60 in.-lbs. (6.78 N·m). Install exhaust pipe and tighten bolts to 20-26 ft.-lbs. (27-35 N·m).

To prime turbocharger, proceed as follows: With oil return tube disconnected, place a container under oil return passage of turbocharger center housing and crank engine with diesel engine stop control out until there is a steady flow of oil from oil return passage of center housing. Connect oil return tube to turbocharger, install hood side panels and muffler. Check all bolts after several hours of operation and retighten, if necessary.

NOTE: Do not start engine until it is certain that turbocharger is receiving lubricating oil.

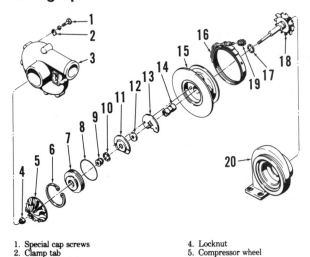

Fig. 128 — Exploded view of Schwitzer turbocharger used on 7000 model tractors. Only special cap screws (1) should be used. Arrangement and identification of parts corresponds for Models 7600 and 7700.

1. Special cap screws
2. Clamp tab
3. Compressor housing
4. Locknut
5. Compressor wheel
6. Snap ring
7. Insert
8. "O" ring
9. Flinger sleeve
10. Seal ring
11. Oil deflector
12. Thrust ring
13. Thrust plate
14. Bearing
15. Center housing
16. Clamp band
17. Seal ring
18. Turbine wheel & shaft
19. Nut
20. Turbine housing

109. OVERHAUL (Schwitzer). Remove turbocharger as outlined in paragraph 108. Before disassembling, mark relative positions of compressor cover, center housing and turbine housing to aid in reassembly. Remove compressor cover and note that bolts are of a special design and same type must be used if bolts are to be renewed. Remove clamp band (16 – Fig. 128) and separate core assembly from turbine housing. It may be necessary to tap lightly on turbine housing to dislodge core assembly. Remove compressor wheel retaining nut (4) being careful not to apply pressure to turbine or compressor wheel fins. Remove compressor wheel (5) and withdraw turbine wheel (18) and shaft assembly. Remove bearing (14) and mark bearing so it can be reinstalled in its original position. Remove center housing snap ring (6) and using two screwdrivers, lift insert (7) and "O" ring (8) from housing. Remove remainder of components in center housing. If sealing rings (10 and 17) are removed, note that they are different in diameter and should be marked to prevent incorrect assembly.

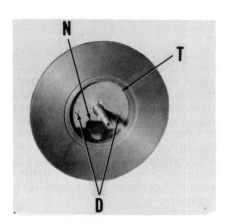

Fig. 129 — Thrust plate (T) is installed on dowel pins (D) with bronze face up. Note cut-out (N) in thrust plate for lip of oil deflector.

CAUTION: Do nut use a wire brush, steel scraper or caustic solution for cleaning as this will damage turbocharger parts.

Inspect turbine wheel and compressor wheel blades. Turbine wheel or compressor wheel should be renewed if any blades are broken, bent or cracked. If any one blade is bent less than 20 degrees, wheel may be used. If more than one blade is bent, discard wheel. Do not attempt to straighten blades.

Inspect all bearing and thrust surfaces for wear, excessive scoring, grooves or heavy scratch marks. Inspect seal rings and grooves. Inspect center housing for cracks and wear in bearing bore. Inspect oil supply holes in housing and be sure they are clear of obstacles.

To assemble, refer to exploded view in Fig. 128. Install copper-coated seal ring (17) on turbine wheel. Be sure correct sealing ring is installed. Lubricate sealing ring and turbine shaft with engine oil and install in center housing being careful not to damage seal ring recess in center housing or bearing bore. Lubricate bearing (14) and install in center housing on turbine wheel shaft. If original bearing is being used, it should be reinstalled in its original position. Install thrust plate (13) on dowel pins with bronze face up as shown in Fig. 129. Lubricate and install thrust ring (12 – Fig. 128). Install oil deflector (11) on dowel pins so lip is in towards housing. Install seal ring (10) on flinger sleeve (9) and install flinger sleeve and seal ring in insert (7) so flat side of sleeve (9) is flush with flat side of insert (7). Lubricate "O" ring and place in groove of insert (7). Install insert (7) in center housing with flat side out and press into counterbore until snap ring groove in center housing is clear. Be sure flinger sleeve remains in place when installing insert. Install snap ring (6) with beveled side out, and place compressor wheel on turbine wheel

shaft. Coat threads of a new locknut with graphite grease and install locknut on turbine wheel shaft and tighten to 156 in.-lbs. (17.63 N·m). Spin compressor wheel and note any binding or rubbing of components. Assembly must spin freely. Check for turbine wheel shaft end play. Normal end play is 0.006 inch (0.152 mm). If end play is excessive, check for abnormal wear on thrust surfaces. If there is no end play, check for carbon build-up on turbine wheel surface.

Reassemble remainder of assembly by reversing disassembly procedure. Tighten clamp band nut to 120 in.-lbs. (13.56 N·m). When installing compressor cover, align cover with reference marks made during disassembly. Tighten compressor cover bolts using a diagonal pattern to 60 in.-lbs. (6.78 N·m). Check that assembly will spin freely and disassemble if any drag is felt. To reinstall turbocharger, refer to paragraph 108.

110. OVERHAUL (AiResearch). Remove turbocharger unit as outlined in paragraph 108. Mark across compressor housing (1 – Fig. 130), center housing (22) and turbine housing (18) to aid alignment when assembling. Unbolt and remove compressor housing and turbine housing from center housing. Use a box end wrench clamped in a vise to hold center of turbine wheel (17) and remove nut (5).

NOTE: A "T" handle wrench is recommended to remove nut so bending of turbine shaft is avoided.

Lift off compressor wheel (6), then remove center housing (22) and shroud (15) from turbine shaft (17). Remove cap screws (23) and separate center housing (22) from backplate (7). Remove seal ring (8) from center housing. Roll pins (12) in center housing should not be removed unless renewal is required. Withdraw thrust collar (10) with thrust bearing (11) from backplate. Slide thrust bearing (11) and piston ring (9) from thrust collar (10). No attempt should be made to remove star spring on center housing side of backplate (7), if renewal is required, spring and backplate can only be renewed as a unit. Remove outer snap rings (13), then withdraw bearings (14) from each side of center housing.

Clean all parts in a cleaning solution which is not harmful to aluminum. A stiff brush and plastic or wood scraper should be used after deposits have softened. When cleaning, use extreme caution to prevent parts from being nicked, scratched or bent. Thoroughly clean oil cavity (C – Fig. 131) and oil squirt hole (H) in center housing. Small passage in oil squirt hole (H) can be

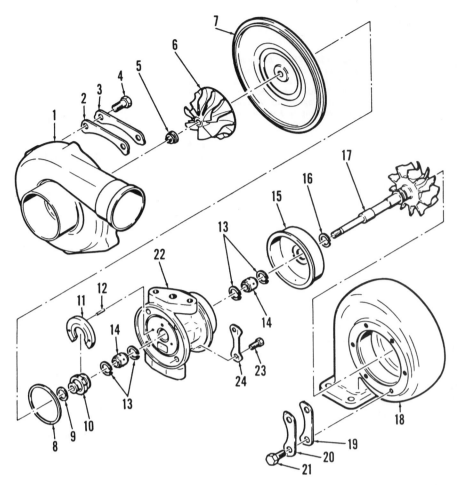

Fig. 130 — **Exploded view of AiResearch turbocharger used on Models 7600 and 7700 produced after January 1978 and all 7610 and 7710 models.**

1. Compressor housing	8. Seal ring	19. Clamp
2. Clamp	9. Piston ring	20. Lockplate
3. Lockplate	10. Thrust collar	21. Cap screw
4. Cap screw	11. Thrust bearing	22. Center housing
5. Nut	12. Roll pin	23. Cap screw
6. Compressor wheel	13. Snap ring	24. Lockplate
7. Backplate	14. Bearing	
	15. Turbine wheel shroud	
	16. Piston ring	
	17. Turbine wheel & shaft	
	18. Turbine housing	

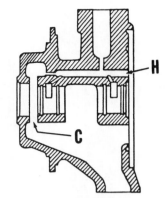

Fig. 131 — **Sectional view of turbocharger center housing.**

C. Turbine end oil cavity   H. Oil squirt hole

pressor housing (1) or the backplate (7). Make certain impeller blades are not bent, chipped, cracked or eroded. Inspect turbine shroud (15) for evidence of turbine wheel rubbing. Turbine wheel (17) should not show evidence of rubbing and vanes must not be bent, cracked, nicked or eroded. Vane tips must not be less than 0.025 inch (0.635 mm) thick.

Turbine wheel shaft must not show signs of scoring, scratching or overheating. Diameter of shaft journals should be 0.3997-0.4000 inch (10.152-10.16 mm) and should not be less than 0.3994 inch (10.145 mm). Groove in shaft for piston ring (16) must not be stepped and width of groove must not exceed 0.0735 inch (1.867 mm). Diameter of hub near piston ring should be 0.682-0.683 inch (17.323-17.348 mm) and must not be less than 0.681 inch (17.297 mm).

Upon reassembly, the following parts should be renewed: Snap rings (13), bearings (14), piston rings (9 and 16), seal ring (8), lockplates (3, 20 and 24) and cap screws (4, 21 and 23).

Install inner snap rings (13) into center housing (22) with rounded edge toward bearings. Oil bearings (14), then insert bearings in each side of center housing. Install outer snap rings with rounded edge toward bearings. Fill piston ring groove with high vacuum silicon grease manufactured by Dow-Corning or equivalent. Install piston ring (16) and shroud (15) on turbine wheel assembly (17), then guide wheel assembly shaft through bearings (14). Slide shaft into center housing (22) as far as it will go. Install new piston ring (9) in groove of thrust collar (10), then install thrust bearing so smooth side of bearing (11) is toward piston ring end of collar. Install thrust bearing and collar assembly over shaft, making certain that pins in center housing engage holes in thrust bearing. Install new seal ring (8), then install backplate (7) making certain that piston ring (9) is not damaged.

cleaned with a 0.053-0.060 inch (1.35-1.52 mm) diameter wire.

Inspect bearing bores in center housing (22 – Fig. 130) for scored or scratched surfaces. Bearing bore diameter must not exceed 0.6228 inch (15.819 mm). Turbine end seal bore may be either stepped or standard depending on production. Turbine end seal bore diameter must not exceed 0.713 inch (18.11 mm) for stepped and 0.703 inch (17.856 mm) for standard. Inside diameter of bearings (14) should be 0.4010-0.4014 inch (10.185-10.195 mm) and outside diameter should be 0.6182-0.6187 inch (15.702-15.715 mm). These bearings and their retaining rings (13) should be renewed each time unit is disassembled for service.

Thrust bearing (11) should be carefully inspected for wear or damage. Thickness should not be less than 0.1716 inch (4.359 mm). Faces of thrust bearing must be flat within 0.0003 inch (0.0076

mm) and diameter of bore for thrust collar must not exceed 0.430 inch (10.922 mm). Make certain oil passages in thrust bearing are clean and free of obstructions. Oil passage in thrust collar (10) must be clean and thrust faces must not be warped or scored. Piston ring groove shoulders must not have step wear. Width of groove for piston ring (9) should be 0.064-0.065 inch (1.626-1.651 mm) and should not exceed 0.066 inch (1.676 mm). Thrust bearing groove width in thrust collar (10) should be 0.1740-0.1748 inch (4.420-4.440 mm) and should not exceed 0.1752 inch (4.450 mm); diameter at bottom of groove should not be less than 0.370 inch (9.398 mm). Inside diameter of backplate (7) bore should be 0.4995-0.5005 inch (12.687-12.713 mm) and must not exceed 0.5010 inch (12.725 mm). Thrust surface and seal contact area must be clean and smooth. Compressor wheel (6) must not show signs of rubbing with either com-

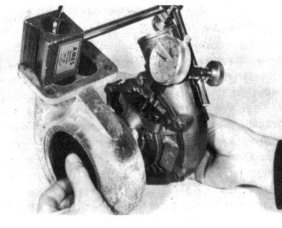

Fig. 132 – View showing method of checking turbine shaft end play. Shaft end play should be checked after unit is cleaned to prevent false reading due to carbon build-up.

Fig. 133 – Turbine shaft radial play is checked with dial indicator through the oil outlet hole.

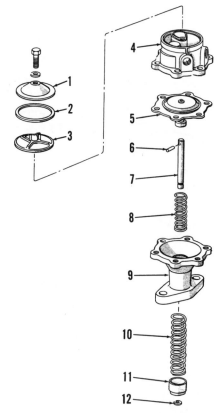

Fig. 134 – Exploded view of gasoline fuel pump assembly. Note filter screen (3). Pump is actuated by eccentric attached to front end of engine camshaft via a push rod.

1. Cover	7. Pump rod
2. Gasket	8. Spring
3. Screen	9. Body, lower
4. Body, upper	10. Spring
5. Diaphragm	11. Spring seat
6. Pin	12. Retainer clip

Install lock plates (24) and cap screws (23). Tighten cap screws (23) to 75-90 in.-lbs. (8.5-10.2 N·m) torque and bend lock plates up around heads of screws. Place serrated end of turbine wheel assembly in a box end wrench clamped in a vise, then install compressor wheel (6) on turbine wheel shaft. After oiling washer face and threads of nut (5), install nut on shaft and tighten to a torque of 18-20 in.-lbs. (2-2.3 N·m), then use a "T" handle to turn nut an additional 90 degrees.

CAUTION: If "T" handle is not used, shaft may be bent when tightening nut (5).

Install turbine housing (18) and compressor housing (1) to center housing (22) while aligning locating marks made before disassembly. Secure turbocharger assembly with clamp plates (2 and 19), lock plates (3 and 20) and cap screws (4 and 21) – coat threads with Fel

Pro high temperature compound. Tighten cap screws to 100-130 in.-lbs. (11.3-14.7 N·m) torque, then bend lock plates up around screw heads.

Check shaft end play and radial play. If shaft end play (Fig. 132) exceeds 0.004 inch (0.10 mm), thrust collar (10 – Fig. 130) and/or thrust bearing (11) is worn excessively. End play of less than 0.001 inch (0.0254 mm) indicates incomplete cleaning (carbon not all removed) or dirty assembly and unit should be disassembled and cleaned. If turbine shaft radial play (Fig. 133) exceeds 0.006 inch (0.152 mm), unit should be disassembled and bearings, shaft and/or center housing should be renewed. Maximum permissible limits of all of these parts may result in radial play which is not acceptable. Fill reservoir with engine oil and protect all openings of turbocharger until unit is installed on tractor.

GASOLINE FUEL SYSTEM

FILTERS AND SCREENS

Models 5000-6600

111. The gasoline fuel system incorporates four separate fuel screens and filters. A filter screen in fuel tank is accessible by disconnecting fuel supply line and unscrewing fuel supply valve from tank. A second screen (3 – Fig. 134) is located in the fuel pump assembly and is accessible after removing cover (1). A disc type filter element is located in sedi-

ment bowl; when reinstalling element, tighten finger tight only. A screen is incorporated in carburetor inlet fitting (29 – Fig. 135). An accessory kit is also available for installing an additional renewable element type filter in line between fuel tank and fuel pump.

FUEL PUMP

Models 5000-6600

112. Refer to Fig. 134 for exploded view of diaphragm type fuel pump used on all models. Fuel pump assembly may be removed after disconnecting lines and unbolting lower body from engine timing gear cover. Fuel pump push rod can be withdrawn from timing gear cover after removing fuel pump. Valves are serviced by renewing upper body assembly (4); all other parts are serviced separately.

CARBURETOR

Models 5000-6600

113. Holley carburetors are used on all gasoline engines. Refer to Fig. 135 for

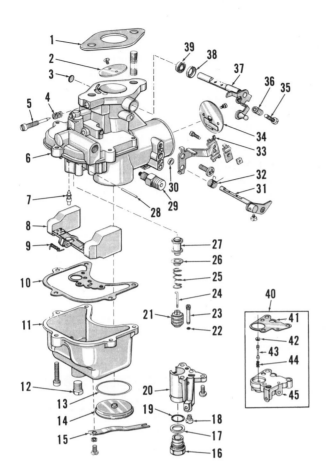

Fig. 135—Exploded view of the Holley carburetor assembly used on non-diesel models. Items shown in inset (40) were used prior to production date 9-66. Note filter screen in inlet fitting (29). For later units, refer to items (16) through (27). Main jet (18) is stamped #57, except for Model 6600, which is stamped #58.

1.	Gasket	23.	Pump discharge valve
2.	Throttle plate	24.	Link
3.	Expansion plug	25.	Spring
4.	Spring	26.	Spring seat
5.	Idle adjusting needle	27.	Vacuum piston
6.	Body	28.	Float hinge pin
7.	Inlet needle	29.	Inlet fitting & screen
8.	Float & lever	30.	Packing
9.	Float spring	31.	Choke shaft
10.	Bowl gasket	32.	Spring
11.	Float bowl	33.	Choke cable bracket
12.	Drain plug	34.	Choke plate
13.	Gasket	35.	Idle speed screw
14.	Air horn plug	36.	Spring
15.	Plug spring	37.	Throttle shaft
16.	Power valve assy.	38.	Seal retainer
17.	Gasket	39.	Seal
18.	Main jet	40.	Metering body assy.
19.	Gasket	41.	Diaphragm
20.	Metering body	42.	Retainer
21.	Accelerating pump piston	43.	Power valve
		44.	Spring
22.	Gasket	45.	Metering body assy.

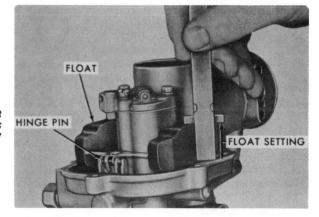

Fig. 137—Measuring float setting Holley carburetor; bend float tang if necessary to correct float level.

Fig. 136—View showing correct installation of float spring (9); float lever is (8).

carburetor exploded view and component parts.

Carburetor for early Model 5000 tractor is equipped with metering cluster cover incorporating a power valve; refer to inset (40). A diaphragm (41) is used between metering cluster cover (45) and carburetor body (6). Power valve (43) can be removed after removing cover (45) and disengaging retainer (42). Use a new diaphragm when reinstalling cover.

Float level is measured as shown in Fig. 137. Float measurement should be from 27/32 to 29/32-inch (21.43-23.04 mm). Adjust float level by bending lever that contacts inlet needle (7 – Fig. 135).

Spring (9) is installed to assist float lift; refer to Fig. 136 for spring installation.

Float valve (inlet needle) seat is an integral part of carburetor body (6 – Fig. 135). As body is not serviced separately, a new carburetor must be installed if seat or body is worn or damaged beyond further use.

DIESEL FUEL SYSTEM

The diesel fuel system consists of three basic components: The fuel filters, injection pump and injection nozzles. When servicing any unit associated with the fuel system, the maintenance of absolute cleanliness is of utmost importance. Of equal importance is the avoidance of nicks or burrs on any of the working parts.

Probably the most important precaution that service personnel can impart to owners of diesel powered tractors is to urge them to use an approved fuel that is absolutely clean and free from foreign material. Extra precaution should be taken to make certain no water enters fuel storage tanks.

TROUBLESHOOTING

All Diesel Models

114. If the engine will not start, or does not run properly after starting, refer to the following paragraphs for possible causes of trouble.

115. **FUEL NOT REACHING INJECTION PUMP.** If no fuel will run from line when disconnected from pump, check the following:

Be sure fuel supply valve is open.
Check the filters for being clogged (including filter screen in fuel supply valve).
Bleed the fuel filters.
Check lines and connectors for damage.

116. **FUEL REACHING NOZZLES BUT ENGINE WILL NOT START.** If, when lines are disconnected at fuel nozzles and engine is cranked, fuel will flow from connections, but engine will not start, check the following:

Check cranking speed.
Check throttle control rod adjustment.
Check pump timing.
Check fuel lines and connections for pressure leakage.
Check engine compression.

117. **ENGINE HARD TO START.** If the engine is hard to start, check the following:

Check cranking speed.
Bleed fuel filters.
Check for clogged fuel filters.
Check for water in fuel or improper fuel.
Check for air leaks on suction side of transfer pump.
Check engine compression.

118. **ENGINE STARTS, THEN STOPS.** If the engine will start, but then stops, check the following:

Check for clogged or restricted fuel lines or fuel filters.
Check for water in fuel.
Check for restrictions in air intake.
Check engine for overheating.
Check for air leaks in lines on suction side of transfer pump.

119. **ENGINE SURGES, MISFIRES OR POOR GOVERNOR REGULATION.** Make the following checks:

Bleed fuel system.
Check for clogged filters or lines or restricted fuel lines.
Check for water in fuel.
Check pump timing.
Check injector lines and connections for leakage.
Check for faulty or sticking injector nozzles.

120. **LOSS OF POWER.** If engine does not develop full power or speed, check the following:

Check throttle control rod adjustment.
Check maximum no-load speed adjustment.
Check for clogged or restricted fuel lines or clogged fuel filters.
Check for air leaks in suction line of transfer pump.
Check pump timing.
Check engine compression.
Check for improper engine valve gap adjustment or faulty valves.

121. **EXCESSIVE BLACK SMOKE AT EXHAUST.** If the engine emits excessive black smoke from exhaust, check the following:

Check for restricted air intake such as clogged air cleaner.
Check pump timing.
Check for faulty injectors.
Check engine compression.

FILTERS AND BLEEDING

122. **MAINTENANCE.** The fuel filter head is fitted with two renewable type elements on models prior to production date 10-69; or one filter element on later models. Water drain plug (cap) should be removed after every 50 hours of operation and any water in sediment bowl drained; drain more often if excessive condensation is noted.

After every 1200 hours (dual filters), or 600 hours (single filter), fuel filter elements should be renewed. Unscrew cap screw at top side of filter head and remove filter element and sediment bowl. Clean sediment bowl and reinstall new element and rubber sealing rings. Bleed diesel fuel system as outlined in paragraph 123.

123. **BLEEDING.** On models with dual filter, open rear bleed screw on filter head and operate lever on fuel lift pump until fuel flowing from bleed screw is free of bubbles. Close rear bleed screw and open front bleed screw on filter head and operate primer lever until bubble free fuel flows from opening. On single filter models, only one bleed screw is involved. Then, close fuel filter bleed screw, open front bleed screw on fuel injection pump and operate primer lever until fuel flowing from bleed screw is free of bubbles. Close pump bleed screw while operating primer lever.

NOTE: Later Simms type injection pumps have a self-bleeding line (L — Fig. 138); and pump bleeding is unnecessary.

Fig. 138 — Later Simms type injection pumps have self-bleeding line (L) to eliminate pump bleeding.

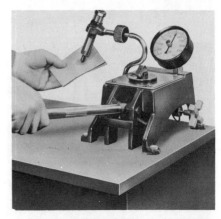

Fig. 139 — A fuel injector tester such as the one shown is necessary for checking and adjusting fuel injector assemblies. Nozzle seat leakage check is illustrated; refer to paragraph 129.

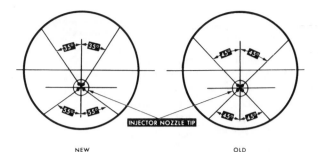

Fig. 140 — Comparative view of old and new diesel injector spray patterns. Refer to text.

Loosen fuel injector lines at injectors and crank engine until fuel appears at all injectors, then tighten fuel injector line connections and start engine.

INJECTION NOZZLES

CAUTION: Fuel leaves injection nozzles with sufficient force to penetrate skin. When testing, use extreme caution to stay clear of nozzle spray.

124. TESTING AND LOCATING A FAULTY NOZZLE. If engine does not run properly and a faulty injection nozzle is indicated, locate faulty nozzle as follows: With engine running, loosen high pressure line fitting on each nozzle holder in turn, thereby allowing fuel to escape at union rather than enter injector. As in checking for malfunctioning spark plugs in a spark ignition engine, a bad injector nozzle is the one which, when its line is loosened, least affects the running of the engine.

125. NOZZLE TESTER. A complete job of testing and adjusting fuel injection nozzle requires use of a special tester such as shown in Fig. 139. The nozzle should be tested for opening preessure, spray pattern, seat leakage and leak back. Also see Fig. 140.

Operate tester until oil flows, then connect injector nozzle to tester. Close tester valve to shut off passage to tester gage and operate tester lever to be sure nozzle is in operating condition and not plugged. If oil does not spray from all four spray holes in nozzle, if tester lever is hard to operate or other obvious defects are noted, remove nozzle from tester and service as outlined in paragraph 131. If nozzle operates without undue pressure on tester lever and fuel is sprayed from all four spray holes, proceed with following tests:

126. OPENING PRESSURE. While slowly operating tester lever with valve to tester gage open, note gage pressure at which nozzle spray occurs. Opening pressure should be 2760 psi (19.04 MPa) on Models 5000 and 7000, 2666-2887 psi (18.395-19.920 MPa) on Models 5600, 5610, 6600, 6610, 6700 and 6710 and

3053-3275 psi (21.065-22.597 MPa) on Models 7600, 7610, 7700 and 7710. If gage pressure is not within these limits, remove cap nut and turn adjusting screw (see Fig. 141) as required to bring opening pressure within specified limits. If opening pressure is erratic or cannot be properly adjusted, remove nozzle from tester and overhaul nozzle as outlined in paragraph 131. If opening pressure is within limits, check spray pattern as outlined in paragraph 127.

127. SPRAY PATTERN. Operate tester lever slowly and observe nozzle spray pattern.

NOTE: Refer to Fig. 140. On Models 5000 and 7000, all four sprays must be similar and spaced at approximately 90 degrees to each other in a nearly horizontal plane. On all other models, all four sprays must be similar and spaced at approximate intervals of 110, 90, 70 and 90 degrees in a nearly horizontal plane.

Each spray must be well-atomized and should spread to a 3-inch (7.6 cm) diameter cone at approximately ⅜ inch (9.525 mm) from nozzle tip. If spray pattern does not meet these conditions, remove nozzle from tester and overhaul nozzle as outlined in paragraph 131. If nozzle spray is satisfactory, proceed with seat leakage test as outlined in the following paragraph.

128. SEAT LEAKAGE. Close valve to tester gage and operate tester lever

quickly for several strokes. Then, wipe nozzle tip dry with clean blotting paper, open valve to tester gage and push tester lever down slowly to bring gage pressure to 150 psi (1035 kPa) below nozzle opening pressure and hold this pressure for one minute. Apply a piece of clean blotting paper (see Fig. 139) to tip of nozzle; the resulting oil blot should not be greater than one-half inch in diameter. If nozzle tip drips oil or blot is excessively large, remove nozzle from tester and overhaul nozzle as outlined in paragraph 131. If nozzle seat leakage is not excessive, proceed with nozzle leak back test as outlined in following paragraph.

129. NOZZLE LEAK BACK. Operate tester lever to bring gage pressure to approximately 2300 psi (15.87 MPa), release lever and note time required for gage pressure to drop from 2200 psi (15.18 MPa) to 1500 psi (10.35 MPa). Time required should be from five to 40 seconds. If time required is less than five seconds, nozzle is worn or there are dirt particles between mating surfaces of nozzle and holder. If time required is greater than 40 seconds, needle may be too tight a fit in nozzle bore. Refer to paragraph 131 for disassembly, cleaning and overhaul information.

NOTE: A leaking tester connection check valve or pressure gage will show up in this test as excessively fast leak back. If, in testing a number of injectors, all show excessively fast leak back, the tester should be suspected faulty rather than the injectors.

130. REMOVE AND REINSTALL INJECTORS. Before removing injectors, carefully clean all dirt and other foreign material from lines, injectors and cylinder head area around injectors. Disconnect injector leakoff line (Fig. 142) at each injector and at fuel return line. Disconnect injector line at pump

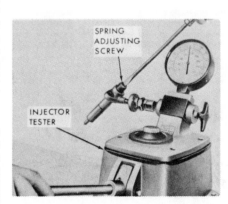

Fig. 141 — Adjusting nozzle opening pressure; refer to paragraph 126 for specifications.

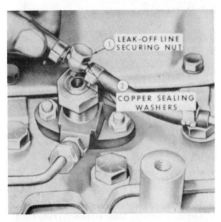

Fig. 142 — View showing injector leak-off line disconnected from injector. Note copper sealing washers placed on each side of banjo fitting.

and at injector. Cap all lines and openings. Remove the two retaining nuts and carefully remove injector from cylinder head (Fig. 143).

Prior to reinstalling injectors, check injector seats in cylinder head to see that they are clean and free from any carbon deposit. Install a new copper washer in seat and a new cork dust sealing washer around body of injector. Insert injector in cylinder head bore, install retaining washers and nuts and tighten nuts evenly and alternately to a torque of 10-15 ft.-lbs. (14-20 N·m). Position new leak-off fitting gaskets below and above each banjo fitting and install banjo fitting bolts to a torque of 8-10 ft.-lbs. (11-14 N·m). Reconnect leakoff line to return line. Check fuel injector line connections to be sure they are clean and reinstall lines tightening connections at pump end only. Crank engine until a stream of fuel is pumped out of each line at injector connection, then tighten connections. Start and run engine to be sure injector is properly sealed and that injector line and leakoff line connections are not leaking.

131. OVERHAUL INJECTORS. Unless complete and proper equipment is available, do not attempt to overhaul diesel nozzles. Equipment recommended by Ford is Kent-Moore or Hartridge Injector Nozzle Tester and Nuday SW 20 Injector Cleaning Kit.

Refer to Fig. 144 and proceed as follows: Secure injector holding fixture in a vise and mount injector assembly in fixture. **Never** clamp injector body in a vise. Remove cap nut and back off adjusting screw, then lift off upper spring disc, injector spring and spindle. Remove nozzle retaining nut using nozzle nut socket and remove nozzle and valve. Nozzles and valves are a lapped fit

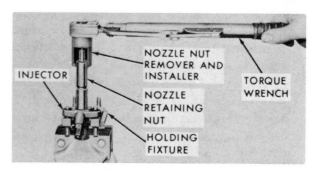

Fig. 144 — View showing injector assembly mounted on holding fixture; nozzle retaining nut is being tightened using special tool and torque wrench.

and must never be interchanged. Place all parts in clean fuel oil or calibrating fluid as they are disassembled. Clean injector assembly exterior as follows: Soften hard carbon deposits formed in spray holes and on needle tip by soaking in a suitable carbon solvent, then use a soft wire (brass) brush to remove carbon from needle and nozzle exterior. Rinse nozzle and needle immediately after cleaning to prevent carbon solvent from corroding the highly finished surfaces. Clean pressure chamber of nozzle with a reamer as shown in Fig. 145. Clean spray holes with proper size wire probe held in a pin vise as shown in Fig. 146. To prevent breakage of wire probe, wire

should protrude from pin vise only far enough to pass through pin holes. Rotate pin vise without applying undue pressure.

Valve seats in nozzle are cleaned by inserting valve seat scraper into nozzle and rotating scraper. Refer to Fig. 147. The annular groove in top of nozzle and pressure chamber are cleaned by using (rotating) pressure chamber carbon remover tool as shown in Fig. 148.

When cleaning is accomplished, back flush nozzle and needle by installing reverse flushing adapter on nozzle tester and inserting nozzle and needle assembly tip end first into adapter and secure with knurled nut. See Fig. 149. Rotate needle in nozzle while operating tester lever. After nozzle is back flushed, seat can be polished by using a small amount of tallow on end of a polishing stick and rotating stick in nozzle as shown in Fig. 150.

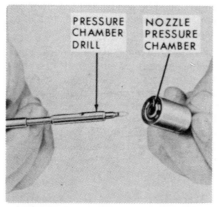

Fig. 145 — Cleaning nozzle tip cavity with pressure chamber drill.

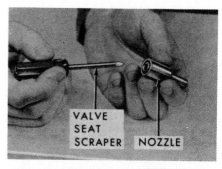

Fig. 147 — Use scraper to clean carbon from valve seat in nozzle body.

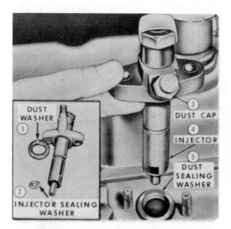

Fig. 143 — Cap or plug all openings when removing injector assembly. Dust washer (1) keeps dirt out of injector bore in cylinder head. Be sure seat in bore is clean and install new sealing washer (2) when reinstalling injector.

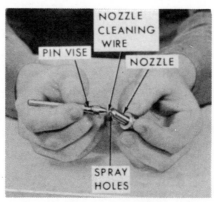

Fig. 146 — Cleaning nozzle spray holes with wire probe held in pin vise.

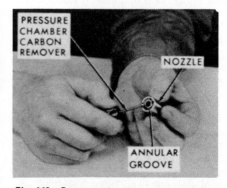

Fig. 148 — Pressure chamber carbon remover is used to clean annular groove as well as to clean carbon from pressure chamber in nozzle.

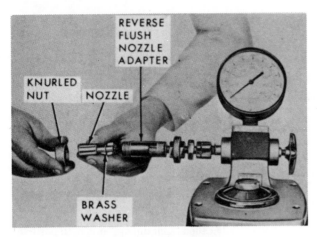

Fig. 149 — A back flush attachment is installed on nozzle tester to clean nozzle by reverse flow of fluid; note proper installation of nozzle in adapter.

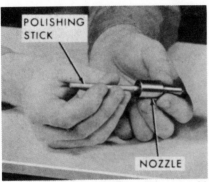

Fig. 150 — Polishing needle valve seat with tallow and polishing stick.

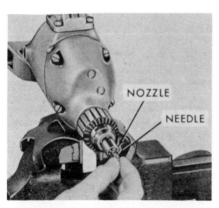

Fig. 151 — Chuck small diameter of nozzle in slow speed electric drill to lap needle to nozzle if leak back time is excessive or if needle sticks in nozzle. Hold pin end of needle with vise grip pliers.

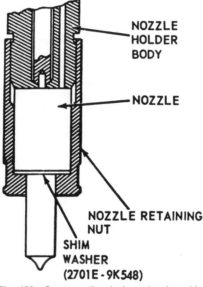

Fig. 153 — Cross-sectional view showing shim washer installed between nozzle body and nozzle retaining nut.

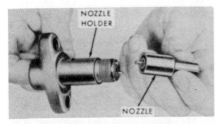

Fig. 152 — Be sure dowel pins in nozzle holder enter mating holes in nozzle body.

If leakback test time was greater than 40 seconds (refer to paragraph 129), or if needle is sticking in bore of nozzle, correction can be made by lapping needle and nozzle assembly. This is accomplished by using a polishing compound (Bacharach 66-0655 is suggested) as follows: Place small diameter of nozzle in a chuck of a drill having a maximum speed of less than 450 rpm. Apply a small amount of polishing compound on barrel of needle taking care not to allow any compound on tip or beveled seat portion, and insert needle in rotating nozzle body. Refer to Fig. 151. It is usually necessary to hold upper pin end of needle with vise-grip pliers to keep needle from turning with nozzle. Work needle in and out a few times taking care not to put any pressure against seat, then withdraw needle, remove nozzle from chuck and thoroughly clean nozzle and needle assembly using back flush adapter and tester pump.

Prior to reassembly, rinse all parts in clean fuel oil or calibrating fluid and assemble while still wet. Position nozzle and needle valve on injector body and be sure dowel pins in body are correctly located in nozzle as shown in Fig. 152. If so equipped, install the ⅜-inch (9.525 mm) shim washer (Fig. 153) and nozzle retaining nut and tighten nut to a torque of 50 ft.-lbs. (68 N·m).

Install spindle, spring, upper spring disc and spring adjusting screw. Connect injector to tester and adjust opening pressure as in paragraph 126. Use a new copper washer and install cap nut. Recheck nozzle opening pressure to be sure installing nut did not change adjustment. Retest injector as outlined in paragraphs 126 through 129; renew nozzle and needle if still faulty. If injectors are to be stored after overhaul, it is rec-

Fig. 154 — View showing filler plug, level plug and drain plug for fuel injection pump cambox and governor lubricating oil.

1. Bleed screws
2. Filler plug
3. Priming lever
4. Level plug
5. Drain plug
6. Excess fuel button
7. Fuel pump
8. High idle stop screw
9. Low idle stop screw

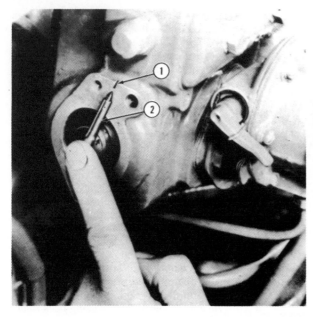

Fig. 155—View of Simms fuel injection pump with proofmeter drive cover removed. With a pointer (2) positioned in "V" notch on rear of pump camshaft, arrow should align with timing mark (1) on housing when No. 1 piston is at recommended degrees BTDC on compression stroke. Refer to text.

ommended that they be thoroughly flushed with calibrating fluid prior to storage.

FUEL INJECTION PUMP

Models 5000-6600-6610-6700-6710-7000-7600-7610-7700-7710

132. **LUBRICATION.** The Simms fuel injection pump is lubricated by an oil sump in pump cambox. After every 300 hours of operation, pump should be drained, cambox breather cleaned and cambox refilled to proper level with new, clean engine oil. Use same weight and type oil as for engine crankcase; refer to Fig. 154 for location of drain plug, oil level plug and filler plug.

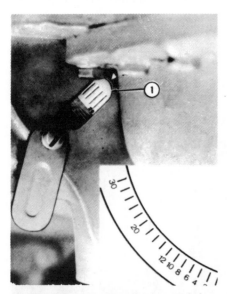

Fig. 156—View with cover plate removed showing timing pointer on engine rear plate and timing marks on flywheel (1).

Whenever installing a new or rebuilt fuel injection pump, be sure cambox is filled with engine oil to level of oil level plug before attempting to start engine. There will be some oil dilution with diesel fuel during engine operation and after engine is stopped, some of fuel oil mixture may run from overflow tube.

133. **PUMP TIMING.** To check and adjust pump timing, remove cover and gasket from rear of pump body and flywheel timing hole cover from right, rear side of engine. With stop cable pulled, turn engine until timing marks are visible through cover opening as shown in Fig. 156; then using a screwdriver through inspection cover opening, continue to turn flywheel until timing mark aligns with arrow. Refer to following table for correct timing setting:

Model	Degrees BTDC
5000, 7000	19
6600, 6700, 7600, 7700	23
6610, 6710	27
7610, 7710	25

Place a pointer in V-notch in rear of pump camshaft as shown in Fig. 155; pointer should align with scribed timing mark (1) if injection pump is correctly timed.

To adjust timing, drain coolant, remove lower radiator hose and injection pump drive gear cover as shown in Fig. 157. Loosen the three gear retaining cap screws (1) and, using a suitable wrench on injection pump shaft nut, slightly turn injection pump camshaft until pump timing marks align as shown in Fig. 155. Tighten gear retaining cap screws to a torque of 20-25 ft.-lbs. (27-34 N·m) when adjustment is correct.

NOTE: If engine is 180 degrees out of time, refer to paragraph 135 for procedure.

134. **ALTERNATE TIMING CHECK.** Factory-calibrated, in-line (Simms) injection pumps are self-timed to a mark on pump mounting flange which sets pump in time with engine (fuel delivery to number one cylinder) with this mark aligned with index on front cover. Steps for timing marked pumps are outlined in preceding paragraph, however, in order to avoid draining coolant, removal of pump drive gear front cover and other operations, the following method is proposed. Commonly called "spill timing," it is both positive and accurate. Follow these steps:

Turn engine over to locate number one piston at bottom of intake stroke, about to begin compression stroke. Remove injector line from number one injector and from pump end. Close fuel shut-off valve at tank. Remove delivery valve holder by use of special wrench, N5447. Remove delivery valve spring and volume reducer, than lift out delivery valve from its guide using needle-nose pliers. Reinstall delivery valve holder and torque to 50 ft.-lbs. (68 N·m). Spill pipe, made up of 12-15 inches (30.5-38.1 cm) of old injector line fitted with a gland nut is now installed on delivery valve holder with open end of pipe aimed at an empty container to catch discharged fuel.

Open fuel tank valve. Fuel will flow from spill pipe. If not, it may be necessary to operate lift pump hand primer a few strokes. Now, turn engine very slowly into its number one cylinder's compression stroke until fuel flow stops at spill pipe, then check position of timing marks on engine flywheel. Refer to table in paragraph 133 for correct timing setting. If correct timing setting does not appear in timing window, repeat cycle. When timing check is complete, reinstall delivery valve parts and reconnect injector line.

Fig. 157—Simms pump drive gear cover is removed to show timing marks on drive gear adapter hub and on pump housing.

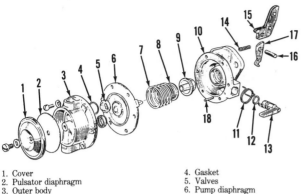

Fig. 158—Exploded view of diaphragm-type fuel lift pump; part of Simms fuel injection pump. Unit shown is used on Models 6610, 6710, 7610 and 7710. Components 4, 8 and 9 are not used on other models.

7. Diaphragm spring
8. Spring seat
9. Seal
10. Inner body
11. Return spring
12. "O" ring
13. Primer lever
14. Cam lever spring
15. Cam lever
16. Pivot pin
17. Diaphragm lever
18. Retaining pin

1. Cover
2. Pulsator diaphragm
3. Outer body
4. Gasket
5. Valves
6. Pump diaphragm

135. R&R FUEL INJECTION PUMP. Thoroughly clean pump, lines and connections and area around pump. Remove cover and gasket from rear of injection pump body as shown in Fig. 155 and timing hole cover from right rear side of engine as shown in Fig. 156. Remove rocker arm cover. Pull fuel stop and turn engine over with starter until number 4 exhaust valve (rearmost valve of engine) is open; then continue turning until timing marks are visible through inspection port as shown in Fig. 156, then, using a screwdriver in starter gear, continue to turn flywheel until timing mark aligns with arrow. Refer to table in paragraph 133 for correct timing setting.

Drain coolant and remove lower radiator hose. Remove injection pump drive cover as shown in Fig. 157. Shut off fuel and remove injector lines, then disconnect injection pump inlet and return fuel lines. Cap all openings as fuel lines are disconnected. Disconnect throttle and shut-off controls. Remove the three cap screws (1) retaining gear to pump hub. Remove gear retaining plate. Remove cap screws retaining pump to engine front plate and withdraw pump assembly. The pump drive gear will remain in timing gear cover and cannot come out of time, however, engine should not be turned with pump removed.

To reinstall fuel injection pump, first make sure number 1 piston is at correct timing setting, refer to table in paragraph 133. Turn injection pump camshaft until V-notch on rear of shaft aligns with scribe mark (1–Fig. 155) on housing and stamped arrow beside notch points upward (toward scribe line). Install pump by reversing removal procedure. Time pump as outlined in paragraph 133. Tighten pump retaining cap screws to 25-30 ft.-lbs. (34-41 N·m), gear retaining cap screws to 20-25 ft.-lbs. (27-34 N·m), and bleed fuel system as outlined in paragraph 123.

136. FUEL LIFT PUMP. Fuel lift pump is mounted on side of Simms fuel injection pump and is driven from a cam on injection pump camshaft. Fuel pump is diaphragm type and component parts are available for service. Refer to exploded view of fuel pump assembly in Fig. 158. Inlet and outlet valves (5) are staked into outer body (3); when renewing valves, insert in body as shown and carefully stake in position. Primer lever retaining pin (18) must be installed with outer end below flush with machined surface of inner body (10). After inserting pivot pin (16) in inner body, securely stake pin in place.

To test pump, operate primer lever with lines disconnected and with fingers closing inlet port; pump should hold vacuum after releasing primer lever. With finger closing outlet port, there should be a well-defined surge of pressure when operating primer lever.

DISTRIBUTOR-TYPE FUEL INJECTION PUMP

Models 5000-5600-5610-6600-6700

137. PUMP TIMING. Refer to Fig. 99 and to Fig. 159 when timing C.A.V. type pump used on these models. As shown in Fig. 99, pump drive gear with dowel engaged must be timed to camshaft gear with number 4 (four-cylinder engine models) matched to "O" index. Scribe line on pump body aligns with "O" mark on engine front cover plate (2–Fig. 159).

NOTE: Engine must not be operated with pump mounting bolts (4) loosened. If injection pump timing is to be altered, as

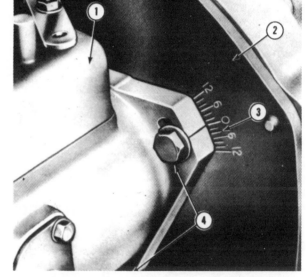

Fig. 159—View of timing marks for distributor-type pump used on Models 5600, 5610, 6600, 6700 and some 5000 series. Refer to text for procedure.

1. Injection pump
2. Engine front cover
3. Timing marks
4. Mounting bolts

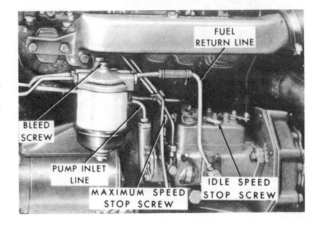

Fig. 160—View of typical distributor-type pump installed to show adjustment points. Note that fuel shutoff cable has been removed.

when timing for maximum horsepower by use of a dynamometer, halt engine, then reset timing and retighten mounting bolts before further testing.

138. LOW IDLE SPEED AND MAXIMUM NO-LOAD SPEED ADJUSTMENT. Whenever required, as when installing new or overhauled injection pump, proceed as follows to set low idle speed:

Engine should be running at normal operating temperature. Disconnect throttle control at injection pump, then loosen locknut and adjust low idle speed stop screw (see Fig. 160). Correct idle speed rpm is 600-700 on Models 5000, 5600, 6600 and 6700 and 600-850 on Model 5610. After correct adjustment is attained, secure locknut. Move operator's control all the way forward to its idle position, then adjust linkage so control rod can be reconnected without affecting idle speed and with no binding in linkage.

For maximum no-load engine speed adjustment, engine must also be at operating temperature. With throttle advanced, check for correct maximum no-load speed reading. Correct rpm is 2325-2375 on Models 5000, 5600, 6600 and 6700 and 2250-2300 on Model 5610. If speed varies out of this range by 50 rpm, proceed as follows:

NOTE: Factory installed sealing wire and locking sleeve must be renewed before adjustment can be made.

Disconnect throttle linkage at injection pump, then loosen locknut on adjuster screw and set engine speed to specification. Reinstall locking sleeve and sealing wire, adjust throttle rod linkage to eliminate binding, reconnect linkage, then check to determine if no-load maximum rpm is within specified range.

139. R&R FUEL INJECTION PUMP. Carefully clean pump exterior,

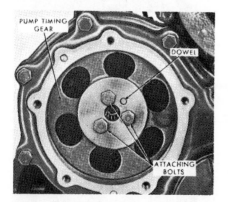

Fig. 161 — View of distributor-type (C.A.V.) injection pump drive gear with cover removed to show drive gear, attaching bolts and dowel pin. Refer to text.

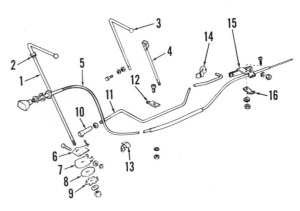

Fig. 162 — Exploded view of typical hand throttle control and linkage system. In earlier production, a coil spring is used in place of friction disc (8).

1. Control lever
2. Grommet
3. Control lever (cab)
4. Extension (cab)
5. Fuel shut-off cable
6. Throttle arm
7. Friction plate
8. Friction disc
9. Spring washer
10. Connector assy.
11. Control rod
12. Cable clamp
13. Cable clamp
14. Control rod clip
15. Cable bracket
16. Cable clamp

lines and connections. Remove pump to injector lines and disconnect fuel inlet and return lines. Cap or otherwise protect all openings from dirt. Disconnect throttle control rod and fuel shut-off control cable. Remove cover plate from front side of engine timing gear cover, then back out three cap screws which retain pump drive gear to its hub. See Fig. 161. Unbolt injection pump from engine front plate and remove pump assembly.

NOTE: Pump drive gear will remain within timing gear cover. Do not turn engine over with pump removed or pump will have to be retimed.

To reinstall pump, reverse removal procedure. Align index mark on pump body with "O" mark on engine front plate as shown in Fig. 159 and then tighten pump retaining cap screws to 26-30 ft.-lbs. (35-41 N·m) and drive gear bolts to 20-25 ft.-lbs. (27-34 N·m). Injector line nuts are tightened to 18-22 ft.-lbs. (24-30 N·m). Follow procedure outlined in paragraph 123 to bleed fuel system.

INJECTION PUMP DRIVE GEAR

All Diesel Models

140. To remove fuel injection pump drive gear, first remove engine timing gear cover as outlined in paragraph 83 or 84. Then, remove the three cap screws, retainer plate (models with Simms injection pump only) and fuel injection pump drive gear.

Prior to installing gear, turn engine crankshaft so timing marks on crankshaft gear and camshaft gear point towards center of idler (camshaft drive) gear. Then, remove self-locking cap screw retaining idler gear to front face of cylinder block, remove idler gear and reinstall with timing marks aligned with marks on crankshaft gear and camshaft gear. Tighten idler gear cap screw to a torque of 100-105 ft.-lbs. (136-143 N·m).

On models with C.A.V. pump, turn pump shaft so pump drive gear can be

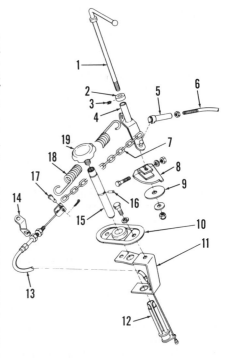

Fig. 163 — Exploded view of typical combination manual and pedal throttle control used on Models 5600, 6600 and 7600 when equipped with cab.

1. Control lever
2. Lever stop
3. Set screw, 10-32 x ¼
4. Throttle arm
5. Connector assy.
6. Control rod
7. Connector chain
8. Friction plate
9. Friction disc
10. Plate & bushing
11. Support bracket
12. Pedal tube
13. Cable assy.
14. Cable bracket
15. Pedal rod
16. Dowel pin, 3/16 x 7/8
17. Clevis pin, 3/16 x 9/16
18. Spring
19. Pedal

installed on dowel pin in pump drive adapter with timing marks on pump drive gear aligned with timing mark on idler (camshaft drive) gear (see Fig. 99). Install and tighten the three pump drive gear retaining cap screws to a torque of 20-25 ft.-lbs. (27-34 N·m).

On models with Simms pump, place pump drive gear on adapter hub with timing mark aligned with timing mark on idler gear. See Fig. 157. Place retainer plate on gear, install socket wrench on nut on front end of pump camshaft and turn pump camshaft until

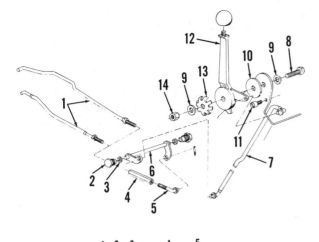

Fig. 164—Exploded view of typical throttle control used on Rowcrop Models 5200, 7200, 6600C and 7600C. Different control rods (1) are used for in-line or distributor-type injection pumps or for gasoline carburetor.

1. Control rod
2. Shaft retainer (2)
3. Flat washer (2)
4. Rod adjuster, ¼-28
5. Control rod
6. Cross shaft assy.
7. Link rod
8. Assy. bolt (⅜-24 x 1¾)
9. Pivot washer (2)
10. Friction disc
11. Lever stop
12. Lever w/knob
13. Spring washer
14. Nut, ⅜-24

Fig. 165—Exploded view of hand and foot throttle linkage used on Models 6700 and 7700.

1. Spring pin
2. Shaft
3. Actuator
4. Hand throttle lever
5. Roll pin
6. Throttle cable
7. Cable support
8. Shut-off cable
9. Snap ring
10. Link
11. Throttle cable
12. Bellcrank
13. Bellcrank
14. Trunnion
15. Rod
16. Lever
17. Bracket
18. Roll pin
19. Pedal
20. Pad
21. Support
22. Bracket
23. Bracket
24. Clamp

To adjust high idle speed, break wire seal, remove cover tube and loosen locknut on adjusting screw; then, turn screw in or out to obtain specified high idle speed, tighten locknut and reseal screw. When reconnecting throttle linkage, refer to paragraph 143 and adjust linkage if necessary.

Simms Pump

142. Start engine and bring to normal operating temperature. Disconnect throttle linkage from fuel injection pump governor arm and hold arm so stop lever contacts slow idle speed stop screw (Fig. 154). Engine slow idle speed should be 600-700 rpm on Models 5000, 6600, 6700, 7000, 7600 and 7700 and 600-850 rpm on Models 6610, 6710, 7610 and 7710. To adjust, loosen locknut on stop screw and turn screw in or out until proper slow idle speed is obtained and tighten locknut. Hold arm so stop lever is against high idle (maximum) speed stop screw; engine high idle speed should be 2325-2375 rpm on all models. If high idle speed is not within specified range, stop screw should be adjusted.

CAUTION: The high idle (maximum) speed stop screw adjustment is sealed with a sealing wire at the factory; this seal should not be broken by anyone other than Ford authorized diesel service personnel if tractor is within factory warranty.

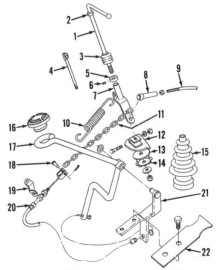

Fig. 166—Exploded view of typical combination manual and pedal throttle control used on Models 5610, 6610 and 7610 when equipped with a cab.

1. Control lever	12. Friction plate
2. Grommet	13. Friction disc
3. Boot	14. Spring washer
4. Extension	15. Boot
5. Lever stop	16. Pad
6. Set screw, 10-32 x ¼	17. Pedal assy.
7. Throttle arm	18. Clevis pin, 3/16 x ¾
8. Connector assy.	19. Cable assy.
9. Control rod	20. Cable assy.
10. Spring	21. Pedal bracket
11. Connector chain	22. Plate

timing mark on adapter is aligned with timing mark (pointer) on pump front plate. See Fig. 155. Then, turn pump slowly in a counterclockwise direction (as viewed from front of engine), if necessary, so cap screws can be installed through retainer plate and drive gear into pump drive adapter. Turn engine until number 1 piston is on compression stroke at correct timing position as noted in paragraph 133. With pump drive gear retaining cap screws loose, turn pump camshaft with socket wrench so pump timing marks are aligned (Fig. 157), then tighten drive gear retaining cap screws to a torque of 20-25 ft.-lbs. (27-34 N·m).

Reinstall timing gear cover as per paragraph 83 or 84.

DIESEL GOVERNOR ADJUSTMENTS
C.A.V. Pump

141. To check idle speed adjustment, proceed as follows: Start engine and

bring to a normal operating temperature. Disconnect throttle linkage from governor arm on fuel injection pump. Hold governor arm against slow idle speed stop screw; slow idle speed should be 600-700 rpm on Models 5000, 5600, 6600 and 6700 and 600-850 on Model 5610. If it is not, loosen locknut on slow idle speed stop screw (Fig. 160) and turn screw in or out to obtain proper slow idle speed, then tighten locknut. Hold injection pump governor arm against high idle (maximum) speed stop screw; engine speed should be 2325-2375 on Models 5000, 5600, 6600 and 6700 and 2250-2300 on Model 5610. If not within specified speed range, high idle speed stop screw should be adjusted.

CAUTION: The high idle speed stop screw is sealed at the factory with a sealing wire and cover tube; this seal should not be broken on tractors within factory warranty by other than Ford authorized diesel service personnel.

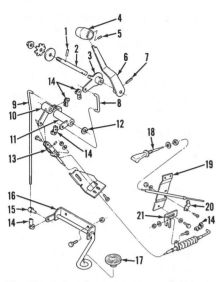

Fig. 167 — Exploded view of hand and foot throttle linkage used on Models 6710 and 7710.

1. Spring pin
2. Shaft
3. Actuator
4. Knob
5. Set screw, 1/4-20 x 5/8
6. Hand throttle lever
7. Roll pin
8. Link
9. Rod
10. Bellcrank
11. Bellcrank
12. Snap ring
13. Cable assy.
14. Clip
15. Trunnion
16. Pedal assy.
17. Pad
18. Shut-off cable
19. Support
20. Clamp
21. Bracket

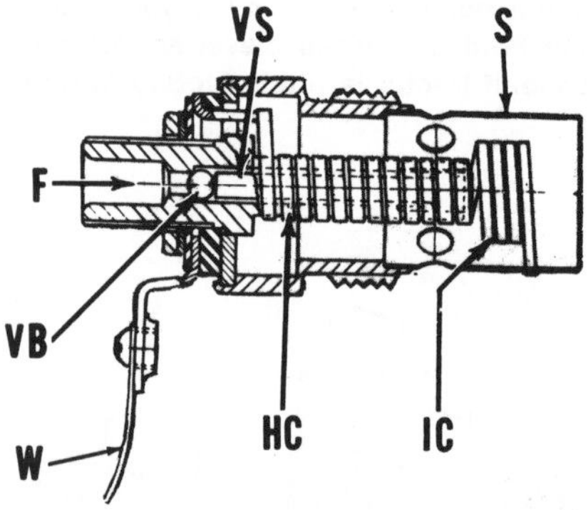

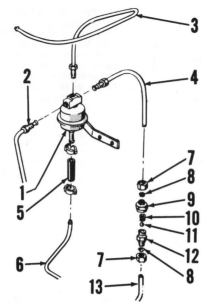

Fig. 169 — View showing typical Thermostart reservoir and full line connections used on early models.

1. Reservoir assy.
2. Injector excess fuel return line
3. Return line to fuel tank
4. Line to pressure valve
5. Flexible connector tube
6. Line to Thermostart fuel inlet
7. Compression nut
8. Compression sleeve
9. Connector, tube to valve assy.
10. Valve spring
11. Valve ball
12. Valve body
13. Line from pressure outlet of fuel filter assy.

Fig. 168 — Cross-sectional view of diesel Thermostart unit (optional) located in intake manifold. Refer to Fig. 169 for fuel reservoir and fuel line connections used on early models.

F. Fuel inlet
HC. Heater coil
IC. Ignition coil
S. Shield

VB. Valve ball
VS. Valve stem
W. Wire to key switch

To adjust high idle speed, break wire seal, loosen locknut and turn adjusting screw in or out until proper speed is obtained; then, tighten locknut and reseal adjusting screw. When reconnecting throttle linkage, refer to paragraph 143 and adjust linkage if necessary.

THROTTLE LINKAGE ADJUSTMENTS

All Models

143. Refer to exploded view of typical throttle linkages shown in Figs. 162 through 167. With control rod disconnected from injection pump governor arm and with engine governed speed adjusted as outlined in paragraphs 141 and 142, proceed as follows. Reference is made to Fig. 162, other models are similar. Move control lever (1) rearward against high idle stop and hold pump governor arm against high idle speed stop screw; forward end of rod (11) should then enter hole in governor arm without binding. If not, loosen locknut and turn rod in or out of end assembly as required. With rod length properly adjusted, reconnect rod to pump governor arm with retainer clip (14) and retighten locknut on rod. With engine running, check operation of throttle; there should be sufficient tension on friction disc (8) so lever will remain in any desired preset position. Some early models have a coil spring instead of spring washer

Fig. 170 — Drawing showing non-diesel governor and throttle linkage.

L. Length of rod
1. Throttle control rod
2. High idle adjustment screw
3. Throttle bellcrank
4. Governor rod
5. Governor arm
6. Throttle rod & governor spring assy.

(9). Throttle lever must not bind. Tighten or loosen lever assembly nut as required. When completed, recheck engine governed speed (paragraphs 141 or 142) with throttle linkage connected.

THERMOSTART COLD WEATHER STARTING AID
All Diesel Models So Equipped

144. Refer to Fig. 168 for cross-sectional view of Thermostart unit which is located in engine intake manifold. The fuel inlet (F) is connected to a small reservoir (see Fig. 169) on early models and to injector leak-off line on later models. In operation, turning key switch to either "HEAT" or "HEAT-START" position connects Thermostart wire (W – Fig. 168) to battery, thus heating heater coil (HC) and ignition coil (IC). The hot coils react to pull valve stem (VS) away from valve ball (VB) allowing ball to unseat and fuel to flow through unit, where it is heated and ignited before entering combustion chamber.

To operate unit in cold weather, turn key-starter switch to "HEAT" position for 30 seconds, then turn switch to "HEAT-START" position to start engine. If engine does not start after 30 seconds, return switch to "HEAT" position for 15 seconds, then back to "HEAT-START." After engine is started, turn switch to "ON" position.

CAUTION: DO NOT attempt to use ether starting fluid and Thermostart unit at same time; use ether starting fluid only if Thermostart unit and intake manifold are cold.

For connecting fuel lines to Thermostart unit and reservoir on early models, refer to Fig. 169.

NOTE: When diesel fuel lines have been disconnected, always make certain that fuel is present at Thermostart fuel inlet (F – Fig. 168) before tightening connection at that point. Any attempt to operate Thermostart unit without fuel available may cause failure of unit allowing fuel oil to leak into intake manifold.

Service of Thermostart unit and/or fuel reservoir consists of renewing defective unit.

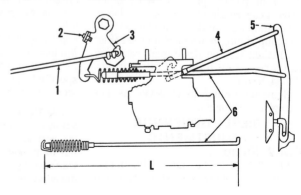

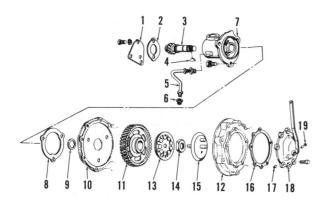

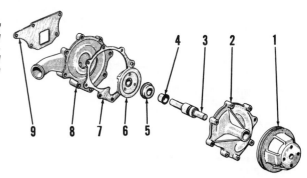

Fig. 171 — Exploded view of non-diesel governor and distributor drive assembly.

1. Rear cover plate
2. Gasket
3. Distributor drive shaft
4. Woodruff key
5. Oil line
6. Fittings
7. Housing
8. Gasket
9. Thrust washer
10. Engine front plate
11. Drive gear
12. Timing gear cover
13. Driver assy.
14. Nut, L.H. thread
15. Outer race
16. Gasket
17. Bushing
18. Governor housing
19. Oil seal

GOVERNOR (Non-Diesel)

SPEED ADJUSTMENT

Non-Diesel Models

145. Refer to Fig. 170 for schematic view of linkage. First make sure linkage is free without excessive looseness. Move hand throttle to high speed position and adjust governor rod (4) if necessary, to fully open carburetor throttle. Adjust length of throttle rod (L) to 17-3/16 inches (43.65 cm).

Start and warm engine and adjust slow idle speed to 650-700 rpm. Adjust high idle speed to 2285-2385 rpm, by turning adjustment screw (2).

R&R AND OVERHAUL

146. Refer to Fig. 171 for an exploded view of engine governor; plus governor drive which also includes distributor drive shaft and housing.

Governor housing (18) can be removed after disconnecting linkage and removing retaining cap screws. See Fig. 172. Outer race (15 – Fig. 171) can be withdrawn after removing governor hous-

ing. Remove drive shaft nut (14), if excessive wear on governor drive is indicated, and withdraw governor driver assembly (13). Nut (14) has left-hand thread.

Governor driver (weight) unit is only available as an assembly.

With governor driver removed, distributor drive housing, shaft and associated parts can be removed after removing ignition coil, distributor, oil feed line and attaching cap screws. Drive gear (11) can only be removed after removing timing gear cover as outlined in paragraph 83.

Distributor drive shaft (3) can be removed from housing (7) after removing rear cover (1), Woodruff key (4) and thrust washer (9).

Renew parts which are worn, damaged or questionable, and assemble by reversing the disassembly procedure. Time ignition as outlined in paragraph 158 and adjust governor linkage if necessary, as outlined in paragraph 145.

COOLING SYSTEM

RADIATOR PRESSURE CAP AND THERMOSTAT

All Models

147. A 7 psi (48.3 kPa) radiator pressure cap is used on all models.

On non-diesel engines, thermostat is located in front end of engine intake manifold. Thermostat in diesel engines is located in front end of cylinder head. On all models, thermostat is accessible after draining coolant from radiator and removing water outlet connection from intake manifold or cylinder head.

Standard thermostat for all 5000 models should start to open at 188°F (86.6°C) and be fully open at 212°F (100°C). Standard thermostat for all other models except Models 6710 and 7710 should start to open at 168°F (75.5°C) and be fully open at 192°F (88.8°C). Thermostat on Models 6710 and 7710 should start to open at 180°F (82°C) and be fully open at 201°F (94°C). Optional thermostats are available; however, use of standard thermostat is recommended for all conditions.

RADIATOR

All Models Except 6700-6710-7700-7710

148. **R&R RADIATOR.** To remove radiator, drain cooling system, disconnect air cleaner hose and headlight wires and remove grille and radiator shell from tractor. Disconnect hoses and transmission oil cooler tubes (Model 5000), or engine oil cooler tubes (Models 6610, 7000, 7600 and 7610) from radiator lower tank. Disconnect engine breather tube from rocker arm cover and fan shroud, then remove tube. Unbolt and remove radiator and fan shroud assembly from tractor. Remove shroud from radiator if necessary.

On 5000 models with transmission oil cooler or 6610, 7000, 7600 and 7610 models with engine oil cooler, radiator lower tank assembly contains a heat exchanger and, on some models, lower tank is available separately from radiator assembly.

NOTE: Some early 5000 model radiators may not have heat exchanger in lower tank.

To reinstall radiator, reverse removal procedure.

Fig. 172 — The governor drive (flyball) assembly is accessible after removing governor housing and outer race from front of timing gear cover. Retaining nut has left-hand thread.

Fig. 173 — Exploded view of water pump assembly. Seal (5) seats against hub of impeller (6). A spacer is used between fan and pulley on some models (not shown).

1. Pulley
2. Housing
3. Shaft & bearing assy.
4. Water slinger
5. Seal assy.
6. Impeller
7. Gasket
8. Rear cover
9. Mounting gasket

Models 6700-6710-7700-7710

149. R&R RADIATOR. To remove radiator, drain cooling system and remove outer radiator grilles and left pulley guard. Disconnect horn wires and remove center radiator grille. Remove hood side panels and hood side support brackets. Disconnect radiator hoses and remove fill tube assembly from radiator. Disconnect oil cooler lines from radiator on Models 6710, 7700 and 7710. Detach and move shroud rearwards. Unscrew retaining bolts and lift out radiator.

Lower tank is available separately. To install radiator, reverse removal procedure.

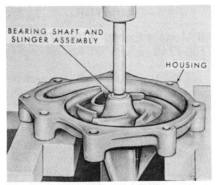

Fig. 174 — View showing shaft and bearing assembly being pressed from impeller and housing.

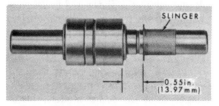

Fig. 175 — Flange of water slinger should be 0.55 inch (13.97 mm) from edge of outer bearing race as shown above.

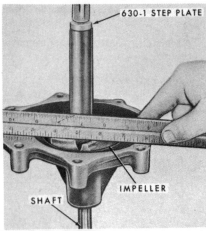

Fig. 176 — Press impeller onto shaft with ¾-inch (1.9 cm) ID pipe so impeller is flush with rear of housing as shown.

WATER PUMP

All Models

150. Water pump can be removed after removing radiator as outlined in paragraph 148 or 149. Refer to exploded view of water pump in Fig. 173 and disassemble pump as follows:

Remove fan (and spacer on models so equipped) from pulley and using standard two-bolt puller, remove pulley front shaft. Remove rear cover (8) and press shaft and bearing assembly (3) out towards front of housing as shown in Fig. 174. Drive seal (5 – Fig. 173) out towards rear of housing.

Using a length of 1-5/16 inch (3.33 cm) ID tubing, press new seal into housing. On models so equipped, check to see that flange on water slinger (4) is located 0.55 inch (13.97 mm) from edge of bearing race as shown in Fig. 175, then press shaft into front of housing until outer bearing race is flush with front end of housing. Using a length of ¾-inch (19 mm) ID pipe, press impeller onto shaft as shown in Fig. 176 so impeller is flush with rear end of housing. Press pulley onto shaft so center of belt groove in pulley is 2½ inches (6.35 cm) from rear face of housing as shown in Fig. 177. Install rear cover with new gasket and

Fig. 177 — Press pulley onto shaft so distance from center of belt groove in pulley is 2½ inches (6.35 cm) from rear face of housing.

tighten retaining cap screws to a torque of 18-22 ft.-lbs. (24-30 N·m).

Reinstall water pump assembly by reversing removal procedure and tighten retaining cap screws to a torque of 23-28 ft.-lbs. (31-38 N·m).

ELECTRICAL SYSTEM

Ford generators and regulators are used on all Model 5000 tractors and on some Model 7000 tractors. Model 7000 tractors which are not equipped with the Ford generator are equipped with a Motorola alternator.

All tractors with "A" or "B" prefixed serial numbers are equipped with a generator and all Model 7000 tractors with "C" prefixed serial numbers are equipped with an alternator. In current production, all tractors are equipped with alternators.

Regulators used with both units are a sealed type and cannot be adjusted.

GENERATOR AND REGULATOR

151. A Ford C7NN-10000-C 22-ampere generator and a Ford D0NN-10505-A regulator are used on Model 5000 and some Model 7000 tractors. See Fig. 178. The generator is a two-pole shunt wound type with type "B" circuit; that is, one field coil terminal is grounded to generator frame and the other field coil terminal is connected to armature terminal through the regulator. Specifications are as follows:

C7NN-10000-C Generator
Max. output (hot) at
1350 rpm and 15 volts ......22 amps.
Renew brushes if
shorter than ....13/32-in.
(1.03 cm)

Fig. 178 — Exploded view of C7NN-10000-C generator used on Model 5000 and some Model 7000 tractors. Note proofmeter (tachometer) drive (9).

1. Drive pulley
2. Fan
3. Drive end plate
4. Bearing
5. Retainer
6. Adjuster arm
7. Mounting bracket
8. Head shield
9. Proofmeter drive
10. Rear plate
11. Brush (2)
12. Brush clip (2)
13. Brush spring (2)
14. Brush holder (2)
15. Housing
16. Pole piece (2)

17. Insulator
18. Field coils
19. Armature bearing
20. Spacer
21. Armature
22. Bearing retainer

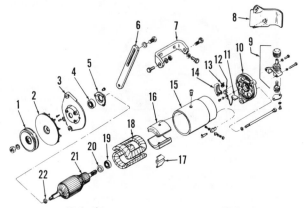

Min. brush spring tension
  with new brushes . . . . .18 ounces
                                                (510.3 gm)
Field coil current . . . . .2 amps.
Field coil resistance . . . . .6 ohms
Commutator min. dia. . . . . . . .1.450 in.
                                                (36.83 mm)
Max. commutator runout . . . . .0.002 in.
                                                (0.05 mm)
Max. armature shaft runout . . .0.002 in.
                                                (0.05 mm)

D0NN-10505-A Regulator
Cutout Relay:
  Cut-in speed (approximate
    engine rpm) 765
  Cut-in voltage 12.3-13.2
  Cut-out voltage 9.5-11.0
  Cut-out current . . . . .4 amps.
  Armature-to-core air
    gap . . . . .0.035-0.045 in.
                                                (0.89-1.14 mm)
  Contact blade
    movement . . . . .0.010-0.020 in.
                                                (0.254-0.508 mm)
Current Regulator:
  No-load setting . . . . .21-23 amps.
  Armature-to-core air
    gap . . . . .0.054 in.
                                                (1.37 mm)
Voltage Regulator:
  Opening voltage –
    50°F (10°C) 14.9-15.5
    68°F (20°C) 14.7-15.3
    86°F (30°C) 14.5-15.1
    104°F (40°C) 14.3-14.9
  Armature-to-core-air gap . . . . .0.053 in.
                                                (1.35 mm)

ALTERNATOR AND REGULATOR

Due to the fact that certain alternator components can be seriously damaged by procedures that would not affect a D.C. generator, the following precautions must be observed.

1. Always be sure that when installing batteries or connecting a booster battery, the negative posts of all batteries are grounded.

2. Never short across any alternator or regulator terminals.

3. Never attempt to polarize alternator.

4. Always disconnect all battery ground straps before removing or replacing any electrical unit.

5. Never operate alternator on an open circuit; be sure that all leads are properly connected and tightened before starting engine.

152. A Motorola 55-amp. alternator (Ford part no. C7NN-10300-B) and transistorized voltage regulator (Ford part no. D1NN-10316-A) are used on some Model 7000 tractors. Service specifications follow:

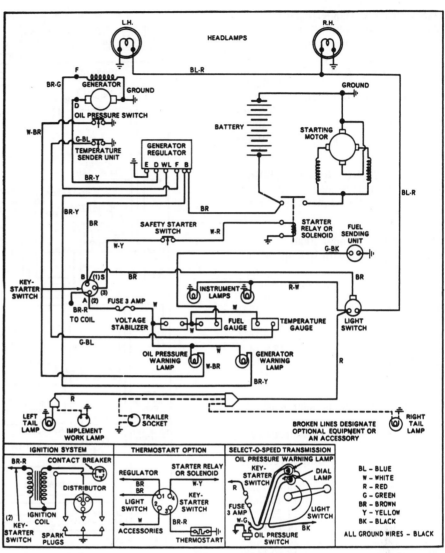

Fig. 179—Wiring diagram for all 5000 models with "C" prefix to tractor serial number. Refer to page two of this manual for serial number location. For tractors with "A" or "B" prefix to tractor serial number, refer to Fig. 180.

C7NN-10300-B Alternator
Max. output at 160°F, 2400
  engine rpm and 14.4 volts . . .47 amps.
Minimum brush length . . . . .3/16-in.
                                                (4.76 mm)
Min. brush spring tension
  with new brush . . . . .4 oz.
                                                (113.4 gm)
Field current . . . . .1.8-2.4 amps.
Field resistance . . . . .6 ohms

153. Alternators used in current production are essentially similar to those used on Model 7000. See Fig. 181. Alternator D5NN-10300-B (32 amp.) is standard for Model 5600 and the 6600 all-purpose model. Alternator D5NN-10300-E (32 amp.) is standard on 5610, 6610, 6700 and 6710 models and replaces alternator D5NN-10300-B. Alternator number D5NN-10300-A (51 amp.) is standard for Model 6600C (Rowcrop), 7600 and 7700 models and all models with a Ford cab. Alternator number D5NN-10300-D is standard on Models 7610 and 7710. When installed in tractor, 32 and 51 amp. alternators can be distinguished from one another by width of gap (exposed stator laminations) between front and rear housings. On 32 amp. alternators, this gap measures 3/8-inch (0.95 cm) and on 51 amp. alternators, this space is 5/8-inch (1.58 cm). All parts are interchangeable between these alternators except stator and rotor assemblies. Service specifications are as follows:

D5NN-10300-B, D5NN-10300-E
Max. output at 160°F (71°C)
  5136 alternator rpm and 14.4
  volts .32 amps.
Minimum brush length . . . . . . .0.187 in.
                                                (4.76 mm)
Min. brush spring tension (new) . . .4 oz.
                                                (113.4 gm)
Field current . . . . .1.8-2.4 amps.
Field resistance . . . . .6 ohms

D5NN-10300-A, D5NN-10300-D
Max. output at 160°F (71°C)
5136 alternator rpm and 14.4
volts .50 amps.
Minimum brush length . . . . . . .0.187 in.
(4.76 mm)
Min. brush spring tension (new) . . .4 oz.
(113.4 gm)
Field current . . . . .1.8-2.4 amps.
Field resistance . . . . .6 ohms

154. VOLTAGE REGULATORS.
Voltage regulators used with alternators have the following service specifications:

D1NN-10316-A (Model 7000)
Voltage regulation with 10 ampere load:

Ambient Temperature	Output Terminal Voltage
40°F (4.4°C)	14.2-15.0
60°F (15.5°C)	14.1-14.9
80°F (26.6°C)	13.9-14.7
100°F (37.7°C)	13.8-14.6
120°F (48.8°C)	13.6-14.4

D3NN-10316A (Models 5600, 6600 and 7600)
Voltage regulation with 10 ampere load:

Ambient Temperature	Output Terminal Voltage
0°F (−18°C)	14.60-15.40
20°F (−7°C)	14.45-15.25
40°F (4°C)	14.26-15.08
60°F (16°C)	14.13-14.92
80°F (27°C)	13.95-14.75
100°F (38°C)	13.80-14.60
120°F (49°C)	13.65-14.45
140°F (60°C)	13.60-14.28

D7NN-10316-B (All Models Except 7000)
Voltage regulation with 10 ampere load:

Ambient Temperature	Output Terminal Voltage
0°F (−18°C)	14.80-15.50
20°F (−7°C)	14.60-15.25
40°F (4°C)	14.26-14.90
60°F (16°C)	13.95-14.60
80°F (27°C)	13.60-14.30
100°F (38°C)	13.15-14.00
120°F (49°C)	12.75-13.70
140°F (60°C)	12.50-13.40
160°F (71°C)	12.30-13.15

IMPORTANT NOTE: Later transistorized regulators (D3NN-10316-A and D7NN-10316-B) do not interchange with regulator used with 55 amp alternator. Alternator, regulator or battery may be damaged if design types are mixed. Newer system uses a completely different method of voltage regulation.

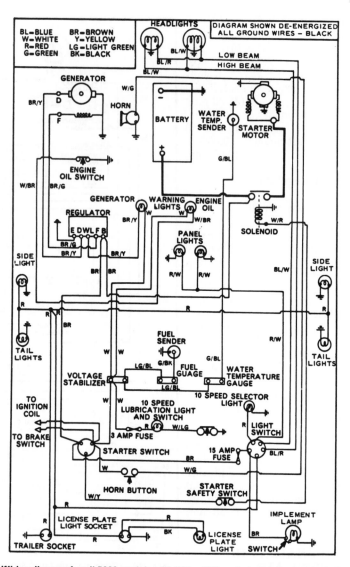

Fig. 180 — Wiring diagram for all 5000 models with "A" or "B" prefix to tractor serial number. Refer to page two of this manual for serial number location. Refer to Fig. 179 for wiring diagram for models with "C" prefix.

STARTING MOTOR

Non-Diesel Starting Motor

155. Ford Model 5000 tractors with non-diesel engine (except Model 5000 with Select-O-Speed transmission) are equipped with a starting motor which utilizes a series-parallel connected field coil arrangement, an integral positive engagement drive assembly and a movable pole piece which, together with one of the field coils, acts as a solenoid to engage drive assembly. When motor is not in use, one of the field coils is grounded through the actuating coil contacts. Closing starter switch completes circuit resulting in movable pole piece being attracted to field coil. A lever attached to movable pole piece engages drive gear pinion with flywheel ring gear. When movable pole piece is fully seated, it opens field coil grounding contacts which applies full field power for normal starting motor operation.

ervice specifications are as follows:
C5NF-11001-B or C7NF-11001-B Starting Motor
Brush spring tension (min.
with new brushes) . . . . .40 oz.
(1134 gms)
Min. brush length . . . . .¼-in.
(6.35 mm)
Commutator min.
diameter . . . . .1.46 in.
(3.7 cm)
Max. armature shaft
end play . . . . .0.058 in.
(1.47 mm)
Max. armature shaft
runout . . . . .0.005 in.
(0.127 mm)
No-load test:
Volts 12
Amps 70
Rpm 6000-9500
Loaded test (with warm engine):
Amps. 250-300
Engine rpm 150-200

Fig. 180A — Wiring diagram of late production Model 7000 tractors. Wire code key follows.

1. Alternator to starter relay (brown)
2. Regulator F to alternator (brown/green)
3. Regulator I to starter switch (brown/white)
4. Starter relay to starter switch (brown)
5. Alternator to starter switch (red)
6. Regulator A to starter switch (brown/yellow)
7. Starter relay to safety switch (white/red)
8. Starter switch to safety switch (white/yellow)
9. Thermostart to starter switch (brown/red)
10. Starter switch to fuse (white)
11. Starter switch to light switch (brown)
12. Fuse to voltage stabilizer & warning lamps (white)
13. Light switch to front & rear lamps (red) Rowcrop only
14. Light switch to head lamps (blue/red)
15. Temperature gage to sending unit (green/blue)
16. Light switch to panel lamps (red/white)
17. Oil pressure switch to warning light (blue)
18. Starter switch to air cleaner restriction vacuum switch (red/yellow)
19. Regulator G to alternator warning light (black)

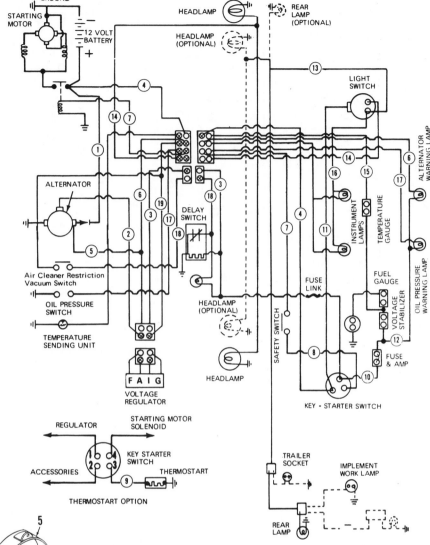

Fig. 181 — Exploded view of alternator typical of that used on current models. Refer to text for model identification and for alternator specifications. Note that isolation diode plate (21) is not used on production models after February 1976 as its function (battery cutout) is now accomplished by higher capacity diodes fitted in positive diode plate (16).

1. Drive pulley & key
2. Fan
3. Spacer
4. Front housing
5. Belt guard
6. Front bearing
7. Spring stop washer
8. Rotor
9. Stator
10. Spacer
11. Brush assy.
12. Brush shield
13. Brush cover
14. Terminal label
15. Rear housing
16. Positive diode plate
17. Diode trio (+)
18. Bearing retainer
19. Rear bearing
20. Negative diode plate
21. Isolation diode plate

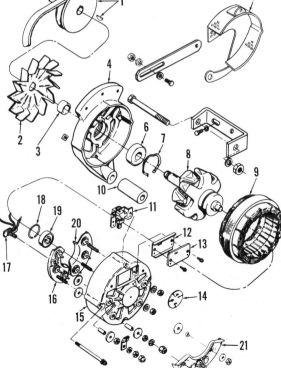

C5NF-11001-A, C6NF-11001-A or C7NF-11001-A Starting Motor

Brush spring tension (min. with new brushes) .... 40 oz. (1134 gms)

Min. brush length .... ¼-in. (6.35 mm)

Commutator min. diameter .... 1.46 in. (3.7 cm)

Max. armature shaft end play .... 0.048 in. (1.21 mm)

Max. armature shaft runout .... 0.005 in. (0.127 mm)

No-load test:
Volts 12
Amps 60
Rpm 5220-9440

Loaded test (with warm engine):
Amps 225-275
Engine rpm 150-200

Diesel Starting Motor

156. Diesel engines (and Model 5000 non-diesel engines with production date

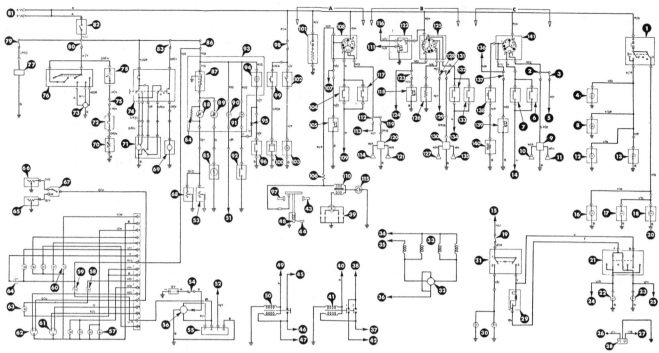

Fig. 182 – General wiring diagram for late-model c-prefix tractors.

B. Black
G. Green
GR. Gray
K. Pink
LG. Light green
N. Brown
O. Orange
P. Purple
R. Red
S. Slate
U. Blue
V. Violet
W. White
Y. Yellow
A. Gas engine ignition circuit
B. Manifold Heater Diesel engine ignition circuit
C. Ether start diesel engine ignition circuit
1. Headlight switch
2. Fuse – 7.5A
3. Fuse – 5A
4. Tail light
5. To voltage regulator – terminal I
7. Safety start switch – flat deck
8. Radio light
9. Radio
10. Speaker – L.H.
11. Speaker – R.H.
12. Instrument lights
13. Headlight – low beam
14. To starter solenoid coil
15. To ignition switch
16. High beam indicator
17. Work lamp
18. Headlight – high beam
19. Fuse – 9A
20. Ground – flat deck only

21. Turn signal switch
22. Turn signal indicator light – R.H.
23. Turn signal indicator light – L.H.
24. To 18 pin connector
25. To 18 pin connector
26. Turn signal indicator light connection – L.H.
27. Turn signal indicator light connection – R.H.
28. Turn signal indicator light assy. – straddle mount
29. Flasher unit
30. Implement lights
31. Toggle switch
32. Starting motor
33. Starting motor field coils
34. To starting motor solenoid
35. To starting motor solenoid
36. To battery negative terminal
37. To nos. 2, 3 & 4 starting motor field coils
38. To battery positive terminal
39. Spark plugs
40. To start ignition switch
41. Starting solenoid
42. To no. 1 starting motor field coil
43. To starting motor
44. To ignition switch start terminal
45. To battery positive terminal

46. To nos. 2, 3 & 4 starting motor field coils
47. To no. 1 starting motor field coil
48. Starter relay – gasoline
49. To start ignition switch
50. Starting solenoid
51. To voltage regulator – terminal I
52. To fuse panel
53. Fuel sender – main tank
54. Fuse link
55. Voltage regulator
56. Alternator
57. Illumination lamps
58. Turn indicator – R.H.
59. Alternator warning light
60. High beam indicator
61. Temperature gage
62. Fuel gage
63. Turn indicator – L.H.
64. Oil pressure warning light
65. Fuel sender – main tank
66. Fuel sender – auxiliary tank
67. Fuel tank selector switch – opt. flat deck only
68. Fuel sender – aux. tank
69. Windshield washer pump motor
70. Compressor clutch solenoid
71. Windshield wiper motor
72. Compressor clutch switch (heat actuated)

73. Blower motor – 3 speed
74. Windshield wiper/ washer switch
75. Fuse – 6A
76. 3 speed blower switch
77. Aux. fuel pump – (As equipped)
78. Thermostatic switch
79. Fuse – 5A
80. Fuse – 30A
81. From 21 pin connector
82. Ignition relay
83. Fuse – 4A
84. Resistor (aux. fuel tank)
85. Temperature sender
86. Fuse – 4A
87. Voltage stabilizer
88. Fuel gage
89. Temperature gage
90. Alternator warning light
91. Oil pressure warning light
92. Oil pressure switch
94. Air cleaner restriction warning light
95. Diode – 3A
96. Air cleaner restriction vacuum switch
97. To battery positive terminal
98. Fuse – 7.5A
99. Horn button
100. Horn
101. Cigar lighter
102. Interior light switch
103. Interior light
104. Safety start switch
105. Ignition relay coil

106. Ignition ballast resistor
107. To flasher/turn signal & implement light
108. Gas ignition switch
109. To starter relay coil
110. Ignition coil
111. To ignition switch terminal 3
112. Fuse – 5A
113. To voltage regulator – terminal I
114. Radio speaker – L.H.
115. Distributor breaker
116. To ignition switch terminal 4
117. Safety start switch
118. Ignition relay coil
119. Fuse – 7.5A
120. Radio
121. Radio speaker – R.H.
123. Fuse – 3A
124. To flasher/turn signal & implement light
125. Diesel ignition switch
126. Manifold heater
127. Radio speaker – L.H.
128. Fuse – 7.5A
129. To voltage regulator – terminal I
130. Radio
131. Fuse – 5A
132. Safety start switch
134. To starter solenoid coil
135. Radio speaker – R.H.
136. To flasher/turn signal & implement light
137. Fuse – 3A
138. Ether start button
139. Ignition relay coil
140. Ether solenoid
141. Diesel ignition switch

6-71 and up), are equipped with a Ford 5-inch (12.7 cm) diameter starting motor and relay assembly. Closing starter switch energizes solenoid; movement of solenoid plunger engages drive pinion and closes a two-stage switch. If drive pinion teeth butt against teeth on flywheel, only first stage of switch is closed which will allow current to flow to one field coil. This will provide enough power to turn starter until drive pinion is in position to engage flywheel ring gear teeth; then, full engagement of

drive pinion will close second stage of switch energizing all four field coils.

When drive pinion is in engaged position, there should be a clearance of 0.010-0.020 inch (0.254-0.508 mm) between drive pinion and thrust collar. To check clearance, first energize solenoid with 6-volt power source, then check clearance with feeler gage as shown in Fig. 184. If clearance is not within 0.010 to 0.020 inch (0.254-0.508 mm), refer to Fig. 186, loosen locknut and turn pivot pin (25 – Fig. 185) as required to obtain

proper clearance. Then, tighten locknut and recheck clearance.

Service specifications are as follows:

Starting Motor & Relay Assembly
Brush spring tension (min. with new brushes) . . . . 42 oz.
(1190 gms)
Min. brush length . . . . 5/16-in.
(7.94 mm)
Commutator min. diameter . . . . 1.53 in.
(3.88 cm)

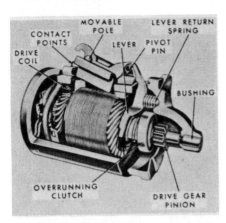

Fig. 183—Cut-away view of a typical starting motor used on non-diesel engine. A similar starting motor is used on Model 5000 non-diesel with Select-O-Speed transmission.

Fig. 185—Exploded view of typical diesel engine starting motor.

1. Tab washer
2. End cover
3. Bearing
4. Brush (2)
5. Spring
6. Brush holder
7. Cover seal
8. Spacer
9. Grommet
10. Shell
11. Relay housing seal
12. Solenoid assy.
13. Thrust washer
14. Thrust washer
15. Drive brake shoe (2)
16. Shaft pin
17. Brake spring (2)
18. Armature & shaft
19. Insulator
20. Field brush (2)
21. Field coils
22. Grommet
23. Thrust washer
24. Lever assy.
25. Eccentric pin
26. End plate bearing
27. Plate
28. Starter drive
29. Solenoid gasket
30. Stop collar
31. Collar retainer
32. Dowel
33. Drive end housing
34. Bearing

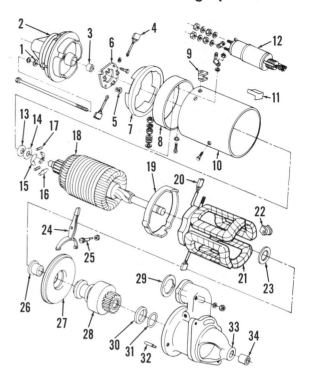

Max. armature shaft
  end play . . . . .0.020 in.
                              (0.508 mm)

Max. armature shaft
  runout . . . . .0.005 in.
                              (0.127 mm)

Drive pinion clearance
  (engaged) . . . . .0.010-0.020 in.
                         (0.254-0.508 mm)

No-load test:
  Volts 12
  Amps 100
  Rpm 5500-7500
Loaded test (with warm engine):
  Amps 250-300
  Engine rpm 150-200
  See Fig. 185 for exploded view of typical diesel engine starter parts.

IGNITION SYSTEM

157. **SPARK PLUGS.** Motorcraft (Autolite) AG5 spark plugs are recommended for all gasoline engines. Set

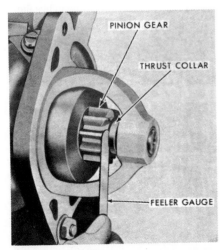

Fig. 184—Measuring drive pinion clearance on diesel engine starter motor. Refer to Fig. 186 for adjustment and to text for measuring and adjusting procedure.

electrode gap to 0.023-0.027 inch (0.584-0.686 mm). Install spark plugs with dry threads and tighten to a torque of 26-30 ft.-lbs. (35-41 N·m).

158. **IGNITION TIMING.** Breaker contact gap is 0.022-0.028 inch (0.559-0.711 mm). Firing order is 1-3-4-2. To install and time distributor, proceed as follows:

Remove No. 1 (front) spark plug and turn engine slowly until air is forced out spark plug hole, then continue turning engine slowly until 0 degree (TDC) flywheel mark is aligned with arrow in inspection opening of engine rear cover plate. Place distributor with dust cover and rotor installed in drive housing with rotor pointing towards No. 1 cylinder distributor cap terminal. This should properly mesh distributor gear with drive shaft gear. Loosen bolt clamping timing arm to distributor base and rotate distributor until breaker points just start to open. Hold distributor housing in this position, center timing arm slot on bolt hole in drive housing and tighten timing arm clamp bolt. Reinstall spark plug, distributor cap and spark plug wire, start engine and set timing with timing light as follows:

159. IGNITION TIMING WITH TIMING LIGHT. A Power Timing Light can be used to check distributor advance mechanism. Stamped timing marks run from 0 degree (TDC) to 30 degrees (BTDC) which is not sufficient to register combined advance. A suggested procedure for checking distributor and adjusting timing is as follows:

Connect timing light to No. 1 spark plug and open timing hole cover on right, front side of engine rear plate, then start engine. Retard ignition timing until 30 degree timing mark aligns with timing pointer, tighten clamp screw and shut off engine. Disconnect and plug vacuum advance line. Restart engine and with engine running at high idle speed, recheck timing which should now be 18 degrees BTDC. Adjust vacuum advance, if necessary, as outlined in paragraph 162.

With engine still running at high idle speed and with vacuum advance line disconnected, reset ignition timing to 28 degrees BTDC.

Fig. 186—Adjusting drive pinion clearance on diesel engine starting motor.

Tighten clamp screw and reduce engine speed to slow idle; timing should retard to 0−4 degrees BTDC. If it does not, overhaul distributor as in paragraph 163 or adjust centrifugal advance mechanism as in paragraph 161.

160. DISTRIBUTOR TEST AND OVERHAUL. The breaker contact gap for all models is 0.022-0.028 inch (0.559-0.711 mm). Cam dwell angle for all models is 35-38 degrees.

Breaker contact points may be either conventional pivoted type or pivotless. On pivoted type points, breaker arm spring tension should be 17-21 ounces (482-595 gms) when measured at end of breaker points contact, or 15-18 ounces (425-510 gms) when measured at center of contact points. To adjust spring tension, loosen nut holding breaker arm spring and move slotted end of spring towards pivot point to decrease tension, or away from pivot point to increase tension. Tension on pivotless point set is nonadjustable. Refer to Fig. 188 for views showing both types of breaker points.

For distributor test stands (synchroscopes), refer to the following test data and vacuum advance test data: (All data is in distributor rpm and distributor degrees.)

Fig. 189 — Exploded view of distributor used on non-diesel tractors.

1. Distributor cap
2. Rotor
3. Retainer
4. Seal
5. Dust cover
6. "O" ring
7. Condenser
8. Breaker points
9. Cam lubricant felt
10. Ground wire
11. Retainer rings
12. Spring washer
13. Upper breaker plate
14. Lower breaker plate
15. Wick
16. Retainer ring
17. Distributor cam
18. Thrust washer
19. Advance weights
20. Secondary advance spring
21. Primary advance spring
22. Rubber sleeve
23. Distributor shaft
24. Bushing
25. Wick
26. Primary wire
27. Distributor base
28. Timing arm
29. "O" ring
30. Spring pin
31. Drive gear
32. Plug
33. Gasket
34. Shim
35. Stop
36. Spring
37. Diaphragm assy.
38. "O" ring
39. Compressing sleeve
40. Vacuum tube
41. Tube adapter

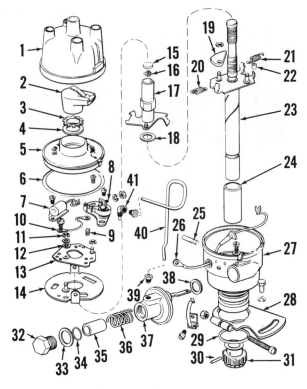

C6NF-12127-A

Centrifugal Advance Data

Distributor Rpm	Degrees Advance
200-450	−0.5 to +0.5
475	0 to 1
800	5.8 to 7.1
1000	9.5 to 10.75
1150	12 to 13.5
1200	12 to 14

Vacuum Advance at 1000 Distributor Rpm

Inches Mercury	Degrees Advance
1	0
3	0 to 0.5
5	0 to 1
7	0 to 1.5
8	0 to 3
10	2.4 to 5.5
12	4 to 7
13.5	5 to 8
15	5 to 8

C7NF-12127-E

Centrifugal Advance Data

Distributor Rpm	Degrees Advance
200 to 400	-0.5 to +0.5
900	6.75 to 9.0
1200	12 to 14

Vacuum Advance at 1000 Distributor Rpm

Inches Mercury	Degrees Advance
1	0
5	0 to 1
9	0 to 3
10	0.5 to 4
13	5 to 7.5

161. ADJUST CENTRIFUGAL ADVANCE. If distributor centrifugal advance did not fall within specifications given in paragraph 160 proceed as follows to adjust or correct centrifugal advance mechanism:

Refer to Fig. 190 and check to be sure sleeve (22) is in place on tang of distributor shaft plate.

NOTE: Top view of distributor in Fig. 190 is with breaker plate (14 — Fig. 189) removed; however, distributor shaft can be rotated and sleeve (22 — Fig. 190) and spring adjustment tabs (T) observed through hole in breaker plate.

If low rpm centrifugal advance is not within specified limits, turn distributor shaft so primary spring (21) adjusting tab (T) is in view through hole in breaker

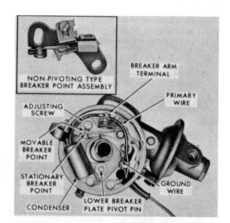

Fig. 188 — View of ignition distributor with cap and rotor removed. Two types of breaker point assemblies are used; refer to inset for non-pivoting type.

NON-PIVOTING TYPE BREAKER POINT ASSEMBLY

BREAKER ARM TERMINAL

ADJUSTING SCREW

PRIMARY WIRE

MOVABLE BREAKER POINT

STATIONARY BREAKER POINT

GROUND WIRE

CONDENSER

LOWER BREAKER PLATE PIVOT PIN

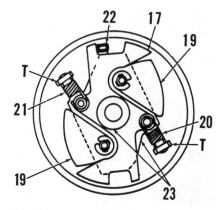

Fig. 190 — Drawing showing top of distributor assembly with breaker plates removed. Refer to Fig. 189 for parts identification. Adjusting tabs for advance springs are (T).

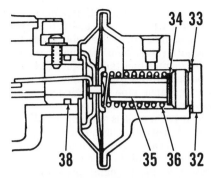

Fig. 191—Cross-sectional view of vacuum advance mechanism showing proper location of vacuum advance stop (35) and adjusting shim washers (34). Refer to Fig. 189 for parts identification.

plate and bend tab in to increase advance or out (away from distributor shaft) to decrease advance.

If high rpm centrifugal advance is not within specified limits, turn distributor shaft so secondary spring (20) adjusting tab (T) is in view through hole in breaker plate and bend tab in to increase advance or out (away from distributor shaft) to decrease advance.

NOTE: Secondary advance spring should be loose on tang when distributor shaft is stationary.

After adjusting centrifugal advance springs, recheck centrifugal advance throughout low, intermediate and high distributor rpm ranges given in test data. Renew advance springs if centrifugal advance is not within specified limits throughout test rpm range and cannot be adjusted. Advance springs have six coils and are color-coded purple.

NOTE: "Primary" and "Secondary" advance springs are identical, only adjustment is different.

162. ADJUST VACUUM ADVANCE. If vacuum advance is not within specified limits as outlined in paragraph 160, remove plug (32—Fig. 191) and add shims (34) between plug and spring to decrease advance, or remove shims to increase advance. Be sure gasket (33) is in good condition or renew gasket when reinstalling plug. Check vacuum unit for leaks after reinstalling plug. Shims are available in four thickness ranges: 0.008-0.010 inch (0.203-0.245 mm), 0.020-0.022 inch (0.508-0.559 mm), 0.040-0.042 inch (1.016-1.067 mm) and 0.080-0.082 inch (2.032-2.083 mm).

163. OVERHAUL DISTRIBUTOR. Refer to Fig. 189 for exploded view of ignition distributor.

It is important to properly lubricate distributor whenever servicing unit. Felts (4, 9 15 and 25) should be lightly saturated with SAE 10W motor oil. When advance unit is disassembled, fill grooves in top of distributor shaft and lubricate pivot pins with multi-purpose lithium base grease.

When installing new breaker points or adjusting breaker point gap, make certain that after breaker point retaining screws are tightened, ground wire (10) is properly positioned as shown in Fig. 188.

To renew distributor shaft and/or shaft bushing, proceed as follows: Remove vacuum advance assembly and breaker plates. Remove felt wick from top of distributor cam (17—Fig. 189) and extract retainer (16) with needle nose pliers. Drive gear retaining pin (30) from gear and press shaft (23) from gear and housing. Press old bushing (24) out towards top of housing with bushing driver (Nuday tool SW503 or equivalent). Lubricate outside of new bushing with motor oil, place flat steel washer against shoulder on driver and press new bushing in with driver until washer seats against top inside surface of housing. Ream bushing to inside diameter of 0.468-0.469 inch (11.887-11.912 mm). Lightly oil shaft, insert shaft into housing and press gear onto lower end of shaft so shaft end play is 0.029-0.042 inch (0.736-1.066 mm). Using pin hole in gear as a guide, drill retaining pin hole through new shaft, then install pin.

After reassembling distributor, check and adjust advance mechanism as outlined in paragraphs 160, 161 and 162.

CLUTCH

Tractors with independent pto will be equipped with a 12- or 13-inch (30.48-33.02 cm) diameter single plate dry type clutch. Tractors with a 10-speed (Select-O-Speed) transmission will have a disc type torque limiting clutch in the flywheel.

TRACTOR SPLIT

Models 5100-7100

164. To split tractor between engine and transmission, first drain cooling system, disconnect battery ground cable and proceed as follows: Remove vertical exhaust muffler if so equipped. Disconnect wiring harness from support clips under engine hood, then unbolt and remove hood. Disconnect proofmeter drive cable. Remove steering gear side covers from under fuel tank. Disconnect steering drag link from steering gear arm.

On models with "C" prefix to serial number, unplug wiring connector located at front of fuel tank. On all other models, disconnect wiring from starter relay terminals, generator, voltage regulator, front lights, oil pressure switch and temperature gage sending unit.

Remove shield from above starting motor and reinstall fuel filter retaining bolts. Unbolt and remove starting motor and flywheel access cover from plate between engine and transmission. Disconnect rear throttle control rod under fuel tank and disconnect diesel shut-off cable or gasoline carburetor choke cable. Shut off fuel supply valve at tank and disconnect fuel supply line and diesel excess fuel return line. Unbolt fuel tank and rear hood (fuel tank cover) from support brackets on rear end of engine. On models equipped wth horizontal exhaust, disconnect exhaust pipe from muffler under left step plate. Disconnect transmission oil cooler lines.

Insert wood wedges between front axle and front support. Place supports under front end of transmission housing and support engine with hoist or rolling floor jack, unbolt engine from transmission and roll front unit away.

To reconnect tractor between engine and transmission, reverse procedure used to split tractor. Refer to paragraph 123 for bleeding the diesel fuel system. Refer to Fig. 192 for engine-to-transmission bolt tightening torque.

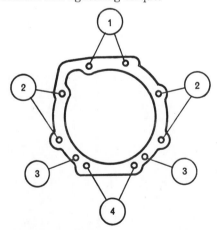

Fig. 192—When reinstalling engine to transmission housing tighten bolts at locations shown to the following torque specifications.

1. 220-300 ft.-lbs. (299-408 N·m)	(engines with stamped steel oil pan)
2. 125-140 ft.-lbs. (170-190 N·m)	4. 220-300 ft.-lbs. (299-408 N·m)
3. 35-50 ft.-lbs. (48-68 N·m)	(engines with case iron oil pan)

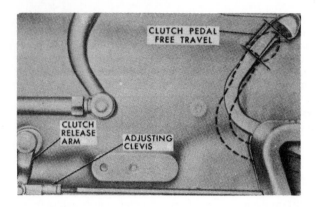

Fig. 193 — View showing typical clutch pedal free travel adjustment except for 6700, 6710, 7700 and 7710 models. Refer to text for adjustment specifications.

Models 5200-7200

165. On these models, follow same procedures as outlined in paragraph 164; however, disconnect the four power steering lines at connections under front of fuel tank and ignore reference to other steering components. Be sure to adequately support tricycle front end when separating tractor.

Models 5600-5610-6600-6610-7600-7610

166. Disconnect battery. Remove vertical exhaust muffler (models so equipped). Remove hood mounting screws and lift hood clear. On models without a cab, disconnect electrical connectors between front and rear wiring harness. On models with a cab, remove front housing rear panel. Disconnect electrical connectors between front wiring harness and safety start switch sender wire, windshield washer motor sender wire and rear wiring harness at multi-connector. Remove any connectors retaining front main harness to hood assembly.

Disconnect proofmeter drive cable from oil pump drive gear on left side of engine. On models so equipped, disconnect power steering pressure and return

lines at pump and reservoir assembly. Remove nut securing drag link to left spindle arm, then separate. On models without a cab, disconnect center and left support struts from rear hood assembly. On models with a cab, remove retaining bolts to withdraw hood rear support assembly. Remove bolts and nuts securing front end of fuel tank. On models without a cab, remove nuts and bolts mounting battery support bracket to rear hood panel. Disconnect wiring from starter motor and remove starter motor.

On models with a cab, close heater water control valves and disconnect hoses and plug. Disconnect fuel shut-off or choke cable. Disconnect throttle con-

trol rod from fuel injection pump or carburetor. Unhook fuel leak-off tube from neck of fuel tank filler tube. Close fuel shut-off valve, then disconnect tube at tank joining fuel tank to fuel filter. On models with horizontal exhaust, disconnect exhaust pipe at engine manifold. Disconnect oil cooler lines from center housing. On models so equipped, disconnect air conditioning quick-release connectors located above rear axle. Remove flywheel inspection cover and cap screws securing engine plate to transmission.

Insert wood wedges between front axle and front support. Place supports under front end of transmission housing and support engine with hoist or rolling floor jack. Unbolt engine from transmission and roll front unit away.

To reinstall engine, reverse removal procedure. Refer to Fig. 192 for engine-to-transmission bolt tightening torques. Refill and bleed systems as previously outlined. On models equipped with air conditioning, system must be evacuated before being recharged.

Models 6700-6710-7700-7710

167. Disconnect battery. Remove vertical exhaust muffler (models so equipped). Remove hood side panels, cowl side panels and cowl top panel (as equipped). On models without a cab, disconnect electrical connectors between

Fig. 194—Loosen locknut (L) and rotate turnbuckle (T) to adjust Model 6700, 6710, 7700 and 7710 clutch linkage. See text.

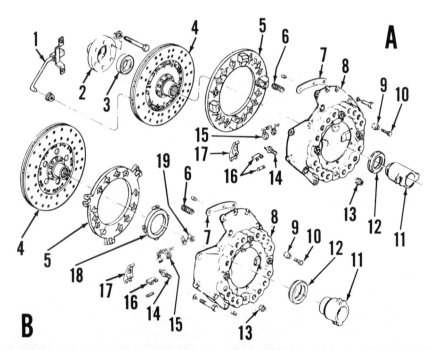

Fig. 195 — Exploded views of (A) 12-inch and (B) 13-inch (30.48-33.02 cm) clutches. Clutch shown in (A) view is used on Models 5600, 5610, 6600 and 6610. Clutch (B) is used on all others and is typical of Model 5000 and 7000 clutches.

1. Lubrication inlet	7. Spacer (A) 0.030-0.035 in. (0.76-0.89 mm) (B) 0.09-0.10 in. (2.28-2.55 mm)	14. Release lever eyebolt
2. Pto drive plate		15. Release lever spring
3. Bearing, pto drive		16. Strut & pin
4. Disc assy.	9. Cover ferrule	17. Release lever
5. Pressure plate	10. Pressure plate bolt	18. Release lever plate
6. Clutch spring(s)	8. Cover assy.	19. Plate retainer spring
	11. Release bearing hub	
	12. Release bearing	
	13. Release lever nut	

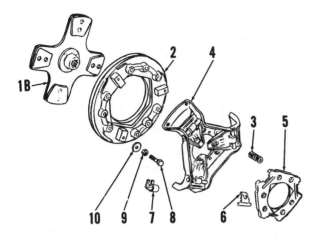

Fig. 196 — Exploded view of Cerametallic single disc clutch used on some Model 5000 tractors.

1B. Clutch disc
2. Pressure plate
3. Clutch spring
4. Cover
5. Retainer
6. Link
7. Clip
8. Adjusting screw
9. Locknut
10. Washer

front and rear wiring harness. On models with a cab, disconnect electrical connectors between front wiring harness and auxiliary fuel tank sender connector, windshield washer motor and rear wiring harness at multi-connector. Remove any connectors retaining front main harness to hood assembly.

Disconnect proofmeter drive cable from oil pump drive gear on left side of engine. Disconnect power steering tubing. Disconnect wiring from starter motor and remove starter motor.

On models with a cab, close heater water control valves then disconnect and plug hoses. Disconnect fuel shut-off cable. Disconnect throttle control rod from fuel injection pump. On models with horizontal exhaust, disconnect exhaust pipe at engine manifold. Disconnect oil cooler lines from center housing. On models so equipped, disconnect air conditioning quick-release connectors located above rear axle. Remove flywheel inspection cover and cap screws securing engine plate to transmission.

Insert wood wedges between front axle and front support. Place supports under front end of transmission housing and support engine with hoist or rolling floor jack. Unbolt engine from transmission and roll front unit away.

To reinstall engine, reverse removal procedure. Refer to Fig. 192 for engine to transmission bolt tightening torques. Refill and bleed systems as previously outlined. On models equipped with air conditioning, system must be evacuated before being recharged.

LINKAGE ADJUSTMENT

All Models

168. The recommended clutch pedal free travel is 1½-2 inches (3.8-5.1 cm) on early 5000 models and 1¼-1½ inches (3.1-3.8 cm) on late 5000 models and 7000 models. Effective 7-72, clutch pedal free travel, all models, is to be set at 1⅜ ± ⅛-inch (3.49 cm ± 3.175 mm). To adjust linkage on all models except 6700, 6710, 7700 and 7710, refer to Fig. 193. Disconnect adjusting clevis from clutch release arm, loosen locknut and turn clevis in or out as required. To adjust linkage on Models 6700, 6710, 7700 and 7710, refer to Fig. 194. Loosen turnbuckle locknut (L), rotate turnbuckle (T) until desired pedal free travel is obtained and retighten locknut.

SAFETY START SWITCH

Models 6700-6710-7700-7710

169. Tractors are equipped with a safety start switch actuated by the clutch pedal. When clutch pedal is depressed, spring plunger in switch closes points thereby completing starter electrical circuit.

R&R AND OVERHAUL

All Models

Early Model 5000 tractors used a full cover clutch with four release fingers and clutch disc facings were full circle fibrous type as shown in Fig. 195. Beginning with 10-69 production and running through 5-71 production, a ceramic button (Cerametallic) clutch was used which has a skeleton type cover with three release fingers

as shown in Fig. 196. Model 5000 tractors from production date 6-71, and all other models use a clutch similar to that shown in Fig. 195.

The two types of clutches are not interchangeable.

170. To remove clutch unit after tractor split, remove retaining cap screws and lift off the unit. Do not lose spacer shims (7–Fig. 195) interposed between flywheel and clutch cover on some units. Shims used on Series 5000 before 1968 were 0.072-0.079 inch (1.82-2.0 mm) thick; later models used shims 0.090-0.100 inch (2.28-2.54 mm) thick. Correct shims must be used. Tighten cover retaining cap screws to a torque of 23-30 ft.-lbs. (31-41 N·m) when reinstalling.

If available in tool stocks, use of Universal Clutch Fixture SW510 is recommended for assembly and disassembly of clutch. For correct release lever height setting on all models except 5000 and 7000, use Tool Gage Spacer SW510-5D on 12-in. (30.48 cm) clutches or Spacer SW510-5C on 13-in. (33.02 cm) clutches.

Place clutch cover on bed of a press or on clutch disassembly tool. On full cover clutches, remove adjusting nuts (13–Fig. 195), thrust plate (18) if used, and the four pressure plate bolts (10).

On button model (Cerametallic) units, carefully depress spring retainer (5–Fig. 196) until it clears clutch cover (4) by approximately ¼-inch (6.35 mm) at nearest point. Using a screwdriver or similar tool, pry spring retainer sideways away from release lever link (Fig. 198) in turn, until all three are disconnected.

NOTE: It may be necessary to thread adjusting screws (8 – Fig. 196) into pressure plate for additional finger clearance.

On all models, slowly release pressure and disassemble cover unit. On button models, DO NOT remove release levers or pivot pins from clutch cover (4–Fig. 196); unit is available only as an assembly. Other parts of all clutches are available individually.

Fig. 197 — When disassembling button type clutch, pry spring retainer to side as shown by heavy arrow, until lever link can be unhooked.

Fig. 198 — Unhooking lever link on button type clutch.

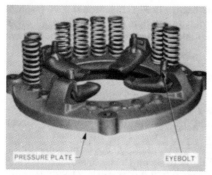

Fig. 199—Partially assembled view of single clutch pressure plate assembly.

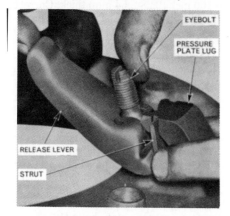

Fig. 200—Installing clutch release lever, strut and eye-bolt.

Various color-coded clutch springs have been used. Renew any springs which are rusted, distorted, heat discolored, cracked, or which fail to meet the test specifications which follow:

NOTE All springs are tested at a length of 1-11/16 inches (4.28 cm).

Given free lengths and test loads are recommended minimum.

Part No.	Color Code
C5NN-7572B	Cream
C7NN-7572B	Cream/Dark Blue
C9NN-7572C	Double Brown Stripe
D0NN-7572A	Double Gray Stripe
D1NN-7572A	Violet/Black Stripe
E1ADDN-7572B	Yellow/Light Green

Free Length	Spring Test
2-21/32 in.	120 lbs.
(6.75 cm)	(54.4 kg)
3.00 in.	125 lbs.
(7.62 cm)	(56.7 kg)
2¾ in.	72 lbs.
(6.98 cm)	(32.6 kg)
2¾ in.	85 lbs.
(6.98 cm)	(38.5 kg)

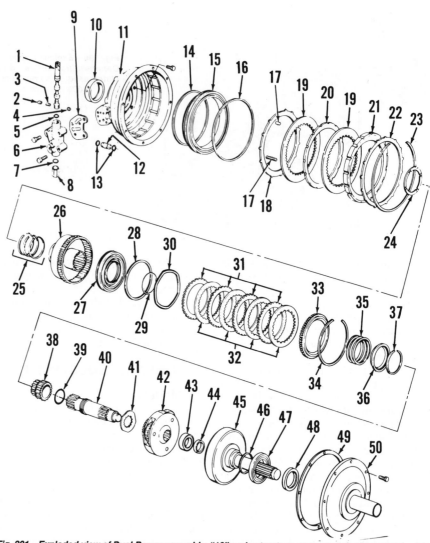

Fig. 201—Exploded view of Dual Power assembly. "10" series tractors use one underdrive clutch plate (19), three steel plates (31) and three friction plates (32); separation plate (20) and feathering spring (30) are deleted. Refer to Fig. 202 for exploded view of control valve assembly used on "10" series tractors.

1. Control valve
2. Valve detent cup
3. Valve spring
4. Valve detent cup
5. Upper "O" ring
6. Valve body
7. Lower "O" ring
8. Valve detent sleeve
9. Body gasket
10. Housing sleeve
11. Housing
12. Lubrication tube
13. Tube "O" rings
14. Rear piston seal
15. Underdrive piston
16. Front piston seal
17. Drive pins (3) Springs (12)
18. Underdrive clutch plate
19. Underdrive clutch plate (2)
20. Separation plate
21. Pressure plate
22. Clutch front plate
23. Retainer ring
24. Thrust washer
25. Seal rings (3)
26. Direct drive housing
27. Drive clutch piston
28. Outer piston seal
29. Inner piston seal
30. Feathering spring
31. Steel plates (4)
32. Friction plates (4)
33. Pressure plate
34. Retaining ring
35. Clutch spring
36. Spring seat
37. Seal retainer ring
38. Sun gear
39. "O" ring
40. Input shaft
41. Carriage washer
42. Planet carrier assy.
43. Input shaft bearing
44. Input shaft seal
45. Shaft & gear assy.
46. Shim*, 0.013 & 0.032 in. (0.33 & 0.81 mm)
47. Bearing
48. Shaft seal
49. Cover gasket
50. Housing cover

*As required

3-13/32 in.	150 lbs.
(8.65 cm)	(68.0 kg)
2-43/64 in.	135 lbs.
(6.78 cm)	(61.2 kg)

Assemble by reversing the disassembly procedure. Use new self-locking adjusting nuts when assembling early models. Finger height should be as outlined in the following table using a new clutch disc and measuring from friction surface of flywheel:

Model 5000 W/12 In.
(30.48 cm) Clutch . . . . . . 2.24-2.26 in.
(5.68-5.82 cm)

Model 5000 W/13 In.
(33.02 cm) Clutch and
Model 7000 . . . . 1.590-1.610 in.
(4.04-4.09 cm)

Models with Cerametallic
Clutch . . . . 1.970-2.030 in.
(5.00-5.16 cm)

NOTE: Lever height settings for all late model tractors are not specified for open adjustment on a shop press bed. Required procedure calls for use of Clutch Fixture SW510 and Spacer Gage SW510-5D on 12 inch (30.48 cm) clutches or Gage SW510-5C on 13 inch (33.02 cm) clutches.

DUAL POWER DRIVE

Dual Power drive is offered as an option on all late model tractors and as an option for later 5000 and 7000 models. Dual Power drive is a hydraulically operated planetary gear set which provides a high range (direct drive) and a low range (underdrive) which will increase torque to rear wheels and reduce ground speed.

Dual Power drive is available on eight-speed non-synchromesh and synchromesh transmissions; option provides a choice of sixteen forward and four reverse speeds on non-synchromesh transmissions and sixteen forward and eight reverse speeds on synchromesh transmissions.

Dual Power unit is mounted in forward compartment of transmission case and consists of a planetary gear set and two clutch assemblies.

On "10" series tractors, Dual Power control valve is engaged or disengaged by use of an electric solenoid valve. On all other models, operator's manual control actuates control valve by means of a cable linkage system. Dual Power unit can be shifted from normal direct drive to power underdrive without reduction of engine speed and without declutching when tractor operating conditions require extra torque.

LUBRICATION

All Models So Equipped

171. Dual Power unit is fitted to pressure lubricated eight-speed synchromesh and non-synchromesh transmissions and its lubricating oil flows directly from transmission oil cooler which is located forward of tractor radiator. Lubricating oil enters Dual Power control valve and is routed through oil passages of Dual Power housing. Oil exits into transmission by tube connecting Dual Power housing and transmission case and through common passages. Lubricating oil pressure for Dual Power should be 5-15 psi (34.5-103.5 kPa).

IMPORTANT NOTE: When air temperatures are below 0°F (−17°C) tractor must be operated with underdrive engaged for approximately fifteen minutes to warm oil and ensure circulation of lubricating oil.

HYDRAULIC PRESSURE

All Models So Equipped

172. Clutch units in Dual Power system are actuated by hydraulic pressure directed by control valve against piston for either direct or underdrive operation. To check hydraulic pressure, tractor must run until operating temperature of hydraulic oil is normal, approximately 120°F (49°C). Stop engine, then connect (tee) a pressure gage into line fittings of pressure line.

NOTE: On Models 5000, 5600, 6600, 7000 and 7600, connect (tee) pressure gage at

Dual Power housing pressure line inlet. On Models 5610, 6610, 6700, 6710, 7610, 7700 and 7710, connect (tee) pressure gage at hydraulic pump end of pressure line.

Start engine and set engine speed at 1000 rpm. Pressure gage reading should be 150-180 psi (1035-1242 kPa) with direct drive engaged and 155-185 psi (1069-1276 kPa) in underdrive. Reading should not differ in excess of 10 psi (69 kPa).

If system pressures are within these limits, hydraulic main system and Dual Power control valve are apparently satisfactory and inspection of Dual Power drive assembly is in order.

If higher pressures are indicated in both direct and underdrive operation, then for all models except "10" series tractors, condition of circuit relief valve (Fig. 203) should be checked. On "10" series tractors, check condition of pressure regulator valve located in pto drive clutch housing (P–Fig. 204). Check for damage to components or for contamination which might cause clogging or sticking.

If low pressures are indicated, hydraulic pump may be defective, control valve spool or "O' rings may be leaking or gasket between control valve body and Dual Power housing may be damaged.

If only direct drive shows low pressure, refer to paragraph 175 for disassembly of Dual Power unit and inspect drive clutch assembly for leaky seals or for a damaged piston.

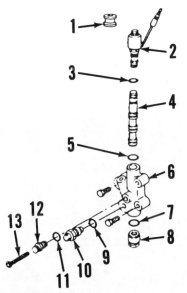

Fig. 202 – Exploded view of Dual Power control valve assembly used on "10" series tractors.

1. Grommet
2. Solenoid valve
3. "O" ring
4. Control valve
5. "O" ring
6. Valve body
7. "O" ring
8. Plug
9. "O" ring
10. Connector
11. "O" ring
12. Connector
13. Filter

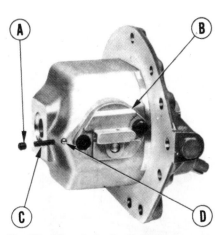

Fig. 203 – View of circuit relief valve located in rear cover of main hydraulic pump used on all models except "10" series tractors.

A. Retaining screw
B. Rear cover
C. Spring
D. Ball

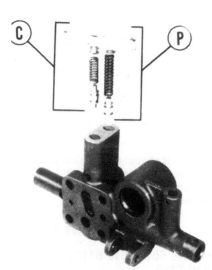

Fig. 204 – View showing pressure regulator valve assembly (P) and cooler/lubrication relief valve assembly (C) contained in pto drive clutch housing.

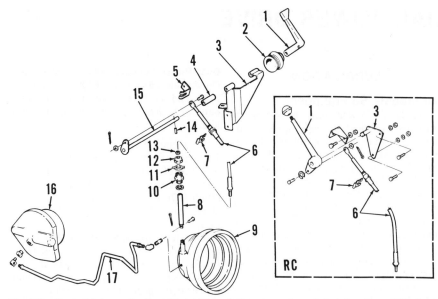

met is pulled inside bell housing. Note filter screen located in pressure line inlet fitting.

Remove pins and disconnect rod between clutch pedal and clutch fork cross shaft. Remove pins from clutch release fork, then remove clutch fork cross shaft, clutch release fork and release bearing. Remove housing cover (50—Fig. 201), shaft and gear assembly (45), planetary gear set (42), shaft (40) and sun gear (38). Remove direct drive housing (26) and thrust washer (24). Back out five retaining cap screws and separate Dual Power housing (11) from rear wall of compartment. Lubrication tube (12), secured by "O" rings (13) is easily pulled out by hand.

Reinstall by reversing the removal procedure, taking care to align lubrication tube (12) in "O" rings (13) when fitting housing (11) to compartment wall. Tighten five retaining cap screws in an alternating diagonal pattern to 77 ft.-lbs. (105 N·m) torque. When all internal parts of drive are reinstalled, tighten eight cap screws retaining housing cover (50) in a diagonal pattern to 35 ft.-lbs. (48 N·m) torque. Install control valve assembly with new gasket and tighten securing cap screws to 32 ft.-lbs. (44 N·m) torque. Lubrication inlet tube is tightened to 8-12 ft.-lbs. (11-16 N·m) and hydraulic pressure tube to 4.2-7.5 ft.-lbs. (5.7-10.2 N·m) when they are reattached to inlet connectors (10 and 12—Fig. 202) on control valve body (6). Use care when routing solenoid wire and protective grommet; reconnect wire in wiring harness connector. Refill transmission/rear axle center housing reservoir with appropriate amount of lubricating oil.

Fig. 205—Control linkages for Dual Power transmission shown in exploded view. Inset (RC) shows linkage for Rowcrop Models 6600C and 7600C.

1. Control handle	10. Cable connector
2. Shaft cover	11. Tab washer
3. Mount	12. Retainer nut
4. Spacer	13. Cable nut
5. Bracket	14. Roll pin, 3/16 in. (4.75 mm)
6. Control cable assy.	15. Control shaft
7. Lock clip	16. Hydraulic pump
8. Valve extension rod	17. Pressure line
9. Housing	

If oil pressure is low for underdrive, check condition of underdrive piston, clutch assembly and seals for possible pressure loss.

CONTROL LINKAGE

All Models So Equipped

173. Engagement of Dual Power in direct or underdrive is by manual linkage shown in Fig. 205. If length of control cable (6) is incorrect, it may not be possible to engage either direct or underdrive operation.

If adjustment of linkage is necessary, disconnect cable (6) at upper end, then manually place control valve to engage underdrive. Set control handle (1) in underdrive (down) position and adjust working length of cable (6) by loosening nuts (12 and 13) and turn end of cable in or out of threaded connector (10) to set cable length correctly. Reconnect cable at upper end, tighten locknuts and check operation of system in both direct and underdrive.

NOTE: If control cable (6) binds slightly while shifting, remove lock clip (7) then reset clip with cable eased to release tension within cable housing.

R&R DUAL POWER

Models 5610-6610-6710-7610-7710

174. First drain lubricating oil from tractor transmission and rear axle center housing. Reservoir capacity is 56 quarts (52.9 liters) on models with single-speed (540 rpm) pto and 60 quarts (56.8 liters) on models with two-speed (540 and 1000 rpm) pto. Split tractor as outlined in paragraph 166 or 167.

Unplug control valve electric solenoid wire from wiring harness located above transmission bell housing. Remove rubber plug located on side of bell housing to expose control valve body and line connections. Remove pressure tube (P—Fig. 207) and lubrication tube (L) from Dual Power control valve. Remove cap screws securing control valve assembly to housing, then withdraw valve assembly and gasket.

NOTE: Take care to prevent damage to grommet and solenoid wire when grom-

All Other Models

174A. First drain lubricating oil from tractor transmission and rear axle

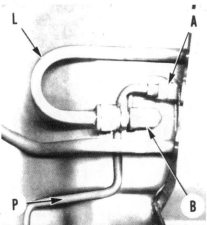

Fig. 206—View showing Dual Power pressure line (P), lubrication line (L) and elbow connectors (A and B) used on all models so equipped except "10" series tractors.

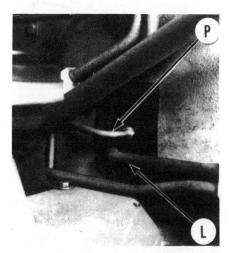

Fig. 207—View showing Dual Power pressure line (P) and lubrication line (L) used on "10" series tractors.

center housing. Reservoir capacity is 56 quarts (52.9 liters) on models with single-speed (540 rpm) pto and 60 quarts (56.8 liters) on models with two-speed (540 and 1000 rpm) pto. Split tractor as outlined in paragraph 164, 165, 166 or 167.

Identify and remove pressure tube (P – Fig. 206) and lubrication tube (L) from Dual Power control valve. Remove pins from clutch release fork, then remove clutch fork cross shaft, clutch release fork and release bearing. Disconnect control valve extension rod (8 – Fig. 205) from control valve.

Remove housing cover (50 – Fig. 201), shaft and gear assembly (45), planetary gear set (42), shaft (40) and sun gear (38). Remove direct drive housing (26) and thrust washer (24). Back out five retaining cap screws and separate Dual Power housing (11) from rear wall of compartment. Lubrication tube (12), secured by "O" rings (13) is easily pulled out by hand.

Reinstall by reversing the removal procedure, taking care to align lubrication tube (12) in "O" rings (13) when fitting housing (11) to compartment wall. Tighten five retaining cap screws in an alternating diagonal pattern to 50-60 ft.-lbs. (68-82 N·m). When all internal parts of drive are reinstalled, tighten eight cap screws for housing cover (50) in a diagonal pattern to 22-27 ft.-lbs. (30-37 N·m). Lubrication inlet tube is tightened to 8-12 ft.-lbs. (11-16 N·m) and hydraulic pressure tube to 4.2-7.5 ft.-lbs. (5.7-10.2 N·m) when they are reattached to control valve body (6). Reconnect linkage to control valve (1), adjusting if necessary as covered in paragraph 173. Refill transmission/rear axle center housing reservoir with appropriate amount of lubricating oil.

DUAL POWER OVERHAUL

All Models So Equipped

175. For overhaul of Dual Power unit, remove as outlined in preceding paragraph 174 or 174A, then proceed as follows with major subassemblies:

Disassemble direct drive clutch housing (26 – Fig. 201) by removing retaining ring (34) and pressure plate (33) followed by clutch plate pack (31 and 32) and feathering spring (30) on models so equipped. Keep clutch plates in assembled order during removal. Set housing up in a shop press and use Nuday tool N775 or equivalent to compress clutch spring (35) by bearing down on spring seat (36) so retainer ring (37) can be removed. Release press with care so as not to damage spring seat or ring groove in hub of clutch housing (26). To remove clutch piston (27), apply compressed air behind clutch piston by inserting a narrow air chuck into hole in hous-

ing hub between second and third housing seals (25). Clean and inspect removed parts, paying special attention to splines of housing (26) and clutch plates (31 and 32) and condition of clutch piston (27). Renew as necessary.

Proceed with disassembly of underdrive clutch by removing retainer ring (23) followed by clutch pack parts (17 through 22) taking care to keep clutch plates in order. Remove piston (15) and seals (14 and 16). Clean all parts thoroughly and inspect for wear or damage. Pay special attention to condition of 12 clutch springs (17). Renew unserviceable parts.

Use a properly fitted puller to remove input shaft bearing (43) from input shaft (40) and slide planetary carrier (42) from splines of shaft. Do not disassemble planetary gear set as it is available only as a complete assembly. Remove bearing (47) from shaft and gear assembly (45) with a puller, taking care not to damage shims (46) which are used to set end play of shaft.

176. Reassemble Dual Power drive unit following this sequence: After parts are lubricated with transmission oil M2C53A or Ford 134, install rear piston seal (14 – Fig. 201), underdrive piston (15) and front piston seal (16) carefully in housing (11). Note assembly order for clutch plates in Fig. 201 and stack plates (18 through 22) in housing with 12 springs in place and each of three drive pins set in every fifth notch of housing. Reset retainer ring (23).

Assemble direct drive clutch parts in housing (26) by reversing removal order, referring to Fig. 201 for placement sequence of steel plates (31) and friction plates (32). Install thrust washer (41) on shaft (40) with planet carrier (42) on shaft splines and press bearing (43) into place.

To set correct end play of 0.004-0.020 inch (0.101-0.508 mm) for shaft and gear

assembly (45), determine make-up of shim pack (46) by this procedure: Assemble all parts except shaft and gear (45) and cover (50) in housing (11). Now, install shaft and gear (45) without shims (46) and bearing (47). Set tool SW-523 on shaft in place of shims and bearing, then place cover (50) on housing. Now, measure gap between cover and housing at three equally spaced points on circumference and figure the average of these measurements. If resulting average gap is 0.046-0.060 inch (1.168-1.524 mm), no shims are needed. If average gap is less than this specification, refer to the following table for thickness of shim pack to be used.

Average Gap Width	Shim Thickness
0.001-0.013 in.	0.041-0.049 in.
(0.025-0.330 mm)	(1.04-1.24 mm)
0.014-0.026 in.	0.030-0.034 in.
(0.355-0.660 mm)	(0.762-0.864 mm)
0.027-0.032 in.	0.022-0.030 in.
(0.685-0.813 mm)	(0.559-0.762 mm)
0.033-0.045 in.	0.011-0.015 in.
(0.838-1.143 mm)	(0.279-0.381 mm)

If tool SW-523 is not on hand, assemble Dual Power unit with bearing (47) and original shim pack (46) on shaft (45), then install cover (50). Measure end play of shaft (45) by using a dial indicator or comparably effective method, then make up a shim pack using 0.013 inch (0.33 mm) and 0.032 inch (0.813 mm) shims to bring end play within limits of 0.004-0.020 inch (0.101-0.508 mm) as required.

When end play adjustment is completed, again strip internal parts from housing (11) and reinstall housing in its place in transmission compartment. Reassemble Dual Power unit as outlined in paragraph 174 or 174A. If necessary, see paragraph 173 for adjustment of control linkage.

8-SPEED TRANSMISSION (NON-SYNCHROMESH)

The 8-speed non-synchromesh transmissions for all models are basically similar in construction, however some differences exist on Model 7000, 7600, 7610, 7700 and 7710 transmissions which will affect service procedures.

Instead of spur type gears used in Model 5000, 5600, 5610, 6600, 6610, 6700 and 6710 transmissions, Model 7000, 7600, 7610, 7700 and 7710 transmissions use helical cut gears on secondary countershaft, 3rd and 7th gear assembly, mainshaft assembly and main countershaft.

To compensate for thrust loads imposed by helical cut gears, tapered roller bearings are used on ends of mainshaft. When mainshaft is serviced, input shaft bearing preload must be checked and adjusted.

A roller bearing is used on main countershaft end of Models 7000, 7600, 7610, 7700 and 7710 instead of ball bearing used on other models.

Except on Models 5000 and 5600 without Dual Power, rear bearing retainer is provided with holes to allow oil to pass be-

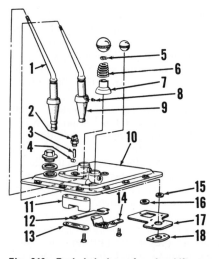

Fig. 208 — Filler plug (1) and dipstick (2) for Model 7000 tractors are located as shown. Same dipstick location applies for late model tractors.

Fig. 210 — Exploded view of early shift cover assembly for 8-speed transmission used on Model 5000 tractor. Refer to Fig. 211 for installation of washers (15 and 16) and baffle plate (17); early production units are not so equipped. Refer to Fig. 212 for late production shift cover.

1. Main shift lever	10. Cover
2. Starter safety switch	11. Plunger retainer
3. Plunger	12. Switch connector
4. Roller	13. Support plate
5. Snap ring	14. Lever stop plate
6. Spring	15. Washer
7. Retainer	16. Washer
8. Pins	17. Baffle
9. High-low shift lever	18. Lever stop plate

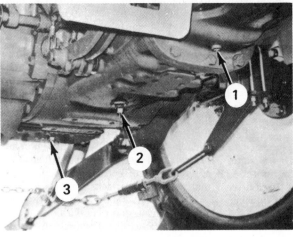

Fig. 209 — Drain plug locations on late model tractors. Plugs (1 and 2) are the same on older models.

1. Transmission drain
2. Rear axle drain
3. Two-speed pto drain

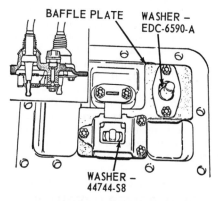

Fig. 211 — Bottom and cross-sectional view of Model 5000 8-speed shift cover showing washer and baffle plate installation; early models are not so equipped. Refer to exploded view in Fig. 210.

tween transmission housing and rear axle housing, thereby providing a common reservoir.

An oil supply tube connecting external oil filter manifold to transmission, provides additional lubrication for transmission when transmission, rear axle and hydraulic system share the same oil reservoir.

LUBRICATION

177. Transmission lubricant capacity for 5000 and 5600 models without Dual Power option is 12 quarts (11.46 liters). Recommended lubricant is SAE 80-EP lubricant (Ford specification No. M2C53A). Recommended lubricant change interval is after each 1200 hours of service. Filler plug is located in transmission shift cover; oil level plug is located on left side of transmission at front end of step plate; drain plug is at rear center of transmission housing.

Transmission lubricant capacity for 7000 models is 55 quarts (52 liters). This includes rear axle housing. Recommended lubricant is Ford specification No. M2C53A or M2C86A (General Specification: SAE 80-EP or SAE 20W/30). Recommended lubricant change interval is after each 1200 hours of service. Filler plug is located in transmission shift cover and oil level dipstick is located on left side of transmission as shown in Fig. 208. The drain plugs are located as shown in Fig. 209.

Lubricant capacity for Model 5600 with Dual Power is 56 quarts (53 liters) which is the same for Models 5610, 6600, 6610, 6700, 6710, 7600, 7610, 7700 and 7710 with or without Dual Power. When these tractors are equipped with two-speed (540 and 1000 rpm) pto, capacity is increased by 3.8 quarts (3.6 liters) On models equipped with front-wheel drive, install an additional 3.8 U.S. pints. Recommended lubricant is M2C134-A, M2C86-A or Ford 134 which correspond to a general specification of SAE 80 EP and SAE 20W/30.

NOTE: The transmission oil level is higher than center housing oil level when tractor is not operating. Therefore, it is essential that the oil level be checked with tractor standing on a level surface. Do not

fill transmission above top mark on dipstick.

REMOVE AND REINSTALL

178. To remove transmission, first split tractor between engine and transmission as outlined in paragraph 164, 165, 166 or 167, then remove steering gear and fuel tank assembly as a unit. Following procedures outlined in paragraph 254, 255 or 256 remove transmission assembly from rear axle center housing.

Reinstall transmission by reversing removal procedure.

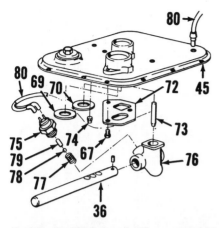

Fig. 212 – Exploded view of shift cover assembly used on late production 5000 model and all other late model transmissions except 6700, 6710, 7700 and 7710 models. Starter safety switch (75) and retainer (76) are mounted inside transmission on high-low shift rail (36) instead of in shift cover.

36. High-low shift rail	74. Dowel pin
45. Shift cover	75. Safety switch
67. Cap screw	76. Switch retainer
69. Washer	77. Spring
70. Washer	78. Steel ball
72. Lever stop plate	79. Switch plunger
73. Locating pin	80. Starter switch wire

Fig. 213 – Exploded view of shift cover components used on Models 6700, 6710, 7700 and 7710.

1. Range lever
2. "O" ring
3. Range shaft
4. Keys
5. Pivot shaft
6. "O" ring
7. Snap ring
8. Pin
9. Bushing
10. Shift rod
11. Snap ring
12. Main shift arm
13. Keys
14. Shaft
15. Spring
16. Ball
17. Boot
18. Seal
19. Snap ring
20. Shift cover
21. Dowel
22. Range shift finger
23. Main shift finger
24. Range shift lever
25. Main shift lever
26. Boot

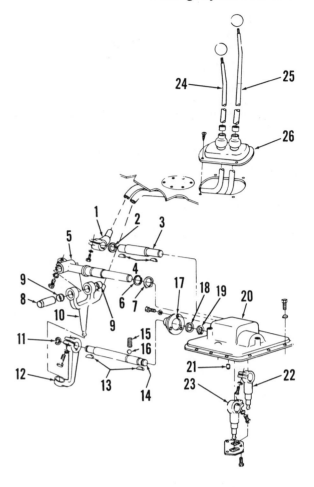

OVERHAUL

179. SHIFT COVER. On early cover (Figs. 210 and 211), disconnect wiring from starter safety switch, then unbolt and remove shift cover assembly. Safety switch plunger (3 – Fig. 210) should measure 1.275-1.280 inch (32.38-32.51 mm). Tighten shift lever stop plate and switch retainer cap screws to torque of 14-17 ft.-lbs. (19-23 N·m). Install dowel cap screw in rear center hole in cover, tighten dowel cap screw first, then tighten remaining cap screws to a torque of 35-47 ft.-lbs. (48-64 N·m). Be sure starter will operate only when main gear shift lever is in neutral position.

On late production cover (Fig. 212), except 6700, 6710, 7700 and 7710, unbolt and lift cover, disconnect starter switch wiring (80) from switch (75), then remove cover assembly. Shift lever units are same as on early cover shown in Fig. 210. Before reinstalling cover, be sure dowel (74 – Fig. 212) and switch retainer locating pin (73) are in place and that wire (80) is connected to switch (75). Then lower shift cover making sure dowel and locating pin enter cover correctly. Tighten cover retaining cap screws to a torque of 35-47 ft.-lbs. (48-64 N·m). Be sure starter will operate only when high-low shift lever is in neutral.

On Models 6700 and 7700, remove Dual Power control pedal if so equipped. On Models 6700, 6710, 7700 and 7710, remove floor mat and transmission access panel in cab floor. Unscrew shift lever knobs. Remove pinch bolt in high/

Fig. 214 – Exploded view of shifter rails, forks and associated parts used on Model 5000 with 8-speed transmission.

1. High speed shift rail
2. 2nd/6th/reverse shift rail
3. High-low shift rail
4. 1st/3rd/7th shift rail
5. Plug
6. Detent plug
7. Spring
8. Ball
9. Interlock plunger
10. Stop plunger
11. 4th/8th shift fork
12. 2nd/6th/reverse shift rail
13. High-low shift arm (early)
13A. High-low shift fork (late)
14. Shift gate
15. 1st/3rd/7th shift arm (early)
15A. 1st/3rd/7th fork (late)

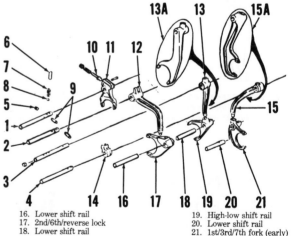

16. Lower shift rail	19. High-low shift rail
17. 2nd/6th/reverse lock	20. Lower shift rail
18. Lower shift rail	21. 1st/3rd/7th fork (early)

low shift rod end (1 – Fig. 213) and detach shift rod from shift shaft (3). Remove pinch bolt in pivot shaft end (5), drive out pin (8), and separate main shift rod end (10) from pivot shaft (5) and main shift arm (12). Remove boot retaining screws and withdraw shift levers. Remove snap ring (11) and main shift arm (12). Unscrew shift cover retaining screws, push Dual Power linkage forward (if so equipped) and remove shift cover. Be sure dowel (21) is in place before installing cover. Tighten shift cover

retaining screws to 35-47 ft.-lbs. (48-64 N·m).

180. SHIFT RAILS AND FORKS. To remove shift rails and forks on early models, first remove transmission from tractor as outlined in paragraph 178, remove top cover (paragraph 179), input shaft and front support plate (paragraph 181) and rear support plate, pto shaft, output shaft and secondary countershaft (paragraph 182). Then, refer to exploded view of shift mechanism in Fig. 214 and to assembled view in Fig. 215 and proceed as follows.

First, be sure all shift rails are in neutral position; then, unscrew locknut and set screw from high-low shift arm (13 – Fig. 214) and push top shift rail (3) and sealing plug out to rear of housing. On models so equipped, remove pin from high-low shift rail (36 – Fig. 212) and remove switch retainer (76) and switch assembly (75) as high-low shift rail is removed. Remove locknut and set screw from 1st-5th/3rd-7th shift connector (14 – Fig. 214) and shift arm (15) and push rail (4) and sealing plug out to front of housing. If necessary, remove sealing plug at rear of rail to gain access to push rail out forward. Unscrew locknut and set screw from Reverse/2nd-6th shift arm (12) and push rail (2) and sealing plug out of rear of housing. Unscrew locknut and set screw from the 4th-8th shift fork (11) and push rail (1) and seal-

ing plug out to rear of housing. Remove the four shift rail detent balls, springs and plungers (Fig. 217). Remove high-low coupling (Fig. 229) from rear of main countershaft, then unscrew locknut and set screw from high-low shift fork (13A – Fig. 214) and remove fork. Unscrew locknut and set screw from Reverse/2nd-6th shift fork (17) and slide rail (16) out to rear pushing sealing plug out with rail. Unscrew locknut and set screw from 1st-5th/3rd-7th shift fork (21) and push rail (20) out to front. Remove interlocking plug (Fig. 218) and interlock plungers (9 – Fig. 214).

On models produced after November 1969, high-low range shift fork (13A) and 1st-3rd shift fork (15A) are one-piece units and lower shift rails are eliminated. Except for 2nd-reverse shift fork (17), shifting components can be removed after splitting tractor between transmission and rear axle center hous-

ing as outlined in paragraph 254, 255 or 256; then removing shift cover as in paragraph 179 and rear support plate as in paragraph 182. Follow the general procedures outlined for earlier models

Fig. 219 — View of transmission front end assembly used on early model tractors.

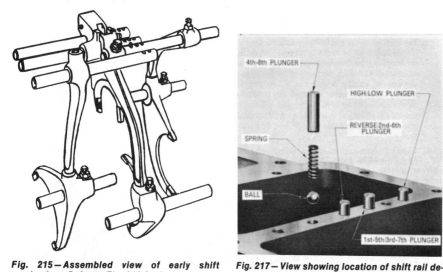

Fig. 215 — Assembled view of early shift mechanism. Refer to Fig. 216 for late units and to Fig. 214 for exploded view and parts identification.

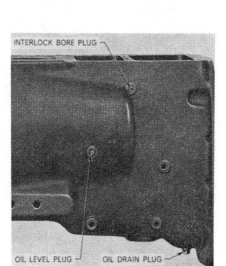

Fig. 217 — View showing location of shift rail detent plungers, springs and balls. Plungers are retained by transmission shift cover.

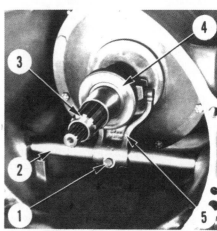

Fig. 220 — View of transmission front end assembly used on late model tractors.

1. Cap screw
2. Clutch release shaft
3. Input shaft
4. Release bearing and hub
5. Release fork

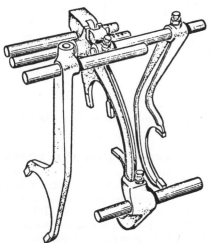

Fig. 216 — Assembled view of late shift mechanism. Refer also to Figs. 214 and 215.

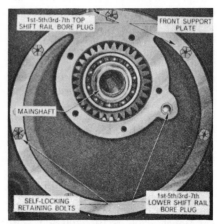

Fig. 218 — View showing location of interlock bore plug, oil level plug and drain plug on typical 8-speed non-synchromesh transmission case.

Fig. 221 — View of transmission front support plate and transmission mainshaft with clutch release bearing support and input shaft removed.

for further disassembly; and for removal of 2nd-reverse shift fork and rail.

Reinstall shift mechanism by reversing removal procedure, renewing damaged or worn parts as necessary. Shift rails must be placed in neutral position as they are installed so interlock pins will allow installation of remaining rails.

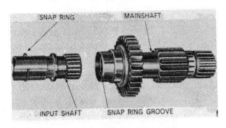

Fig. 222 — Input (clutch) shaft is retained in transmission mainshaft by a snap ring. Refer to items 7, 8 and 12 in Fig. 223.

Tighten fork and connector set screws and locknuts to a torque of 20-25 ft.-lbs. (27-34 N·m). When installing top high-low shift rail (36 – Fig. 212) in late production transmissions, be sure starter safety switch retainer (76) is properly positioned.

NOTE: There is no sealing plug on bottom high-low shift rail.

The plug sealing the 1st-5th/3rd-7th shift rail bore at front of transmission is ¼-inch (6.35 mm) thick; other bore sealing plugs are 5/16-inch (7.93 mm) thick.

181. INPUT SHAFT, CLUTCH RELEASE BEARING SUPPORT AND FRONT SUPPORT PLATE. The transmission input shaft, clutch release bearing support (input shaft housing) and

front support plate can be removed after splitting tractor between engine and transmission housing as outlined in paragraph 164, 165, 166 or 167. Then, proceed as follows:

On early model tractors, refer to Fig. 219 and disengage clutch release fork return spring from stop inside transmission housing. Remove the two clevis pins from fork and slide release shaft from housing.

On late model tractors, refer to Fig. 220. Remove cap screw (1) securing release fork (5) to release shaft (2), then slide release shaft from housing.

Remove release bearing and hub assembly from support, then unbolt and remove bearing support from front support plate. Remove snap ring (Fig. 222) retaining input shaft in transmission main shaft and remove input shaft.

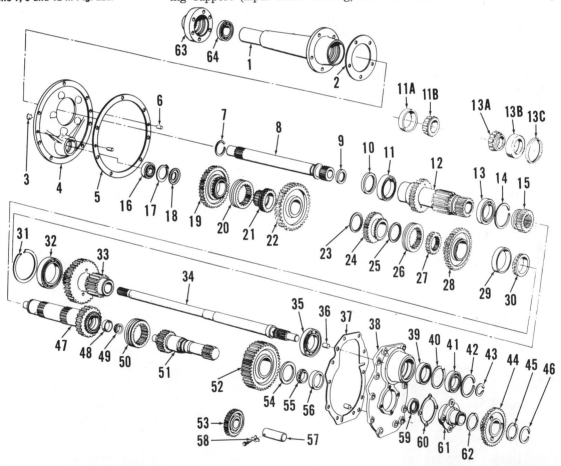

Fig. 223 — Exploded view of Model 5000, 5600, 5610, 6600, 6610, 6700 and 6710 eight-speed non-synchromesh transmission gears, shafts and related parts. Model 7000, 7600, 7610, 7700 and 7710 transmission is similar except for tapered bearings (11B and 13A) used on mainshaft, gears (12, 19, 33 and 47) are helical cut and a roller bearing is used in place of ball bearing (16). Component (2) is used as a gasket on Models 5000, 5600, 6600, 6610, 6700 and 6710 and as a selective shim (0.003, 0.005 and 0.012 inch) for adjusting input shaft bearing preload on Models 7000, 7600, 7610, 7700 and 7710. Oil seal (59) is not used on models which share transmission and center housing as a common reservoir.

1. Clutch release bearing support	11. Ball bearing	18. Thrust washer	30. Bearing cone
2. Gasket	11A. Bearing cup	19. 3rd gear	31. Snap ring
2. Shims	11B. Bearing cone	20. Shift collar	32. Ball bearing
3. Plug	12. Main shaft	21. Connector	33. Secondary counter-shaft
4. Front support plate	13. Ball bearing	22. 1st gear	34. Pto drive shaft
5. Gasket	13A. Bearing cone	23. Thrust washer	35. Ball bearing
6. Dowel pins	13B. Bearing cup	24. Reverse gear	36. Dowel pins
7. Snap ring	13C. Retaining ring	25. Thrust washer	37. Gasket
8. Input (clutch) shaft	14. Snap ring	26. Shift collar	38. Rear support plate
9. Oil seal	15. Sliding gear	27. Connector	39. Oil seal
10. Oil seal	16. Ball bearing	28. 2nd gear	40. Snap ring
	17. Snap ring	29. Bearing cup	41. Ball bearing
			42. Snap ring
			43. Snap ring
			44. Hydraulic pump idler gear
			45. Thrust washer
			46. Snap ring
			47. Countershaft
			48. Bearing cup
			49. Bearing cone
			50. Shift collar
			51. Output shaft
			52. Output shaft gear
			53. Reverse idler
			54. Thrust washer
			55. Bearing cone
			56. Bearing cup
			57. Reverse idler shaft
			58. Lock plate
			59. Oil seal
			60. Shims
			61. Bearing & idler support
			62. Thrust washer
			63. Pto drive plate
			64. Ball bearing

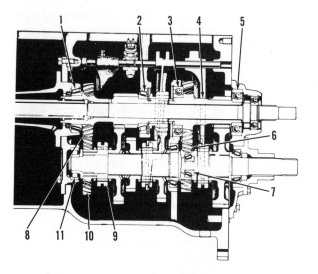

Fig. 224 – Cross-section view of transmission for Models 7000, 7600, 7610, 7700 and 7710 to show differences from basic 8-speed transmission shown in Fig. 223. Note helical-cut gears and tapered roller thrust bearings. See paragraph 181 for bearing preload instructions for these models.

1. Thrust bearing
2. Thrust bearing
3. Front bearings, secondary countershaft
4. Secondary countershaft assy.
5. Rear bearing, secondary countershaft
6. Main countershaft assy.
7. Output shaft bearing
8. Mainshaft assy.
9. 1-3, 5-7 coupling
10. 3-7 gear assy.
11. Front countershaft bearing

Remove self-locking cap screws retaining front support plate to housing and remove plate. If difficulty in removing plate is encountered, remove shift cover assembly and drive plate out with a suitable drift.

Remove input shaft seal (10 – Fig. 223) from support (1) and install new seal with lip to rear. Renew any excessively worn or damaged parts and reassemble by reversing disassembly procedure. The main countershaft front bearing (16) can be renewed at this time (Fig. 225); however, this is not recommended due to other possible related wear or damage which would not be discovered without disassembly of transmission. Refer to overhaul procedure for main countershaft. Install front support plate with new gasket and tighten self-locking cap screws to a torque of 23-30 ft.-lbs. (31-41 N·m). Use new snap ring (7 – Fig. 223) when reconnecting input shaft to mainshaft if old snap ring is distorted in any way. On 5000, 5600, 5610, 6600, 6610, 6700 and 6710 models, install release bearing support to support plate (lubricate seal lip) with new gasket and tighten self-locking cap screws to a torque of 35-47 ft.-lbs. (48-64 N·m).

When installing input shaft in 7000, 7600, 7610, 7700 and 7710 models, input shaft bearing preload must be checked and adjusted as follows: Use two 0.012 inch (0.30 mm) shims (2 – Fig. 223) between support (1) and front plate (4), install support on front plate and tighten retaining cap screws alternately to a torque of 35-47 ft.-lbs. (48-64 N·m). Remove snap ring which retains pto drive shaft rear bearing and move shaft rearward. Place transmission in neutral, then adapt an inch-pound (N·m) torque wrench to input shaft. Turn input shaft until it rotates steadily and record reading at which this occurs. Repeat this operation at least twice and average results.

Loosen support retaining cap screws and repeat operation already outlined to obtain a no-preload torque reading.

Now, subtract no-load (second) value from preload (first) value. The shaft rolling torque should be 10-20 in.-lbs. (1.130-2.260 N·m). If shaft rolling torque is not as stated, add shims to decrease or subtract shims to increase bearing preload. Shims are available in thicknesses of 0.003, 0.005 and 0.012 inch (0.08, 0.13 and 0.30 mm).

After initial bearing preload has been obtained, add another 0.005 inch (0.13 mm) shim between support and front plate.

Use a small box end wrench to hook release fork spring over stop in transmission housing on models so equipped.

182. REAR SUPPORT PLATE, PTO DRIVE SHAFT, OUTPUT SHAFT AND SECONDARY COUNTERSHAFT. With transmission removed from tractor as outlined in paragraph 178, proceed as follows:

Remove snap ring (46 – Fig. 223), hydraulic pump idler gear (44) and the two thrust washers (45 and 62) from output shaft retainer (61). Cut locking wire as needed and remove cap screws retaining rear support plate (38) to transmission housing. With a heavy soft-faced mallet, drive pto shaft (34) rearward to remove shaft and rear support plate as a unit. Pull secondary countershaft (33) rearward until front bearing (32) is free of bore in housing, lift rear end of shaft up and remove output shaft (51) and gear (52) as an assembly as shown in Fig. 227. Then, remove secondary countershaft assembly.

Disassemble components as follows: Remove snap ring (42 – Fig. 223) and drive or press pto shaft and bearing out to rear of support plate. Unbolt and remove output shaft bearing retainer and shims (60) from support plate. Re-

Fig. 225 – Removing transmission mainshaft front bearing assembly using special puller (Nuday tool SW-501).

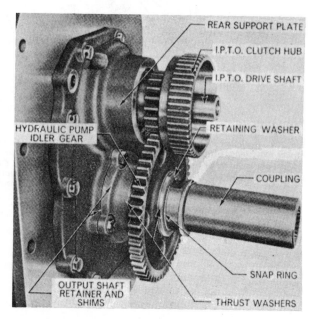

Fig. 226 – View of transmission rear support, pto clutch hub, hydraulic pump idler gear and drive shaft coupling installed. Note locking wire in rear support retaining bolts used on early models.

move bearing (41) from pto shaft. Drive seal (39) from support plate. Drive oil seal (59), then bearing cup (56) from output shaft retainer. Remove bearings (32 and 35) from secondary countershaft; refer to Fig. 228 for front bearing removal. Press output shaft from gear (52—Fig. 223), thrust washer (54) and bearing cone (55). Remove bearing cone (49) from front end of output shaft. Carefully inspect all parts and renew any that are excessively worn or damaged.

Reassemble components and reinstall in transmission by reversing removal and disassembly procedure and by observing the following: Install oil seals with lips to inside of transmission (forward). Be sure that high-low coupling (15) is engaged on teeth on rear end of mainshaft (12) and that snap ring (14) is in groove in bore of housing. Place secondary countershaft in top of rear compartment in transmission and hold shaft up while installing output shaft and gear assembly; refer to Fig. 227.

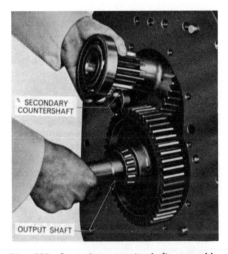

Fig. 227—Secondary countershaft assembly and output shaft assembly must be removed together as shown. Refer to text.

Fig. 228—Secondary countershaft front bearings can be removed by pressing against steel pins inserted through shaft against inner bearing race as shown.

Then, align front bearing of countershaft with bore in housing and bump shaft forward until bearing seats against snap ring (14—Fig. 223). Install rear support plate with new gasket, but without output shaft retainer or pto shaft. Align support plate on dowels and tighten retaining cap screws to a torque of 24-30 ft.-lbs. (33-41 N·m). Insert pto shaft and bearing assembly into transmission from rear and bump shaft forward until snap ring (42) can be installed. Install output shaft retainer with shim (60) thickness of at least 0.060 inch (1.524 mm) and tighten retaining cap screws to a torque of 24-30 ft.-lbs. (33-41 N·m). Measure end float of output shaft with a dial indicator as shown in Fig. 230, then remove shims (60—Fig. 223) in thickness equal to measured end float plus 0.002 inch (0.05 mm) for proper bearing preload. Shims are available in thicknesses of 0.003, 0.005 and 0.012 inch (0.08, 0.13 and 0.30 mm). Reinstall output shaft retainer and install locking wire (as needed) through heads of rear support plate retaining cap screws. Install hydraulic pump idler gear, thrust washers and retaining snap ring on output shaft retainer.

183. REVERSE IDLER. With shift mechanism removed as outlined in paragraph 180, bend down locking tabs from idler gear shaft retaining cap screw head and remove cap screw. Drive reverse idler shaft out towards front of transmission and remove reverse idler gear.

Inspect gear and bushing assembly and shaft for excessive wear or other damage. Gear and bushing are serviced as an assembly only. Renew parts as necessary and reinstall by reversing removal procedure. Use new lock tab washer, tighten retaining cap screw to a

Fig. 229—View showing rear of transmission with output shaft assembly and secondary countershaft assembly removed. Refer to text for further disassembly procedure.

torque of 15-18 ft.-lbs. (20-24 N·m) and bend tabs of washer against cap screw head.

184. MAINSHAFT. With shift mechanism removed as in paragraph 180 and reverse idler removed as outlined in paragraph 183, use suitable step plate and drift to drive mainshaft (12—Fig. 223) out towards front of transmission and remove the 4th-8th shift coupling (15). Remove power take-off shaft seal (9) from bore in front end of mainshaft and remove ball bearings (11 and 13) from shaft if worn or damaged and shaft is otherwise serviceable.

Install new seal (9) in front end of shaft with seal lip to rear of shaft. Press new ball bearings onto shaft until seated against shoulders. Be sure snap ring (14) is in place in bore of transmission housing, place shift coupling (15) on rear end of shaft with shift fork groove forward and drift the assembly rear-

Fig. 230—Checking transmission output shaft end play with dial indicator.

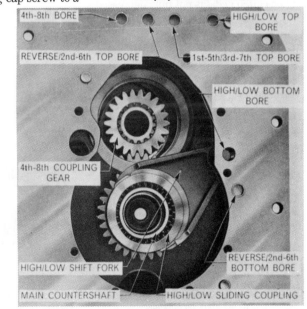

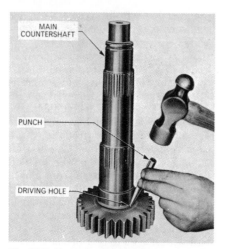

Fig. 231—Removing bearing from rear end of transmission main countershaft.

ward into the transmission housing until bearing (13) is seated against snap ring (14).

185. MAIN COUNTERSHAFT. With transmission mainshaft removed as outlined in paragraph 184, remove main countershaft (47–Fig. 223) as follows:

On Models 7000, 7600, 7610, 7700 and 7710, front end of main countershaft rides in a roller bearing which is removed with front plate (4–Fig. 223). Remove snap ring (17) and using a suitable puller, remove roller bearing race. Remove thrust washer (18). On all other models, front of main countershaft is supported by a ball bearing. Using a suitable puller, remove bearing (16) as shown in Fig. 225. Remove snap ring (17–Fig. 223) and thrust washer (18).

Pull main countershaft out from rear of transmission housing and remove gears, couplings, connectors and thrust washers as shaft is being withdrawn.

NOTE: Connectors (21 and 27) and couplings (20 and 26) are serviced in matched sets only and connectors should not be turned or interchanged on couplings.

Carefully inspect all parts and renew any that are excessively worn or damaged. If a bushing in a gear is worn or damaged beyond further use, gear and bushing must be renewed as an assembly. Bearing cup (48) can be driven out of rear bore of shaft with a punch inserted through holes in shaft (Fig. 231) after first removing bearing cone (30–Fig. 223).

Refer to exploded view in Fig. 223 and reinstall main countershaft by reversing removal procedure.

EIGHT-SPEED TRANSMISSION (SYNCHROMESH)

Transmission is fully synchronized. Speed and range shifting may be performed while tractor is moving. Eight forward and four reverse speeds are manually selected by two levers. Models equipped with Dual Power have an option of sixteen forward and eight reverse speeds. On Models 5610, 6610 and 7610, a safety start switch is provided to prevent engine from starting when speed selection lever is not placed in neutral (N) position. On Models 6710 and 7710, a safety start switch is provided to prevent engine from starting when clutch pedal is not fully depressed.

NOTE: Tractor should only be towed with both selector levers in neutral (N) position.

Models 6710 and 7710 are equipped with an interlock system to prevent speed selector lever movement unless clutch pedal is fully depressed.

LUBRICATION

186. Transmission and rear axle center housing share a common reservoir. Capacity for models with single-speed pto (540 rpm) is 56.0 U.S. quarts and for models with dual-speed pto (540/1000 rpm) is 60.0 U.S. quarts. On models equipped with front-wheel drive, install an additional 3.8 U.S. pints. Maintain fluid level at full mark on dipstick located in rear axle center housing (Fig. 232).

NOTE: Fluid level should be checked with tractor on a level surface, with hydraulic lift arms raised and with remote cylinders retracted.

Recommended lubricant is Ford 134 or a suitable equivalent. Recommended lubricant change interval is every 1200 hours of service. Filler plug is located at top backside of rear axle housing (Fig. 233). Transmission and rear axle drain plugs are located as shown in Fig. 234 for models without front-wheel drive and in housing of transfer case (Fig. 235) on models equipped with front-wheel drive.

Fig. 232—View showing location of dipstick (D) in rear axle center housing. Maintain fluid level to full mark on dipstick.

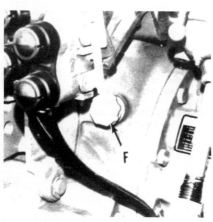

Fig. 233—Fill reservoir through filler plug (F) located in top backside of rear axle housing.

Fig. 234—View showing location of reservoir drain plugs on models not equipped with front wheel drive.

P. Pto (Dual Speed)
R. Rear axle
T. Transmission

REMOVE AND REINSTALL

187. To remove transmission unit, first split tractor between engine and transmission housing as outlined in paragraph 166 or 167. Remove components as needed from transmission assembly to expose only the transmission unit. Support transmission and rear axle center housing using suitable fixtures, then separate assemblies as outlined in paragraph 254, 255 or 256.

Reinstall transmission by reversing removal procedure.

OVERHAUL

Models 5610-6610-7610

188. **GEAR SHIFT ASSEMBLY.** Place gear shift levers in neutral (N) position. Disconnect battery ground cable. Remove knob and indicator from speed shift lever. Remove screws securing protective cover at top of steering

Fig. 235 — View showing location of drain plug (D) located in housing of transfer case on models equipped with front wheel drive.

Fig. 236 — View of lower shift control components used on Models 5610, 6610 and 7610 equipped with eight-speed synchromesh transmission.

1. Locking pin
2. Selector assy.
3. Shift control sleeve
4. Inner control shaft
5. Outer control shaft
6. Roll pin

column and gear shift plate. On models without a cab, loosen protective boot clamps and slide boot up outer control shaft. On cab equipped models, raise floor mat and remove screws securing lower column cover to front cowl panel. Disengage locking pin (1 – Fig. 236) from selector assembly (2). On cab equipped models, remove screws securing transmission cover to cab floor and withdraw cover. Loosen shift control sleeve cap screw (3), then slide sleeve up outer control shaft (5). Using a suitable punch and hammer, drive roll pin (6) from inner control shaft (4) and selector shaft (7). While supporting weight of gear shift assembly, remove cap screws and clamp securing plate assembly to

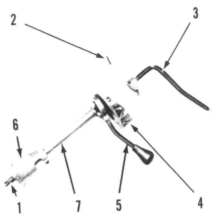

Fig. 237 — View showing gear shift assembly used on Models 5610, 6610 and 7610 equipped with eight-speed synchromesh transmission.

1. Inner control shaft	5. Range shift lever
2. Roll pin	6. Shift control sleeve
3. Speed shift lever	7. Outer control shaft
4. Plate assy.	

Fig. 238 — Exploded view of shift cover assembly used on Models 5610, 6610 and 7610 equipped with eight-speed synchromesh transmission.

1. Wiring harness connectors
2. Safety start switch
3. Oil seal
4. Cover
5. Oil seal
6. Selector assy.
7. Washer
8. Spring
9. Selector shaft
10. Spring
11. Snap ring
12. Washer
13. Shaft
14. Lever
15. Washer
16. Lever
17. Washer
18. Support
19. Roll pin (5/32 x 7/16 in.)
20. Snap ring
21. Pin
22. Nut
23. Lockwasher
24. Washer
25. Bellcrank
26. Bellcrank
27. Washer
28. Seal
29. Pin

steering column. Pull gear shift assembly away from dowel pin located in steering column. Withdraw gear shift assembly from selector shaft.

Scribe a reference mark on speed shift lever and inner control shaft relative to position for correct reassembly. Using a suitable punch and hammer, drive roll pin (2 – Fig. 237) from speed shift lever (3) and inner control shaft (1). Withdraw speed shift lever and inner control shaft. Lift plate assembly (4) off range shift lever (5). Remove roll pin securing range shift lever (5) to outer control shaft (7), then separate components. Slide shift control sleeve (6) off outer control shaft (7).

Inspect all components for excessive wear, cracks or any other damage.

NOTE: Examine roll pin and locking pin locating holes closely, a loose fitting pin could become dislodged during operation.

Renew all parts as needed.

Reassembly is reverse of disassembly. Use new roll pins when assembling components. Be sure reference marks on speed shift lever and inner control shaft are aligned before installing roll pin. Tighten clamp cap screws securing steering column to plate assembly to 32 ft.-lbs. (43 N·m) torque. Tighten shift control sleeve cap screw to 32 ft.-lbs. (43 N·m) torque.

189. **SHIFT COVER.** Remove gear shift assembly as outlined in paragraph 188. Disconnect wiring harness connectors (1 – Fig. 238) from safety start switch (2). Remove cap screws securing cover to transmission housing and withdraw cover (4).

Remove safety start switch (2) as needed. Remove nut (22) and lock-

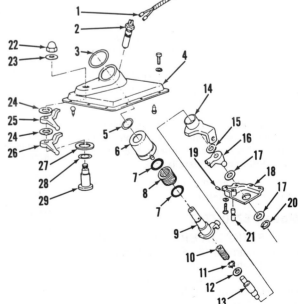

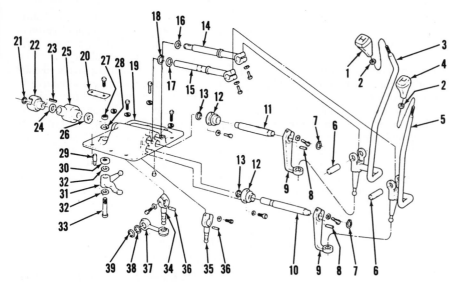

Fig. 239 — Exploded view of gear shift and cover assembly used on Models 6710 and 7710 equipped with eight-speed synchromesh transmission.

1. Knob	21. Snap ring
2. Locknut	22. Pivot rod stop
3. Shift lever (Speed)	23. Key
4. Knob	24. Washer
5. Shift lever (Range)	25. Guide assy.
6. Shaft	26. Washer
7. Snap ring	27. Nut
8. Key	28. Lockwasher
9. Shift arm	29. Pin
10. Shaft	30. Oil seal
11. Shaft	31. Bellcrank
12. Boot	32. Washers
13. Oil seal	33. Pin
14. Pivot shaft (Speed)	34. Shift finger (Speed)
15. Pivot shaft (Range)	35. Shift finger (Range)
16. Oil seal	36. Key
17. Oil seal	37. Speed shift link
18. Snap ring	38. Bearing halves
19. Cover	39. Circlip
20. Strap	

washer (23), then using a suitable mallet, drive pin (29) along with bellcrank assembly from cover (4). Drive dowel pins from support (18) and cover (4) using a suitable hammer and punch. Remove cap screws securing support to cover. Remove selector assembly, spring (8) and washers (7) from cover. Remove snap rings (11 and 20), then separate components to complete disassembly.

Inspect cover (4) for cracks or any other damage. Examine all other components for excessive wear or any other damage and renew as needed. Renew seals and gasket for reassembly.

Reassembly is reverse of disassembly. Lubricate all parts with clean Ford 134 lubricant or a suitable equivalent during reassembly. Tighten bellcrank retaining nut (22) to 77 ft.-lbs. (105 N·m) torque. Tighten support (18) cap screws to 32 ft.-lbs. (43 N·m) torque.

NOTE: Be sure ball ends on selector assembly levers and ball ends on bellcrank levers correctly engage gear shift gates during cover installation.

Tighten cover retaining cap screws to 48 ft.-lbs. (65 N·m).

Models 6710-7710

190. **SHIFT LEVERS AND COVER.** Loosen locknuts (2 – Fig. 239) and unscrew knobs (1 and 4). Loosen cap screws which secure pivot pins (6) in pivot shafts (14 and 15), then slide out pins (6). Withdraw shift levers (3 and 5)

from shift arm sockets. If cover is being removed without transmission removal, remove cab or platform as outlined in paragraph 415. Disconnect clutch interlock lever from guide assembly (25). Re-

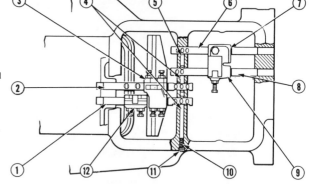

Fig. 240 — Gear shift mechanism used on Models 5610, 6610 and 7610.

1. Mainshaft synchronizer rail
2. Countershaft synchronizer rail
3. Countershaft synchronizer gate & fork assy.
4. Locking pin
5. Interlock
6. Reverse/low rail
7. Reverse/low gate & fork assy.
8. High range rail
9. High range gate & fork assy.
10. Spring
11. Plug
12. Mainshaft synchronizer gate & fork assy.

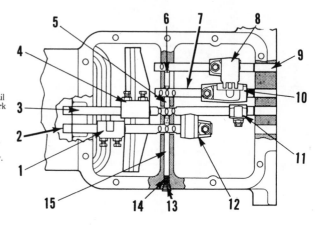

Fig. 241 — Gear shift mechanism used on Models 6710 and 7710.

1. Mainshaft fork
2. Mainshaft synchronizer rail
3. Countershaft synchronizer rail
4. Countershaft synchronizer fork
5. Locking pin
6. Interlock
7. Reverse/low rail
8. High range gate & fork assy.
9. High range rail
10. Reverse/low gate & fork assy.
11. Countershaft synchronizer gate
12. Mainshaft synchronizer gate
13. Plug
14. Spring
15. Locking pin

move cap screws securing cover to transmission housing and withdraw cover (19).

To disassemble proceed as follows: Remove cap screw and snap ring securing range shift finger (35) to shaft and arm assembly. Slide shaft and arm assembly from cover and withdraw shift finger from shaft. Remove cap screw and snap ring securing speed shift finger (34) to shaft and arm assembly. Remove nut (27) and lockwasher (28), then using a suitable mallet, drive pin (33) along with bellcrank assembly from cover (19). Slide speed shift finger (34) from shaft as an assembly with shift link (37) and bellcrank assembly (31). Remove snap ring (21), pivot rod stop (22), key (23), washer (24), guide assembly (25) and washer (26) from pivot shaft (14).

NOTE: It may be necessary to drive pivot shaft (14) from pivot rod stop (22).

Extract pivot shaft (14) from cover (19). Remove snap ring (18) and withdraw range pivot shaft (15) from cover (19). Remove cap screws securing boot (12) to cover, then remove boot.

Inspect cover (19) for cracks or any other damage. Examine all other components for excessive wear or any other damage and renew as needed. Renew seals and gasket.

Reassembly is reverse order of disassembly. Lubricate all parts with clean Ford 134 lubricant or a suitable equivalent during reassembly. Tighten bell-

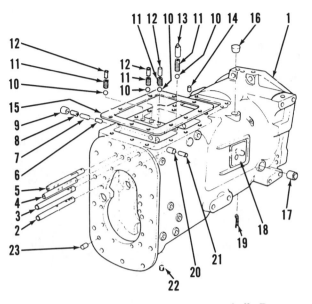

Fig. 242 – Exploded view of transmission housing and shift rail components used on all models with eight-speed synchromesh transmission.

1. Housing
2. Shift rail
3. Shift rail
4. Shift rail
5. Shift rail
6. Interlock
7. Locking pin
8. Spring
9. Plug
10. Ball (⅜ in.)
11. Spring
12. Pin (⅜ x 1.26 in.)
13. Pin (⅜ x 0.965 in.)
14. Pin (⅜ x 11/16 in.)
15. Gasket
16. Cover (Grommet W/DP)
17. Bushing
18. Cover
19. Cotter pin (5/32 x 1.0 in.)
20. Locking pin
21. Interlock
22. Plug
23. Pin (½ x 13/16 in.)

crank retaining nut (27) to 77 ft.-lbs. (105 N·m) torque.

NOTE: Be sure ball ends on shift fingers and ball end on bellcrank lever correctly engage gear shift gates during cover installation.

Tighten cap screws retaining cover to 48 ft.-lbs. (65 N·m).

Models 5610-6610-7610

191. **GEAR SHIFT MECHANISM.** With transmission removed as outlined in paragraph 187 and shift cover separated from housing as outlined in paragraph 189, proceed as follows for disassembly.

Remove all pins (12 and 13 – Fig. 242) and springs (11) from their housing bores. Loosen all locknuts on gearshift forks and gates. Remove cap screw securing reverse/low range gate and fork. Ensure high range rail (8 – Fig. 240) is in neutral position, then slide reverse/low range rail (6) rearward from transmission case.

NOTE: Reassemble gate and shift fork on rail. Lay rails out in sequence to ensure correct reassembly. Be sure to collect ball (10 – Fig. 242) from bore as shift rail is being removed.

Remove cap screw securing high range gate and fork. Slide high range rail (8 – Fig. 240) rearward from transmission case. Collect ball and reassemble gate and shift fork on rail, then lay to the side in correct sequence.

Remove cap screws securing countershaft synchronizer gate and fork assembly (3). Ensure mainshaft synchronizer rail (1) is in neutral position, then rotate countershaft synchronizer rail to push locking pin (4) away from rail. Slide countershaft synchronizer rail (2) rearward from transmission case and collect

ball. Remove plug (11), spring (10) and locking pin (4).

Remove cap screws securing mainshaft synchronizer gate and fork (12), then slide mainshaft synchronizer rail (1) rearward from transmission case. Collect ball and reassemble countershaft synchronizer gate and fork assembly (3) on countershaft synchronizer rail (2), then lay to the side in correct sequence. Reassemble mainshaft synchronizer gate and fork (12) on mainshaft synchronizer rail (1), then lay to the side in correct sequence.

Using a suitable magnet, withdraw remaining interlock pins and locking pin from transmission cross-drilling bore. Note correct position of pins.

Examine shift rails, gates and forks for excessive wear or any other damage and renew all parts as needed.

Reassemble in reverse order of removal. Tighten gearshift fork and gate cap screws and locknuts to 23 ft.-lbs. (31 N·m) torque.

NOTE: With mainshaft synchronizer rail (1) and countershaft synchronizer rail (2) in neutral position, ensure that shift fork pads are centered in synchronizer slot with synchronizer in neutral position.

NOTE: Slots of countershaft synchronizer gate and fork assembly (3) and mainshaft synchronizer gate (12) should align within 0.020 inch (0.508 mm) of each other when countershaft synchronizer rail (2) and mainshaft synchronizer rail (1) are in neutral position.

Models 6710-7710

192. **GEAR SHIFT MECHANISM.** With transmission removed as outlined in paragraph 187 and shift cover separated from housing as outlined in paragraph 190, proceed as follows for disassembly.

Remove all pins (12 and 13 – Fig. 242) and springs (11) from their housing bores. Loosen all locknuts on gearshift forks and gates. Remove cap screw securing high range gate and fork. Ensure reverse/low rail (7 – Fig. 241) is in neutral position, then slide high range rail (9) rearward from transmission case.

NOTE: Reassemble gate and shift fork on rail. Lay rails out in sequence to ensure correct reassembly. Be sure to collect ball (10 – Fig. 242) from bore as shift rail is being removed.

Remove cap screw securing reverse/low gate and fork assembly (10 – Fig. 241). Slide reverse/low rail (7) rearward from transmission case. Collect ball and reassemble gate and shift fork on rail, then lay to the side in correct sequence.

Remove cap screw securing countershaft synchronizer gate (11) and cap screws securing countershaft synchronizer fork (4). Ensure mainshaft synchronizer rail (2) is in neutral position, then rotate countershaft synchronizer rail to push locking pin (5) away from rail. Slide countershaft synchronizer rail (3) rearward from transmission case. Remove gate (11) and collect ball. Remove plug (13), spring (14) and locking pin (15).

Remove cap screw securing mainshaft synchronizer gate (12) and cap screws securing mainshaft fork (1). Slide mainshaft synchronizer rail (2) rearward from transmission case. Collect ball and reassemble countershaft synchronizer gate (11) and fork (4) on countershaft synchronizer rail (3), then lay to the side in correct sequence. Reassemble mainshaft synchronizer gate (12) and fork (1) on mainshaft synchronizer rail (2), then lay to the side in correct sequence.

Using a suitable magnet, extract remaining interlock pins and locking pin

Fig. 242A – Remove snap ring and thrust washer to withdraw hydraulic pump idler gear.

from transmission cross-drilling bore. Note correct position of pins.

Examine shift rails, gates and forks for excessive wear or any other damage and renew all parts as needed.

Reassemble in reverse order of removal. Tighten gearshift fork and gate cap screws and locknuts to 23 ft.-lbs. (31 N·m) torque.

NOTE: With mainshaft synchronizer rail (2) and countershaft synchronizer rail (3) in neutral position, ensure that shift fork pads are centered in synchronizer slot with synchronizer in neutral position.

All Models

193. **SUPPORT HUB, FRONT SUPPORT PLATE AND INPUT SHAFT.** Remove clip and withdraw pin securing clutch release fork to shaft and slide shaft from transmission case. Withdraw release bearing and fork from support hub. On models equipped with Dual Power, disassemble unit as outlined in paragraph 174. On models without Dual Power, remove cap screws retaining support hub (1 – Fig. 243) and separate hub from front support plate (9). Withdraw input shaft (11).

Remove cap screws securing front support plate (9) to transmission case and extract support plate.

NOTE: Support plate may be driven from transmission case by using a suitable punch and hammer inserted through shift cover opening.

Inspect support hub (1) and front support plate (9) for cracks or any other damage. Examine bearing (14) located in front support plate (9) for roughness, corrosion, excessive wear or any other damage. Inspect input shaft and splines for excessive wear or any other damage. Inspect oil seal lips for hardness, excessive wear or any other damage. Renew all parts as needed using suitable tools.

Reassemble in reverse order of removal, renewing gasket (10), oil seals (2 and 3) and "O" ring (8) as needed. Lubricate all working components and seal lips with clean transmission fluid during reassembly. Tighten front support plate (9) and support hub (1) cap screws to 35 ft.-lbs. (48 N·m) torque.

194. **REAR SUPPORT PLATE, PTO SHAFT, OUTPUT SHAFT ASSEMBLY AND RANGE CLUSTER GEAR ASSEMBLY.** Remove snap ring (Fig. 242A) securing hydraulic pump idler gear on bearing retainer (72 – Fig. 243), then withdraw thrust washer and gear. Remove cap screws securing rear support plate (67), then using a suitable soft-faced mallet, drive input end of pto shaft (37) rearward to separate rear

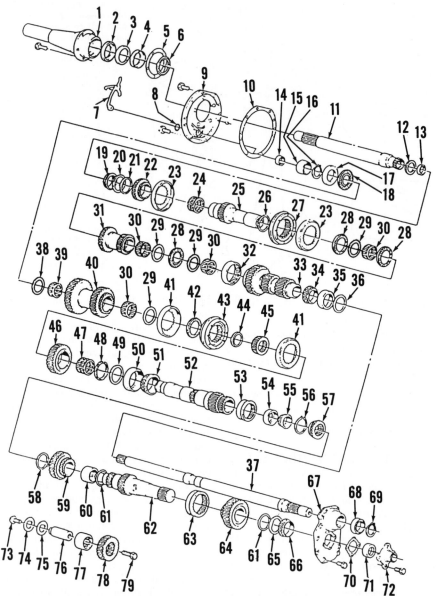

Fig. 243 – Exploded view of typical eight-speed synchromesh transmission gears, shafts and related parts. Some parts will differ on models equipped with Dual Power.

1. Support hub	22. Mainshaft gear	41. Synchronizer cup
2. Seal	23. Synchronizer cup	42. Thrust bearing
3. Seal	24. Needle roller bearing	43. Synchronizer assy.
4. Bearing cup	25. Mainshaft	44. Thrust washer
5. Gasket	26. Thrust washer	45. Inner coupling
6. Shield	27. Synchronizer assy.	46. Countershaft gear
7. Tube assy.	28. Thrust bearing	47. Needle roller bearing
8. "O" ring	29. Thrust washer	48. Thrust washer
9. Front support plate	30. Bearing	49. Washer
10. Gasket	31. Mainshaft cluster gear	50. Bearing cup
11. Input shaft	32. Bearing	51. Bearing cone
12. Seal	33. Range cluster gear	52. Countershaft
13. Bushing	34. Bearing cone	53. High range coupling
14. Bearing	35. Bearing cup	54. Bearing cone
15. Race	36. Shim	55. Bearing cone
16. Snap ring	37. Pto shaft	56. Snap ring
17. Shim	38. Thrust washer	57. Connector
18. Thrust bearing	39. Bearing	58. Thrust washer
19. Bearing cone	40. Countershaft cluster	59. Reverse gear
20. Washer	gear	60. Bushing
21. Thrust washer		

61. Thrust washer
62. Output shaft
63. Coupling
64. Output shaft gear
65. Washer
66. Bearing cone
67. Rear support plate
68. Bearing
69. Snap ring
70. Shim
71. Bearing cup
72. Bearing retainer
73. Cap screw
74. Washer
75. Spacer
76. Shaft
77. Bearing
78. Idler gear
79. Cap screw

support plate (67) from transmission case. Withdraw rear support plate (67) and pto shaft (37) as an assembly. Remove output shaft assembly (62 – Fig. 244) from rear of transmission. Slide high range coupling (53) from countershaft end (52). Pry range cluster

gear (33) rearward until front stub of gear disengages roller bearing (32 – Fig. 243) located in center web of transmission case, then withdraw range cluster gear assembly from transmission case. Disassemble output shaft assembly with reference to Fig. 243. Remove snap ring (69–Fig. 243) and separate bearing (68) and pto shaft (37) from rear support plate (67).

Examine all gears and splines for chipped or missing teeth, cracks, excessive wear or any other damage. Inspect bearings for corrosion, roughness, cracks, excessive wear or any other damage. Inspect thrust washers and rear support plate for cracks, excessive wear or any other damage. Examine shim (36) located behind range cluster gear rear bearing cup (35). Tapered bearing cone and cup must be serviced as a matched set. Renew all parts as needed using suitable tools.

If range cluster gear parts are renewed, end play between mainshaft and range cluster gear must be checked and adjusted as needed. Reassemble front end components as outlined in paragraph 193. Reassemble range cluster gear to mainshaft assembly. Install rear support plate and tighten securing cap screws to 35 ft.-lbs. (48 N·m) torque. Position dial indicator as shown in Fig.

245 and place indicator gage needle on range cluster gear, parallel to mainshaft. Move cluster gear forward and backward while observing indicator needle movement. End play should be 0.008-0.011 inch (0.20-0.28 mm). To adjust end play, remove rear support plate (67–Fig. 243) and withdraw bearing cup (35). Increase or decrease thickness of shim (36) until correct end play is attained. Shims (36) are available in thicknesses of 0.005, 0.012 and 0.020 inch (0.127, 0.305 and 0.510 mm).

A dial indicator mounted as shown in Fig. 246 is used to measure end play between output shaft and countershaft. End play should be 0-0.004 inch (0-0.102 mm) for models equipped with reduction gear (Paragraph 251) and 0-0.0025 inch (0-0.064 mm) for models without reduction gearbox. Adjust end play by removing retainer (72–Fig. 243) and adjusting shim (70) until correct end play is attained.

Reassemble in reverse order of disassembly. Lubricate all working components with clean transmission fluid

during reassembly. Tighten cap screws securing rear support plate (67) and bearing retainer (72) to 35 ft.-lbs. (48 N·m) torque.

195. **COUNTERSHAFT ASSEMBLY AND MAINSHAFT ASSEMBLY.** Remove front end components as outlined in paragraph 193. Remove rear end components as outlined in paragraph 194. Remove countershaft front bearing race (15–Fig. 247) from countershaft end. Remove snap ring (16) and withdraw shim (17–Fig. 243) and thrust bearing (18). Slide countershaft (52) rearward enough to allow countershaft cluster gear (40–Fig. 247) to lie on bottom of transmission case.

Remove bearing (19–Fig. 243) from mainshaft end (25). Use a suitable bearing puller and shaft end protector and withdraw bearing (19). Remove washer (20), thrust washer (21), mainshaft gear (22) and synchronizer assembly (27). Lift mainshaft (25) with remaining components from transmission case.

Withdraw countershaft (52) from the rear and remove countershaft cluster gear (40), synchronizer assembly (43), inner coupling (45), countershaft gear (46), thrust bearings, thrust washers and washer as they become free from shaft.

Remove cap screws securing reverse idler gear shaft (76) and withdraw idler shaft, idler gear (78), spacer (75) and washer (74).

Complete disassembly of components as needed with reference to Fig. 243.

Examine all gears and splines for chipped or missing teeth, cracks, excessive wear or any other damage. Inspect bearings for corrosion, roughness, cracks, excessive wear or any other damage. Inspect thrust washers for cracks, excessive wear or any other damage. Renew all parts as needed using suitable tools.

Countershaft end play must be checked and adjusted as needed before

Fig. 244—Removal of output shaft assembly (62) from rear of transmission. Slide high range coupling (53) from countershaft end (52). Note range cluster gear (33).

Fig. 246—View showing correct mounting position of dial indicator (D) for checking end play between output shaft (S) and countershaft. Refer to text.

Fig. 245—View showing correct mounting position of dial indicator gage (D) for checking end play between mainshaft and range cluster gear (R). Refer to text.

Fig. 247—View showing countershaft front bearing race (15), mainshaft assembly (25), countershaft cluster gear (40) and snap ring (16).

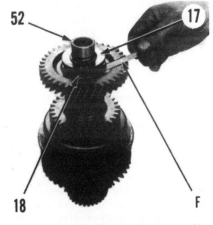

Fig. 248—View showing procedure for checking countershaft (52) end play by measuring gap between shim (17) and thrust bearing (18) using feeler gage (F). Correct end play is 0.0005-0.003 inch (0.013-0.076 mm).

reinstalling assembly in transmission. Reassemble components as shown in Fig. 248. Use a feeler gage (F) and measure the gap between shim (17) and thrust bearing (18) at three locations around shaft (52). Correct end play is 0.0005-0.003 inch (0.013-0.076 mm). Adjust end play by changing shim (17) until correct gap is attained.

Install in reverse order of removal. Check and adjust end play between mainshaft and range cluster gear as outlined in paragraph 194. Lubricate all working components with clean transmission fluid during reassembly. Tighten rear end components as outlined in paragraph 194. Tighten front end components as outlined in paragraph 193. Check and adjust end play between output shaft and countershaft as outlined in paragraph 194.

SELECT-O-SPEED (10-SPEED) TRANSMISSION (Model 5000)

The Select-O-Speed transmission is a planetary gear drive unit providing 10 forward and two reverse speeds. Desired gear ratio is selected by moving a control lever and starting, stopping or changing gear ratios is accomplished without operation of a conventional clutch. A foot-operated feathering valve is provided for interrupting gear train in case of emergency or for close maneuvering such as hitching or unhitching of implements.

To better understand operation of Select-O-Speed transmission, fundamental operating principles of a planetary gear system are outlined in following paragraphs 196 through 203.

PLANETARY GEAR POWER FLOW

196. Refer to Fig. 249; the three "elements" of a planetary system are sun gear, pinion carrier and ring gear. When any element is rotated, the other two elements will also turn unless one is held by an external force. Depending upon which element is held, power can be applied or taken out at the sun gear, pinion carrier or ring gear. The possible means of obtaining different gear ratios from a planetary gear set are as follows:

197. Turning sun gear and holding ring gear forces pinions to turn within ring gear moving pinion carrier with them. Thus, carrier turns in the same direction as sun gear but at a slower speed (Underdrive gear ratio).

198. Turning ring gear and holding sun gear forces pinions to turn within ring gear moving carrier with them; pinion carrier turns in the same direction as ring gear, but at a slower speed (Underdrive gear ratio).

199. Turning pinion carrier and holding ring gear forces pinions to turn within ring gear causing sun gear to turn in the same direction as carrier, but at a higher speed (Overdrive gear ratio).

200. Turning pinion carrier and holding sun gear forces pinions to turn around sun gear causing ring gear to turn in the same direction as carrier, but at a higher speed (Overdrive gear ratio).

201. Turning sun gear and holding pinion carrier forces pinions to act as idlers turning ring gear in opposite direction from sun gear and at a slower speed (Underdrive reverse gear ratio).

202. Turning ring gear and holding pinion carrier forces pinions to act as idlers turning sun gear in opposite direction from ring gear and at a higher speed (Overdrive reverse gear ratio).

203. Locking any two units of a planetary system together results in a solid drive unit and if any element is turned, all three elements turn in the same direction at the same speed (Direct drive gear ratio).

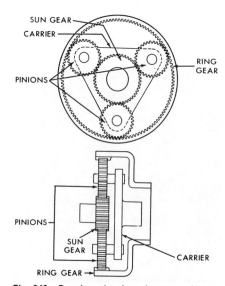

Fig. 249—Drawing showing elements of planetary gear system. Refer to paragraphs 196 through 203 for different gear ratios obtainable from such a gear set.

TRANSMISSION OPERATION

204. **PLANETARY SYSTEMS.** The "Select-O-Speed" transmission utilizes three planetary systems designated, from front of transmission to rear, as "A", "B" and "C" planetary units. Elements of each unit are designated as "A" sun gear, "A" carrier, "A" ring gear, etc.

Six basic speed ratios, five forward and one reverse, are obtained by various combinations of holding or applying power to the elements of the "B" and "C" planetary units. The "A" (front) planetary unit is used as a direct drive-overdrive unit to double the basic speed ratios providing ten forward and two reverse speeds.

205. **PLANETARY CONTROLLING UNITS.** Three brake bands and four multiple disc clutches are used to control the "A", "B" and "C" planetary units to provide the different speed ratios.

The three brake bands are designated, from front to rear, as Band 1, Band 2 and Band 3. The bands are actuated by hydraulic servos designated Servo 1, Servo 2 and Servo 3 to correspond with band numbers. The servos contain springs that work in the opposite direction from hydraulic pressure. The servos apply the bands with spring pressure and release the bands utilizing hydraulic pressure. The bands, when applied, hold planetary elements stationary as follows:

Band Applied	Planetary Element Held
Band 1	"A" Sun Gear
Band 2	"B" Ring Gear and "C" Sun Gear
Band 3	"C" Carrier

Multiple disc clutch packs are designated, from front to rear, as Direct Drive Clutch, Clutch 1, Clutch 2 and Clutch 3. (Clutch 2 and Clutch 3 are con-

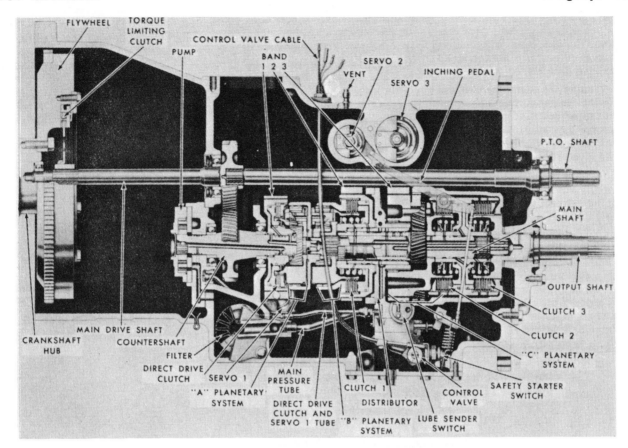

Fig. 250 — Cross-sectional view of Model 5000 Select-O-Speed transmission.

tained in the same housing.) Clutches are engaged by hydraulic pressure against a piston within clutch unit and are disengaged by spring pressure returning piston to disengaged position. Clutches, when engaged, lock planetary elements together as follows:

Clutch Engaged	Planetary Elements Locked
Direct Drive	"A" Sun Gear to "A" Carrier
Clutch 1	"B" Carrier to "B" Ring Gear
Clutch 2	"B" Carrier to "C" Carrier
Clutch 3	"B" Carrier to Output Shaft

SELECT-O-SPEED HYDRAULIC SYSTEM

206. **HYDRAULIC CIRCUITS.** Refer to Fig. 251 for diagram of the "Select-O-Speed" hydraulic circuits.

Pressure is supplied directly to Band 1, Band 2 and Band 3 control valves; indirectly to Clutch 1, Clutch 2 and Clutch 3 control valves via transmission sequencing valve and transmission feathering valve (indirect transmission circuit) and indirectly to transmission cooling and lubrication circuit via transmission regulating valve.

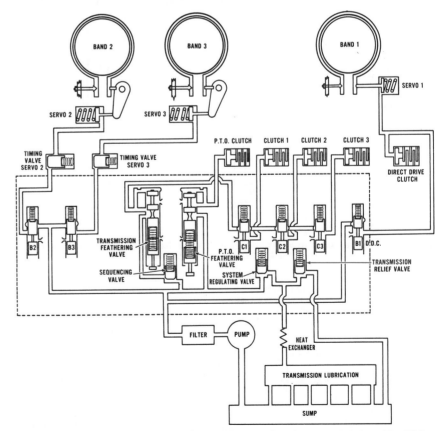

Fig. 251 — Schematic diagram of Select-O-Speed transmission hydraulic system. Refer to Fig. 252 for diagram showing operation of feathering valve and to Figs. 253 and 254 for operating principles of the Servo 2 and 3 timing valves. Pto feathering valve is not used in Model 5000 transmission.

Refer to Fig. 252 for diagrams showing operation of transmission and feathering valves. With inching pedal (transmission, feathering valve control pedal) up, pressure is directed to indirect transmission circuit. With inching pedal fully depressed, feathering valves block pressure from indirect transmission circuit. With inching pedal partially depressed, pressure in indirect transmission circuit can be "feathered" to provide smooth starting of tractor. This is similar to slipping a conventional clutch. Transmission should not be operated for any period of time with the control in "feathering" position.

The transmission **sequencing valve** separates the direct and indirect hydraulic circuits so oil is supplied only to direct circuit when system oil pressure drops below valve setting of 125 psi (862.5 kPa). This prevents Bands 2 and 3 from locking up transmission when large quantities of oil are required such as during the 4-5 and 5-4 shifts.

The system **regulating valve** controls pressure within the transmission hydraulic system to 200-210 psi (1380-1449 kPa) at 800 engine rpm and oil temperature of 80°-120°F (26.6°-48.8°C). When pressure in system exceeds regulating valve pressure setting, valve opens and oil passes into transmission cooling and lubrication circuit. When system pressure is below regulating valve setting, oil pressure in lubrication circuit drops to zero and the oil pressure warning light on models so equipped will come on.

The transmission **relief valve** limits maximum pressure in the hydraulic system to 220-230 psi (1518-1587 kPa). Oil by-passing relief valve is returned to sump via return tube attached to control valve assembly.

Timing valves are incorporated in Servo 2 and Servo 3 circuits to permit gradual application of bands and to pre-

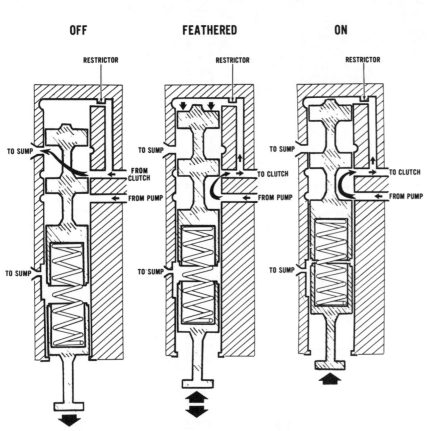

Fig. 252—Cross-sectional views showing transmission feathering valve assembly in "OFF," "FEATHERED" and "ON" positions.

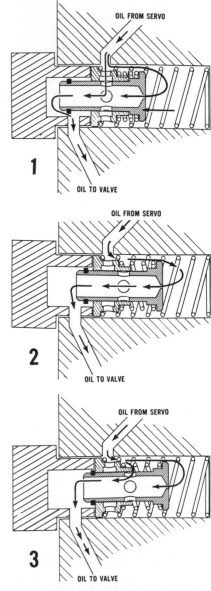

Fig. 254—Sequence views of Servo 2, Servo 3 timing valve during time servo hydraulic pressure is being released, and servo spring pressure is applying band. In view 1, valve position allows a fast initial pressure drop in servo to begin band application. As servo pressure decreases, timing valve spring moves valve to permit a relatively slow continued servo pressure drop as shown in view 2 to permit gradual band application. In view 3, servo hydraulic pressure has decreased far enough to allow timing valve spring to move valve so a final dumping of oil from servo can take place, completely applying band by servo spring pressure.

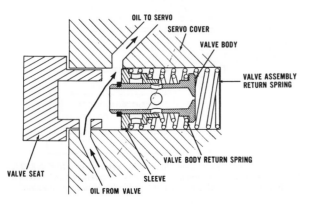

Fig. 253—Cross-sectional view of Servo 2, Servo 3 timing valve. Valve position during time when hydraulic pressure is being applied to servo is shown. Refer to sequence views in Fig. 254 for valve positions during time servo hydraulic pressure is being released.

vent both Band 2 and Band 3 from being engaged at the same time during the 4-5 and 5-4 shifts. During these shifts, one servo is pressurized to release its corresponding band and pressure is released from other servo to apply its corresponding band. Oil flow to servo (Fig. 253) to release band is not restricted as valve assembly return spring will be compressed and oil will bypass valve. On the other servo, return oil will first compress valve body return spring (Step 1–Fig. 254) and oil will flow through both valve orifice and the four annular ports. As servo oil pressure begins to drop, valve body return spring expands moving valve to close off the four annular ports (Step 2) which slows down the flow of oil from servo and allows band to gradually apply. When servo pressure drops further, valve body spring expands to full length again opening the four annular ports (Step 3) to provide a final dumping of oil to permit the band to be completely applied.

When Band 1–Direct Drive Clutch spool in control valve assembly is closed to pressurize circuit, oil pressure will apply Direct Drive Clutch as Servo 1 spring is compressed to release Band 1. Conversely, as spool is opened to release circuit pressure, spring in Servo 1 will apply Band 1 as Direct Drive Clutch is released.

207. CONTROL POSITIONS (GEAR SELECTION). Fourteen control positions are provided on the gear selector. Units of the transmission that are applied in each position of the gear selector are shown in the Application of Bands and Clutches Chart (Fig. 255).

LUBRICATION AND FILTERS

208. As the Select-O-Speed transmission incorporates a hydraulic control system, use of correct lubricating oil and keeping the oil clean is of utmost importance.

Recommended fluid is Ford Specification No. M2C41-A transmission and hydraulic oil. Capacity is 11.8 quarts (11.2 liters). Fluid should be maintained to bottom of oil level plug (Fig. 256). Add oil through oil level plug openings.

To remove pump intake screen, first drain oil from transmission, remove cover plate from bottom of transmission, then remove filter screen as in Fig. 257.

The pleated paper filter element should be renewed each 600 hours of

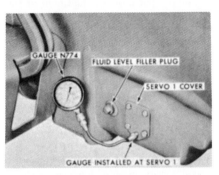

Fig. 256—View showing transmission fluid (oil) level-filler plug on Model 5000. View above also shows installation of test gage at servo 1.

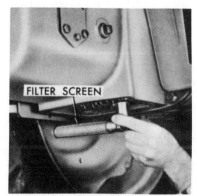

Fig. 257—View showing transmission pump filter screen being removed from Model 5000 transmission.

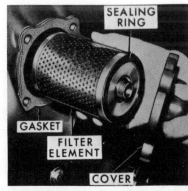

Fig. 258—Removing Model 5000 transmission oil filter element; element is threaded into cover.

POSITION OR GEAR	DIRECT DRIVE CLUTCH	BAND 1 (B1)	BAND 2 (B2)	BAND 3 (B3)	CLUTCH 1 (C1)	CLUTCH 2 (C2)	CLUTCH 3 (C3)
PARK (P)		A	A	A			
R2		A		A	A		
R1	A			A	A		
NEUTRAL (N)	A						
1ST	A			A			A
2ND	A			A		A	
3RD		A		A			A
4TH		A		A		A	
5TH	A		A				A
6TH	A		A			A	
7TH		A	A				A
8TH		A	A			A	
9TH	A				A	A	
10TH		A			A	A	

Fig. 255—Chart showing transmission units applied at each gear selector position. Bands are applied by servo spring pressure and are released by hydraulic pressure. Clutches are applied by hydraulic pressure and are released by spring pressure. Hydraulic pressure is indicated by shaded area.

service and the fluid changed every 2400 hours to coincide with each 4th oil filter change. To remove oil filter, unbolt filter cover (Fig. 258) and unscrew filter from cover. When reinstalling filter cover, tighten retaining bolts to a torque of 35-40 ft.-lbs. (48-54 N·m).

SYSTEM ADJUSTMENTS

Malfunction of the "Select-O-Speed" transmission could be from a number of causes, the most common of which is maladjustment of one or more units of the transmission. Therefore, the first step in correcting troubles would be a complete operational adjustment of the three transmission bands and pressure control valves within control valve assembly and correcting any misalignment of the transmission.

As some adjustments are made with engine running and gear ratio selector lever in an operational position, the first step is to disengage traction coupling as follows:

209. TRACTION COUPLING. All tractors equipped with a "Select-O-Speed" transmission have a traction coupling sleeve which can be shifted to disengage transmission output shaft from differential pinion. The traction coupling shift lever is located as shown in Fig. 259 for Model 5000 tractors. Move lever forward to disengage traction coupling.

210. FLUID LEVEL CHECK. Before an attempt is made to start or service the tractor, first check transmission fluid level. To check fluid level, remove pipe plug (Fig. 256) located on right side of transmission housing. Fluid level should be even with bottom of plug opening with tractor standing level. If fluid level is below bottom of plug opening, add fluid as necessary through opening. Recommended fluid is Ford M2C41-A transmission and hydraulic oil.

211. PRESSURE CHECK. Provisions are made for installation of pressure gages in Servos 1, 2 and 3 hydraulic circuits as shown in Figs. 256 and 260. However, pressure adjustments and system diagnosis can usually be made with gage installed at Servo 2 location only. With a 0-300 psi (0-2070 kPa) gage installed at Servo 2, proceed as follows:

Start engine and operate transmission until fluid temperature is 80°-120°F (26.6°-48.8°C). Disconnect traction coupling as outlined in paragraph 209. With engine running at 800 rpm, place gear selector in Neutral (N) position. Gage reading should then be 200-210 psi (1380-1449 kPa). A gage reading higher than 210 psi (1449 kPa) would indicate need of adjusting transmission regulating valve as outlined in paragraph 212. If pressure gage reading is below 200 psi (1380 kPa), proceed with hydraulic circuit checks as outlined in paragraph 206. If, during the checks, the gage reading remained evenly low, readjust transmission regulating valve as outlined in paragraph 212.

If pressure gage reading with engine running at 800 rpm was in the specified range of 200-210 psi (1380-1449 kPa), increase engine speed to 2400 rpm. Pressure gage reading should then be 220-230 psi (1518-1587 kPa). If pressure is not within the range specified at 2400 engine rpm, readjust transmission relief valve as outlined in paragraph 212.

If transmission lubrication light (when used) remains on, or if pressure gage reading at Servo 2 location is within specified range, check lubrication circuit pressure as follows:

Obtain a spare oil line adapter bolt (Ford part No. C5NN-7D192-B), drill through head of bolt and thread to connect a hydraulic pressure gage. Install drilled adapter bolt and gage in bottom heat exchanger line (B – Fig. 261). With engine running at 800 rpm, gage reading should be above 22 psi (151.8 kPa). If gage reading is above that specified, renew lubrication light sending switch (when used).

If lubrication pressure is below specified pressure, a leak in transmission lubrication circuit is indicated.

212. CONTROL VALVE ADJUSTMENT. To adjust control valve, valve assembly must first be removed from transmission (refer to paragraph 223) and adjusting screw retainers (11 and 12 – Fig. 272) be removed from upper valve housing. If valve assembly has been disassembled, turn adjusting screws in flush with upper valve housing.

NOTE: On Model 5000 tractor, control valve assembly cannot be adjusted with valve mounted in transmission due to valve location. To adjust the 5000 control valve, remove valve, make trial adjustment, then reinstall valve and recheck pressure.

With transmission fluid at operating temperature of 80°-120°F (26.6°-

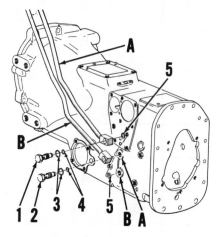

Fig. 260 – Hydraulic test gages installed at Servo 2 and Servo 3 pressure check ports in servo cover; refer to Fig. 256 for installing gage in Servo 1.

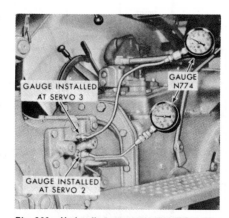

Fig. 261 – View showing method of attaching transmission oil cooler tubes to Model 5000 transmission housing.

A. Return tube & port	3. "O" rings
B. Pressure tube & port	4. "O" rings
1. & 2. Adapter bolt	5. "O" rings

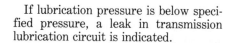

Fig. 259 – View showing location of drive line (shaft) disengagement lever on Model 5000.

48.8°C), a 0-300 psi (0-2070 kPa) gage installed at Servo 2 location (Fig. 260) and with traction coupling disengaged, place valve camshaft in Park position (all control valves out and aligned), ground starter safety switch wire and start engine. Set engine speed to 800 rpm and turn valve camshaft to Neutral position (three detent positions away from Park; Servo valves will be pushed in, clutch valves will be out). With sequencing valve adjusting screw retainer, turn transmission regulating valve adjusting screw so gage reading is 200-210 psi (1380-1449 kPa). Set engine speed at 4200 rpm and with retainer, turn transmission relief valve adjusting screw so gage reading is 220-230 psi (1518-1587 kPa). Final adjustment for sequencing valve is adjusting screw flush with upper valve housing.

After final adjustments are made, remove control valve assembly and reinstall adjusting screw retainers. If slots in adjusting screws are not aligned for retainer installation, turn screw in as required (not more than ¼-turn). Reinstall control valve assembly and transmission cover as outlined in paragraphs 223 and 221.

213. **TRANSMISSION BAND ADJUSTMENT.** Before adjusting transmission bands, operate transmission until fluid temperature is 80°-120°F (26.6°-48.8°C). Disengage traction coupling as outlined in paragraph 209, refer to Figs. 262 and 263; then, proceed as follows:

BAND NO. 1: Adjust Band 1 with engine stopped. Loosen adjusting screw locknut at least two full turns while holding adjusting screw from turning. Tighten adjusting screw to a torque of 19-21 ft.-lbs. (26-29 N·m). Check to be sure locknut did not turn down tight

against sealing washer; if so, back nut off farther and retighten adjusting screw to specified torque. Then, back adjusting screw out exactly 1¼ turns, hold adjusting screw stationary and tighten locknut to a torque of 20-25 ft.-lbs. (27-34 N·m).

BAND NO. 2: Adjust Band 2 with engine running at 800 rpm. Move selector lever to Park position to aid in holding adjusting screw and back off locknut at least two full turns while holding screw from turning. Move selector lever to neutral position and tighten adjusting screw to a torque of 110-130 inch-pounds (12.429-14.689 N·m). Check to be sure locknut did not turn down tight against sealing washer; if so, back nut off farther and retighten adjusting screw to specified torque. Then, back screw out exactly ¾-turn, move

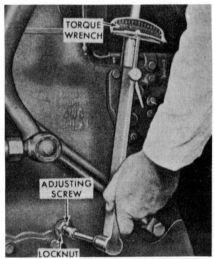

Fig. 263—Adjusting transmission bands with torque wrench. Refer also to Fig. 262 and to text for procedure.

selector lever to Park position to aid in holding adjusting screw stationary and tighten locknut to a torque of 20-25 ft.-lbs. (27-34 N·m), while holding adjusting screw from turning.

BAND NO. 3: To adjust Band 3, follow the same procedure as outlined for Band 2.

214. **SELECTOR DIAL ADJUSTMENT.** The individual speed (gear ratio) identifications on selector dial should be in alignment with pointer in selector housing. If not, readjust dial as follows:

Remove selector shaft end cover from left side of selector unit to expose shaft and hex nut as shown in Fig. 264. If control lever is mounted on left end of shaft, remove lever and reinstall it on right end of shaft while making adjustment; refer to paragraph 215. Move selector lever to Neutral (N) position and hold lever firmly in detent notch while loosening hex nut with a deep well socket. The dial can now be moved in either direction to align Neutral (N) identification on dial with pointer. Hold lever in detent and tighten hex nut to a torque of 25-35 ft.-lbs. (34-48 N·m). Recheck dial alignment; if correct, move lever back to left side if desired and reinstall shaft end cover.

215. **SELECTOR LEVER POSITION, LEVER STOPS AND NEUTRAL BY-PASS PLATE.** The selector lever is normally placed on right side of selector housing for right-hand operation; however, if operator prefers, lever can be installed on left side as shown in Fig. 265.

Stop screws can be placed at R2 or R1 and 3rd, 5th or 7th shift positions if desired. When not in use, stop screws are installed on opposite side of housing from shift lever position.

A neutral by-pass plate can be installed over the neutral notch in housing for shuttle operation. The plate is retained to housing by stop screws. Install

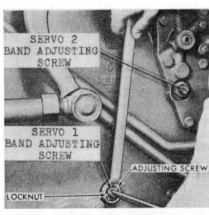

Fig. 262—Hold band adjusting screw from turning while loosening locknut prior to band adjustment. Refer also to Fig. 263. View shows Model 5000 transmission band 1 and band 2 adjusting screw locations. Band 3 adjusting screw is at rear of band 2 screw location.

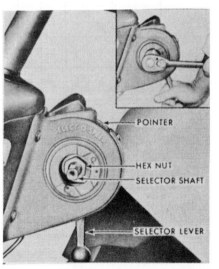

Fig. 264—View showing left end cover removed from gear selector assembly. Inset shows hex nut being loosened to adjust dial pointer.

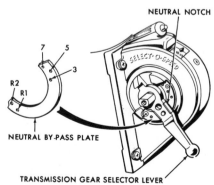

Fig. 265—Selector lever can be mounted on left side of housing if desired. A neutral by-pass plate is available for shuttle work. Numbers indicate possible stop screw locations.

one screw in the R1 or R2 position and the second screw in 3rd, 5th or 7th position as shown in Fig. 265.

216. ADJUST INCHING PEDAL (MODEL 5200 ONLY). Refer to Fig. 266; the length of the inching pedal rod assembly must be adjusted to give a 0.010-0.060 inch (0.254-1.524 mm) gap between pedal return stop and pedal shank. Incorrect adjustment may prevent full closing of transmission feathering valve, thus resulting in improper operation and/or damage to transmission components.

TROUBLESHOOTING

217. OPERATIONAL (MECHANICAL) CHECK. If the system adjustments outlined in previous section fail to correct transmission malfunction, the next step in trouble diagnosis would be an operational check. To perform this check, proceed as follows:

Place traction coupling in engaged position, start engine and set engine speed to 800 rpm and depress inching pedal. Shift transmission selector to each of the 14 positions in turn starting with Park (P), gradually release inching pedal at each position and record reaction when pedal is released. One of the five following conditions will be encountered.

(1) The transmission will operate properly for the control position selected.

(2) The transmission will operate in a different speed ratio than that selected.

(3) The transmission will go into a "neutral" condition in control position other than neutral.

(4) The transmission will go into park condition in control position other than park.

(5) The transmission will lock up and stall the engine.

If transmission seems to operate properly in all 14 control positions, proceed with TORQUE LIMITING CLUTCH CHECK as in paragraph 218, then the HYDRAULIC CIRCUIT CHECKS as outlined in paragraph 219. If any of the malfunction conditions (2), (3), (4) or (5) are encountered, compare the recorded reactions for each of the control positions with the columns in the operational troubleshooting chart in Fig. 267. The matching column in the troubleshooting chart will indicate the malfunctioning unit and whether the trouble is caused by the unit being continually applied ("A" column) or released ("R" column).

218. TORQUE LIMITING CLUTCH CHECK. A defective or worn torque limiting clutch should be suspected if transmission malfunction exists only under heavy loads and in higher speed ratios, especially if transmission operates properly in lower speed ranges.

To check torque limiting clutch, operate tractor in 8th, 9th or 10th speed position at wide open throttle and quickly apply both brakes. If tractor forward motion can be halted without pulling engine down below 1000 rpm or without stalling engine, renew torque limiting clutch as outlined in paragraph 225.

219. HYDRAULIC CIRCUIT CHECKS. Leakage in any of the hydraulic circuits to servos or clutches can be detected by performing the following pressure checks.

With a 0-300 psi (0-2070 kPa) hydraulic gage installed at Servo 2 location (Fig. 260), transmission fluid temperature at 80°-120°F (26.6°-48.8°C), traction coupling disconnected as outlined in paragraph 209 and with engine running

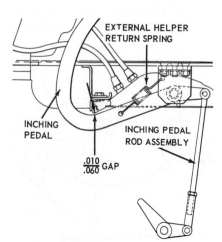

Fig. 266 — On Model 5200, inching pedal rod must be adjusted to obtain a 0.010-0.060 inch (0.508-1.524 mm) gap between foot pedal arm and platform at point shown.

GEAR RATIO	DIRECT DRIVE CLUTCH				BAND 1		BAND 2		BAND 3		CLUTCH 1		CLUTCH 2		CLUTCH 3	
	A_h	R_h	A_m	R_m	A	R	A	R	A	R	A	R	A	R	A	R
PARK	P	P	L	P	P	P	P	N	P	N	L	P	L	P	L	P
R2	R1	R2	L	R2	R2	N	L	R2	R2	N	R2	N	L	R2	L	R2
R1	R1	P	R1	N	L	R1	L	R1	R1	N	R1	N	L	R1	L	R1
NEUTRAL	N	P	N	N	L	N	N	N	N	N	N	N	N	N	N	N
1st	1	P	1	N	L	1	L	1	1	N	L	1	L	1	1	N
2nd	2	P	2	N	L	2	L	2	2	N	L	2	2	N	L	2
3rd	1	3	L	3	3	N	L	3	3	N	L	3	L	3	3	N
4th	2	4	L	4	4	N	L	4	4	N	L	4	4	N	L	4
5th	5	P	5	N	L	5	5	N	L	5	L	5	L	5	5	N
6th	6	P	6	N	L	6	6	N	L	6	L	6	6	N	L	6
7th	5	7	L	7	7	N	7	N	L	7	L	7	L	7	7	N
8th	6	8	L	8	8	N	8	N	L	8	L	8	8	N	L	8
9th	9	P	9	N	L	9	L	9	L	9	9	N	9	N	9	9
10th	9	P	L	10	10	N	L	10	L	10	10	N	10	N	10	10

Fig. 267 — Operational troubleshooting chart for Select-O-Speed transmissions for Model 5000 tractors.

Ah. Applied hydraulically & will not release
Rh. Released hydraulically & will not apply

Am. Applied mechanically & will not release
Rm. Released mechanically & will not apply

A. Applied & will not release
R. Released & will not apply
L. Lock-up condition

at 800 rpm, record hydraulic gage reading at each selector position shown in the hydraulic troubleshooting chart in Fig. 268 except 5th gear. Normal gage reading for each position is 200-210 psi (1380-1449 kPa) with a maximum variation of 3 psi (20.7 kPa). A leak in any one circuit will show up as one of the following conditions; leakage in more than one circuit will show up as a combination of conditions.

A. Low Gage Reading In All Positions—Move gage to Servo 3 location (Fig. 260) and observe pressure gage reading with selector at 5th gear. A normal gage reading 200-210 psi (1380-1449 kPa) would indicate leakage in Servo 2 circuit; a continued low reading would indicate incorrectly adjusted transmission regulating valve (refer to paragraph 212), clogged filter and/or worn transmission pump.

B. Low Gage Reading In Positions R2 and R1—Normal reading in other positions and returns to normal in R2 and R1 when inching pedal is fully depressed would indicate leakage in Clutch 1 circuit.

C. Low Gage Reading In Positions R2, N, 1st and 2nd—Normal reading in other positions and depressing inching pedal does not change gage readings would indicate leakage in Servo 1/ Direct Drive Clutch circuit.

D. Low Gage Reading In Position N—Normal reading in other positions would indicate leakage in Servo 3 circuit.

E. Low Gage Reading In Positions 1st and 3rd—Normal readings in other positions and gage reading returns to normal in 1st and 3rd when inching pedal is depressed indicates leakage in Clutch 3 circuit.

F. Low Gage Reading In Positions 2 and 4—Normal readings in other positions and gage reading returns to normal in 2 and 4 when inching pedal is depressed indicates leakage in Clutch 2 circuit.

OVERHAUL TRANSMISSION

All Models

CAUTION: The Select-O-Speed transmission is a hydraulically controlled unit and merits the same degree of care and cleanliness as for any hydraulic system. Disassembly or service should be attempted only in a clean, dust-free shop. Use only lint-free paper shop towels to wipe internal transmission parts; lint from cloth shop towels or rags will clog the oil filter screen, etc., and possibly cause serious damage or transmission malfunction.

220. **R&R GEAR SELECTOR ASSEMBLY.** To remove gear selector assembly, refer to Fig. 269 and proceed as follows:

Place gear selector in Park (P) position, loosen conduit retaining nut at top of transmission and remove the four screws retaining selector housing to rear hood panel. Shift selector to 10th gear position while lifting up on selector housing. Disconnect exposed control cable at lower end of conduit, disconnect selector lamp and oil pressure warning light wires and remove selector assembly from tractor.

Prior to turning or removing lower selector cable, measure distance (D—Fig. 270) that cable protrudes from conduit fitting and record this measurement for reassembly. Lower cable can now be removed by turning it out of control valve camshaft wheel, then pulling it out of transmission.

To reinstall selector assembly, first insert lower cable into conduit fitting and thread it into control valve camshaft wheel until cable protrudes distance (D) measure on disassembly. If measurement was not made on disassembly, or if selector assembly has been disassembled, thread cable into camshaft wheel until measurement (D) is 2¾ inches (6.98 cm). Place selector lever in 10th gear position and connect cables at lower end of conduit. Move selector lever to Park position while lowering selector assembly into opening in rear hood panel. Tighten conduit retaining nut, reconnect light wires and install the four selector housing retaining screws.

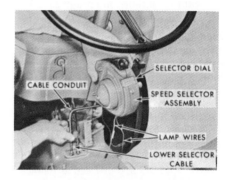

Fig. 269—Removing speed (gear ratio) selector assembly; refer to text.

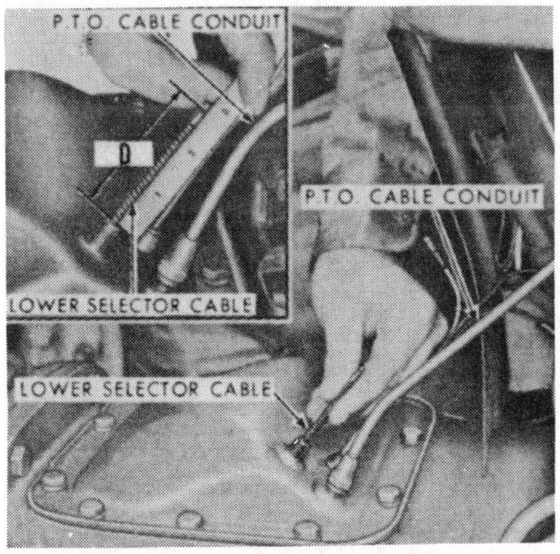

GEAR RATIO	DIRECT CIRCUIT				INDIRECT CIRCUIT		
	D.D.C.	B1	B2	B3	C1	C2	C3
R₂			P		P		
R₁	P	P	P		P		
NEUTRAL	P	P	P	P			
1st	P	P	P				P
2nd	P	P	P			P	
3rd			P				P
4th			P			P	
5th	P	P		P			P

Fig. 268—Hydraulic troubleshooting chart for use with gage installed at one or more servo pressure ports. "P" indicates hydraulic pressure applied to unit for different gear ratio selector positions. Refer to text for hydraulic circuit troubleshooting procedure.

Fig. 270—Measure distance lower selector cable extends from conduit fitting (distance "D"-see inset) before removing cable.

Check to be sure that selector dial correctly indicates gear ratios; refer to paragraph 214 if adjustment of dial is required.

221. R&R TRANSMISSION BOTTOM COVER. First, drain transmission fluid, then unbolt and remove cover from bottom of transmission housing. Refer to Fig. 271 for view of transmission with bottom cover removed.

Reinstall cover with new gasket and tighten retaining cap screws to a torque of 20-23 ft.-lbs. (27-31 N·m). Refill transmission with proper fluid as outlined in paragraph 208.

222. R&R TRANSMISSION ASSEMBLY. To remove transmission, first split tractor between engine and transmission as outlined in paragraph 164 or 165. Remove gear selector assembly as outlined in paragraph 220. Disconnect rear light wires, transmission lube warning light wire (when used) and starter safety switch wire, then remove steering gear assembly and fuel tank as a unit from top of transmission. Refer to paragraph 254 or 256 and remove transmission from rear axle center housing.

Reinstall transmission by reversing removal procedure. Refill with proper lubricant as outlined in paragraph 208.

NOTE: Although complete transmission overhaul requires removal of transmission, most work can be completed after splitting tractor between engine and transmission or between transmission and rear axle center housing.

223. R&R CONTROL VALVE ASSEMBLY. First, remove selector assembly as outlined in paragraph 220 and bottom cover as in paragraph 221. Then, refer to Fig. 271 and proceed as follows:

Disconnect wires from starter safety switch, remove cap screws retaining switch bracket and remove switch and bracket assembly. Tie inching pedal down and remove the four remaining cap screws retaining control valve assembly.

NOTE: Do not remove cap screw (C) at corner of valve assembly.

It may be necessary to pry or bump valve assembly to free it from gasket, then remove assembly from bottom of transmission.

Reinstall control valve assembly with new gasket and the four end cap screws only; tighten cap screws snug. Turn camshaft to Park position (all valves out and cam for starter safety switch to rear) and install safety switch and bracket assembly. Push switch and bracket forward and with switch button centered on cam, tighten cap screws to a torque of 6-8 ft.-lbs. (8-11 N·m), then tighten the four end cap screws to same torque. Install selector assembly as in paragraph 220, reconnect switch wires and reinstall bottom cover as in paragraph 221.

224. OVERHAUL CONTROL VALVE. With control valve assembly removed as outlined in paragraph 223, refer to exploded view of assembly in Fig. 272 and proceed as follows:

Remove the two cap screws that hold the halves of control valve assembly together, then separate valve body half (27) from camshaft body half (16). The six control valve spools (43) can now be removed. Remove feathering valve retaining plate (39) and withdraw feathering valve (41), spring (38) and valve plunger (34). On Model 5000, a passage blocking spool (42) is substituted for pto feathering valve used in other control valves. The retainer plate (20) can now be removed from valve body half (27) for better access to clean spool bores. Remove adjusting screw retainers (11 and 12), adjusting screws (9 and 13), springs (8, 10 and 14) and valves (7 and 15) from camshaft body half (16).

Check cam followers (rocker arms) (21 and 22) and shaft (19) for wear or damage; to disassemble, remove snap rings (18) and slide shaft from body.

Check valve camshaft (4), detent plunger (5), spring (6) and camshaft cable wheel (17) for wear or damage. To disassemble, thread a ¼-inch diameter, 20 thread 3-inch (7.62 cm) long socket head screw (or other hardened steel ¼-inch x 20 bolt with at least 1¾ inches [4.44 cm] threaded) into cable wheel (17), hold camshaft with wrench and turn screw to force cable wheel and trunnion assembly from camshaft. Remove trunnion (3) from opposite end of camshaft using same screw.

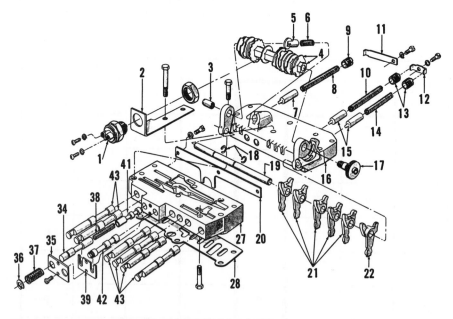

Fig. 272 — Exploded view of Model 5000 control valve assembly. Valve is shown inverted from actual position in transmission.

1. Starter safety switch	10. Regulating valve spring	18. Snap rings
2. Bracket	11. Retainer	19. Rocker arm shaft	35. Spring retainer
3. Trunnion	12. Retainer	20. Valve cover plate	36. Snap rings
4. Camshaft	13. Adjusting screw	21. Rocker arms	37. Return springs
5. Detent	14. Relief valve spring	22. Rocker arm, end	38. Valve spring
6. Spring	15. Relief & regulating	27. Lower valve body	39. Valve retainer
7. Sequencing valve	valves	28. Gasket	41. Feathering valve
8. Valve spring	16. Upper valve body	34. Feathering valve	42. Passage blocking spool
9. Adjusting screw	17. Cable wheel	plunger	43. Control valves

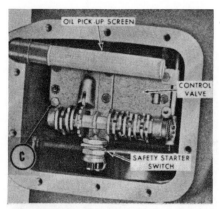

Fig. 271 — View of Model 5000 transmission control valve and oil pick-up screen with transmission bottom cover removed. Do not remove cap screw (C) when removing control valve from distributor.

NOTE: DO NOT use a common steel or rethreaded bolt for a forcing screw; this type of bolt or screw will twist off in cable wheel or trunnion.

The camshaft, detent plunger and spring can now be removed.

Carefully clean all parts in solvent, air dry and check against the following values:

Control Valve Spool
Diameter, Model
5000.............0.3743-0.3747 in.
(9.50-9.52 mm)

Feathering Valve Spool
Diameter, All......0.3743-0.3747 in.
(9.50-9.52 mm)

Valve Bore Diameter,
All ....0.3751-0.3758 in.
(9.527-9.545 mm)

Spring Free Lengths:
Camshaft Detent Plunger ....0.75 in.
(1.9 cm)
Control Spool Return........1.24 in.
(3.15 cm)
Sequencing Valve...........2.89 in.
(7.34 cm)
System Regulating Valve ......2.47 in.
(6.27 cm)
Transmission Relief Valve .....2.06 in.
(5.23 cm)

Transmission Feathering Valve:
Return....................1.22 in.
(3.098 cm)
Plunger ....1.40 in.
(3.55 cm)

To reassemble, proceed as follows: Lubricate all parts prior to assembly. Insert detent spring (6) and detent plunger (5) in bore of body half (16) and position camshaft (4) so detent cam is aligned with detent; hold camshaft in place with a ⅝-inch (1.58 cm) bolt inserted in trunnion (3) end of shaft.

NOTE: The bolt must be long enough to bottom in shaft.

Place unit in arbor press as shown in Fig. 273 and press cable wheel into camshaft so there is 0.010 inch (0.254 mm) clearance between wheel and valve body. Thread the forcing screw that was used in disassembly into cable wheel until bottomed, then turn unit over in arbor press and pressing trunnion (3—Fig. 272), chamfered end first, into camshaft until flush with valve body camshaft support boss.

Insert shaft (19) through hole in boss of body (16) and install cam followers (21 and 22) on shaft as it is pushed into place.

NOTE: Follower on cable wheel end (22) is constructed with cam arms placed opposite cam arms of other five followers (21).

Secure shaft with snap rings (18) inserted in grooves of shaft.

Insert valves (7 and 15) (all three valves are alike) into bores of body (16). Insert spring (8) (2.89 inches – 7.34 cm – long) in bore farthest from cable wheel end. Insert spring (14) (2.06 inches – 5.23 cm – long) in bore nearest cable wheel end and insert spring (10) (2.47 inches – 6.27 cm – long) in middle bore. Thread adjusting screws (9 and 13) (all screws are alike) in flush with valve body. Do not install retainers (11 and 12) until valve has been adjusted as outlined in paragraph 212.

Install retaining plate (20) on valve body (27). Install snap ring (36) in groove of feathering valve plunger (34), insert valve through return spring (37) and plate (35). Compress return spring and install notched plate (39) over plunger (34) and blocking spool (42). Insert feathering valve (41), with hollow end out, into bore towards end of body having two control valve spool bores. Insert plunger spring (38) in hollow end of feathering valve. Install assembled plunger, spring and plate assembly so plunger (34) is in feathering valve bore and in contact with spring (38), then secure with plate retaining cap screws.

Lubricate control valve spools (43) and carefully insert them in their bores with a twisting motion to avoid scraping bores with sharp edges of valve lands. Turn valve spools so notches in outer ends of spools are positioned as shown in Fig. 272. Place the two valve body halves together so ends of followers engage notches in valve spools. Install the two cap screws that hold body halves together and tighten cap screws to a torque of 6-8 ft.-lbs. (8-11 N·m). Refer to paragraph 212 for adjustment of the

transmission regulating and relief valves, then install adjusting screw retainers (11 and 12).

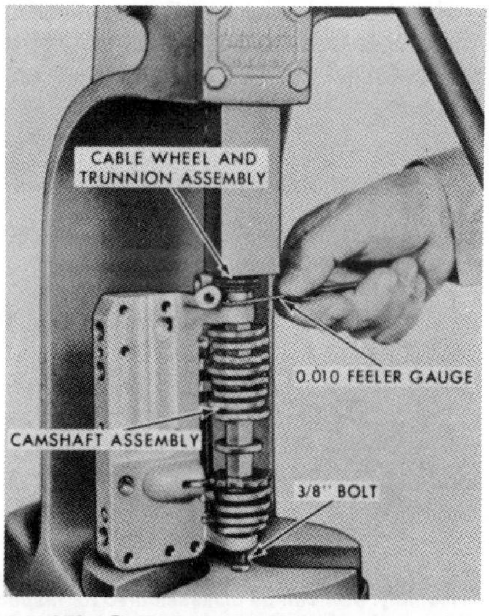

225. TORQUE LIMITING CLUTCH. To remove torque limiting clutch, first split tractor between engine and transmission as outlined in paragraph 164 or 165. Then, unbolt and remove clutch assembly from flywheel. Refer to Fig. 274 for exploded view of the clutch assembly.

Renew clutch disc facings or clutch disc assembly if facings are glazed, oil soaked or excessively worn. Thickness of new disc with facings is 0.329-0.353 inch (8.35-8.96 mm).

Renew Belleville (spring) washer if discolored, cracked, or if it fails to meet the following specifications: Thickness, 0.111-0.114 inch (2.82-2.89 mm); free height, 0.198 inch (5.03 mm); color code, blue. Free height is measured as dish in the spring washer. Color code is ⅛-inch (3.18 mm) wide paint mark on outer edge of the spring washer.

Drive plate thickness should be 0.278-0.282 inch (7.06-7.16 mm). Renew drive plate if cracked, scored, warped or excessively worn.

Check splines in clutch disc hub and on transmission input shaft for rust or excessive wear. Renew clutch disc assembly if backlash of disc on input shaft, measured at outer diameter of disc, exceeds ½-inch (12.7 mm). Renew input shaft if backlash of new disc on input shaft exceeds ¼-inch (6.35 mm). Before reassembly, be sure splines on input shaft and in clutch disc are clean and free of rust; then, apply a thin film of light silicone grease (Ford part No. M1C-43) to splines of both parts.

When reassembling, be sure that concave side of Belleville washer (2) is towards clutch cover (1) as shown in Fig. 275. Use a suitable pilot to center clutch disc in flywheel. Tighten clutch cover retaining cap screws to a torque of 25-30 ft.-lbs. (34-41 N·m).

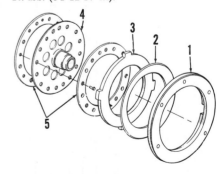

Fig. 273—Press cable wheel and trunnion into camshaft so there is 0.010 inch (0.254 mm) clearance between cable wheel and upper valve body. Note ⅜-inch bolt support opposite end of camshaft.

Fig. 274—Exploded view of typical torque limiting clutch assembly. Refer to Fig. 275 for cross-sectional view showing proper installation of the Belleville (spring) washer (2).

1. Clutch housing
2. Spring
3. Pressure plate
4. Clutch disc
5. Facings

226. **TRANSMISSION PUMP.** To remove transmission pump, first split tractor between engine and transmission as outlined in paragraph 164 or 165, then unbolt and remove pump assembly. Refer to Fig. 276 for exploded view of pump. The pump is driven by transmission countershaft.

The transmission pump, except for rear plate retaining screws, "O" ring (2) and plug (6) is serviced as a complete assembly only. Disassemble, clean and inspect pump paying particular attention to cam surface in pump body (5). If cam surface is worn or pitted, renew pump assembly. If pump is serviceable, reassemble using a new "O" ring. Install plug (if removed), using a light application of "Loctite" sealer grade "C".

Reinstall pump with new mounting gasket. Be sure pump is correctly positioned and install two retaining cap screws opposite each other. Tighten these screws to a torque of 3-5 ft.-lbs. (4-7 N·m), then install two remaining screws to a torque of 3-5 ft.-lbs. (4-7 N·m). Finally tighten all four cap screws alternately and evenly to a torque of 15-18 ft.-lbs. (20-24 N·m).

227. **TRANSMISSION END PLAY.** Transmission end play is controlled by bronze thrust washers placed between each of the rotating members. Variation in total length of transmission components is compensated for during assembly by providing two selective fit thrust washers, one at the rear of each transmission section, to hold end play within specified limits. Cumulative wear of thrust washers and of thrust surfaces on other parts will require that end play be checked at each overhaul, then renewing selective fit thrust washers with ones of greater thickness to provide correction for wear.

End play should be checked as outlined in paragraphs 228 and 229 before disassembly of the involved transmission components. The overhaul procedures outlined in this manual are based on the supposition that the two selective fit thrust washers are the only ones which will be renewed. However, if any other thrust washers or other transmission parts are renewed, the difference in thickness of the new parts will affect thickness of the selective fit washers to be installed. Thus, it is important to check end play during reassembly to be sure the proper thickness of selective fit thrust washers have been installed.

228. **TRANSMISSION FRONT END PLAY.** To check transmission front end play, all drive components to rear of "C" sun gear and distributor must first be removed and if transmission has not been removed from tractor, the steering gear assembly or front compartment cover plate must be removed to provide access to the front transmission components. Loosen Band 1 and Band 2 adjusting screws. Then, mount dial indicator against rear end of "C" sun gear as shown in Fig. 277. Pry "C" sun gear forward and set dial indicator to zero while holding slight pressure against sun gear. Insert a screwdriver between the "B" carrier and "A" ring gear and pry "B" carrier rearward.

The resulting dial indicator reading is transmission front end play which should be 0.005 to 0.015 inch (0.127-0.381 mm). Record the end play measurement for reassembly.

Front end play is adjusted by selecting the proper thickness of thrust washer to be installed between distributor and clutch 1. This thrust washer is available in thicknesses of 0.062 to 0.122 inch (1.574-3.098 mm) in steps of 0.010 inch (0.254 mm).

229. **TRANSMISSION REAR END PLAY.** Transmission rear end play can be checked after splitting tractor between transmission and rear axle center housing and removing bottom cover. To check rear end play, proceed as follows:

Loosen Band 3 adjusting screw so band is completely released. Mount a dial indicator as shown in Fig. 278. Push in transmission output shaft and zero dial indicator while holding pressure on shaft. Then pry between "C" carrier and "C" ring gear with a screwdriver; resulting dial indicator reading will be transmission rear end play which should be 0.005 to 0.015 inch (0.127-0.381 mm).

Transmission rear end play is adjusted by selecting the proper thickness of thrust washer to be installed between the shoulder on transmission output shaft and rear support. This washer is available in thicknesses of 0.102 to 0.142 inch (2.59 to 3.60 mm) in steps of 0.010 inch (0.254 mm).

230. **SERVO 1 COVER AND SERVO 1 ASSEMBLY.** To remove Servo 1 cover and Servo 1 assembly, first drain transmission fluid, then proceed as follows:

While holding Band 1 adjusting screw from turning, remove adjusting screw locknut and flat washer. Then, slowly back adjusting screw out until servo spring tension is relieved.

CAUTION: Do not back adjusting screw out farther than necessary as this will allow Band 1 strut to fall out of place making disassembly of transmission necessary.

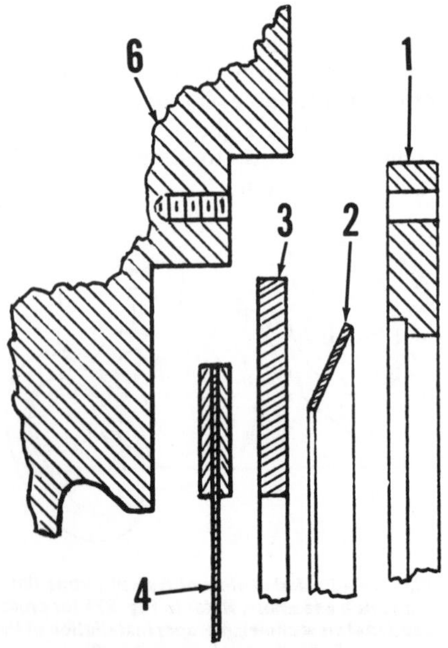

Fig. 275 — Cross-sectional view showing placement of Belleville spring washer (2). Flywheel is (6); refer to Fig. 274 for parts identification.

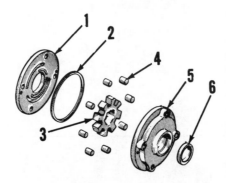

Fig. 276 — Exploded view of transmission pump assembly. Unit can be disassembled for inspection as shown; however, only "O" ring (2) and plugs (6) are available for service.

1. Cover	4. Rollers
2. "O" ring	5. Body
3. Rotor	6. Plug

Fig. 277 — View showing dial indicator set against "C" Sun Gear to check transmission front end play on Model 5000 tractor.

Remove the four cap screws retaining Servo 1 cover to transmission and carefully remove cover. Slowly turn adjusting screw in, forcing servo assembly out, until screw is tight. Then, withdraw servo assembly from transmission housing. Refer to Fig. 279.

The servo sealing "O" rings (7 and 10 – Fig. 280) can now be renewed and servo assembly reinstalled. However, if inspection of unit reveals excessive wear or other defect, proceed as follows to disassemble and overhaul the servo:

NOTE: On early production units (prior to 4-68), Servo 1 piston rod pilot diameter was 0.730-0.734 inch (18.54-18.64 mm), and piston outer diameter was 1.993-1.997 inch (50.62-50.72 mm). If these early parts are encountered, servo operation can be improved by renewing both the rod and piston; service and late production piston rod pilot diameter is 0.734-0.736 inch (18.64-18.69 mm), and piston outer diameter is 1.981-1.985 inch (50.317-50.419 mm). The larger rod pilot diameter and the smaller piston diameter reduce the possibility of piston cocking and wedging in bore of transmission housing. At the same time, a check valve assembly was installed in port directly below servo piston bore. See Fig. 280.

231. To disassemble servo, place assembly in a press using sleeve and step plate as shown in Fig. 281. Compress servo spring and remove nut and washer from piston rod (9 – Fig. 280). Slowly release spring tension, remove unit from press and remove retainer (3) and spring (4). Remove piston retaining snap ring (6), then push or press piston rod from piston. To reassemble, reverse disassembly procedure. When reinstalling piston, be sure sharp edge of snap ring is placed away from piston. Install new sealing "O" rings (7 and 10) on piston rod and piston.

232. To reinstall servo, first lubricate "O" rings with petroleum jelly, then pro-

ceed as follows: Insert assembled servo in bore of housing with notch in end of servo aligned with strut. Push servo against strut, then slowly back band adjusting screw out while pushing servo into housing. When servo contacts inner end of bore and stops moving inward, immediately stop backing adjusting screw out and reinstall servo cover with new gasket. Tighten cover retaining cap screws to a torque of 20-25 ft.-lbs. (27-34 N·m). Clean adjusting screw threads, coat threads lightly with a non-hardening plastic lead sealer (Crane Packing Company "Plastic Lead Sealer No. 2" or equivalent); then, turn adjusting screw in to remove free play in linkage. If adjusting screw is equipped with a sealing type nut and flat copper washer, discard nut and washer and install a ⅝-inch (1.7 cm) flat steel washer (Ford part No. 351502-S) and a plain ⅝-inch x 11 hex nut (Ford part No. 33849-S8). Refill transmission and adjust Band 1 as outlined in paragraph 213.

233. **SERVO 2 AND 3 COVER, TIMING VALVES AND SERVOS 2 AND 3.** The servo cover and servos can be removed with transmission installed in tractor. The servo timing valves are

Fig. 279—Removing Servo 1 assembly and cover.

located in the servo cover; refer to exploded view in Fig. 282. Refer to appropriate following paragraphs for removal, overhaul and reinstalling procedure.

234. **R&R SERVO 2 AND 3 COVER.** Drive retaining pin out of inching pedal and pivot (do not remove) pedal back out of way. Hold Band 3 adjusting screw from turning and remove locknut and flat washer from screw. Slowly back adjusting screw out until Servo 3 spring tension is relieved.

CAUTION: Do not back adjusting screw out any farther than necessary to relieve spring tension; if screw is excessively loosened, the band struts may fall out which would require transmission disassembly.

Similarly remove Band 2 adjusting screw locknut and flat washer, then back Band 2 adjusting screw out, observing same caution as for Band 3 adjusting screw, until Servo 2 spring tension is relieved. Remove cap screws retaining Servo 2 and 3 cover to transmission, insert pry bar in relief behind front to rear edge of cover and pry cover loose.

CAUTION: Do not attempt to remove the Servos 2 and 3 without first referring to procedure outlined in paragraph 237.

235. Before reinstalling servo cover, clean threads of Band 2 and 3 adjusting screws and apply a light coat of non-hardening plastic lead sealer (Crane Packing Company "Plastic Lead Sealer No. 2" or equivalent) to screw threads. Install servo cover with new gasket and

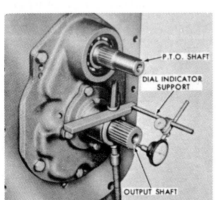

Fig. 278—View showing dial indicator set against output shaft to check transmission rear end play on Model 5000.

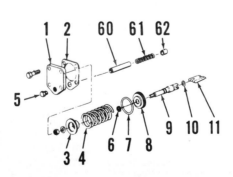

Fig. 280—Exploded view of the Servo 1 assembly.

1. Cover	8. Piston
2. Gasket	9. Piston rod
3. Retainer	10. "O" ring
4. Spring	11. Strut
5. Plug	60. Pin
6. Snap ring	61. Spring
7. "O" ring	62. Check valve

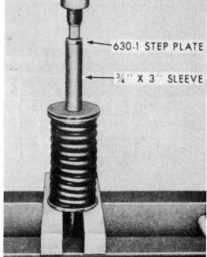

Fig. 281—Compressing servo assembly in hydraulic press to remove nut and washer from piston rod.

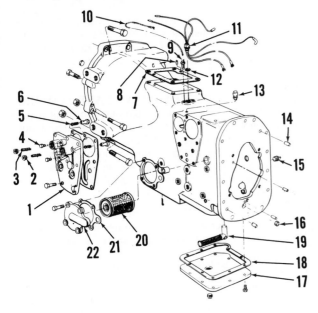

Fig. 282 — Exploded view showing Servo 2 and 3 cover, transmission case and related parts.

1. Servo cover
2. Plug
3. Springs
4. Plugs
5. Timing valves
6. Valve seats
7. Gasket
8. Dowel
9. Adapter
10. Control valve cable assy.
11. Selector cable harness
12. Retainer plug
13. Vent plug
14. Dowels
15. Oil level plug
16. Expansion plug
17. Cover
18. Gasket
19. Inlet screen
20. Oil filter
21. Gasket
22. Filter cover

taining plugs securely after installing timing valve assemblies (V – Fig. 283), snap ring end first, and springs (S).

237. R&R AND OVERHAUL SERVOS 2 AND 3. With Servo 2 and 3 cover removed as outlined in paragraph 234, proceed as follows: Turn Band 3 adjusting screw in slowly until tight while pushing against outer end of Servo 3. When Band 3 adjusting screw is tight, withdraw Servo 3 from housing (refer to Fig. 284). With Servo 3 removed, withdraw piston rod seal gland with hooked wire tool as shown in Fig. 285. Follow same procedure as outlined for Servo 3 and seal gland to remove Servo 2 and seal gland.

CAUTION: If either Servo 2 or Servo 3 is removed from housing without first turning the appropriate band adjusting screw in tight, the band struts may fall out of place requiring disassembly of transmission for reinstallation.

238. To disassemble Servo 2 or Servo 3, refer to procedure outlined in paragraph 231 for Servo 1, and also to Fig. 281. Exploded view of Servo 2 assembly is shown in Fig. 287 and Servo 3 assembly in Fig. 286. Note that Servo 3 has an inner spring (4) and an outer spring (5). Check servo springs against the following specifications:

Model 5000
Servo 2 spring:
Free Length . . . . .7.22 inches
(18.34 cm)
Lbs. at 4.36 inches 375-423
Kg at 11.07 cm 168.7-190.3
Servo 3 Inner Spring:
Free Length . . . . .7.22 inches
(18.34 cm)
Lbs. at 4.36 inches 375-423
Kg at 11.07 cm 168.7-190.3
Servo 3 Outer Spring:
Free Length . . . . .10.31 inches
(26.13 cm)
Lbs. at 4.88 inches 505-556
Kg at 12.30 cm 227.2-250.2

alternately and evenly tighten retaining cap screws to a torque of 35-40 ft.-lbs. (48-54 N·m). If adjusting screws are equipped with sealing (self-locking) type nuts and copper washers, discard nuts and washers and install plain steel 5/8-inch (1.7 cm) flat washers and plain 5/8-inch x 11 hex nuts. Tighten band adjusting screws to remove free travel in linkage. Refill transmission with proper fluid as outlined in paragraph 208, then adjust the bands as outlined in paragraph 213.

236. SERVO TIMING VALVES. The Servo 2 and 3 cover assembly contains the Servo Timing Valves and Valve Seats; refer to exploded view in Fig. 282. To remove timing valves, remove socket head plugs (2) from outer side of cover, then remove springs (3) and timing valve assemblies (5).

With timing valves removed, remove valve seats by using a 1/4-inch (6.35 mm) diameter steel rod to press them out towards inside of cover. Refer to cross-sectional view of cover, timing valve and seat in Fig. 283. When installing new valve seats, press them into servo cover so hole in side of seat is aligned with passageway (P) in cover.

The timing valves are serviced as complete assemblies only. Renew valve assembly if any part is damaged in any way; refer to paragraph 206 for timing valve operation. The color of the valve assembly spring (AS – Fig. 283) correctly identifies the timing valve for Servo 2 or Servo 3 application as follows:

On models having transmission with serial number prior to 5H18C, the Servo 2 timing valve spring was coded green and the Servo 3 timing valve spring was coded red. On models having transmission with serial number 5H18C and up, both Servo 2 and Servo 3 timing valve springs are color coded green.

After selecting proper timing valve (spring) color code, install timing valves in servo cover. Tighten socket head re-

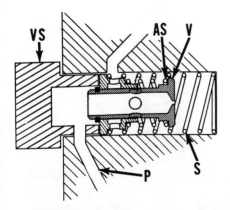

Fig. 283 — Cross-sectional view of Servo 2 and 3 cover showing timing valve (V), valve return spring (S), valve assembly spring (AS), oil passage (P) and timing valve seat (VS). Note that hole in side of valve seat must align with oil passage.

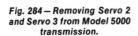

Fig. 284 — Removing Servo 2 and Servo 3 from Model 5000 transmission.

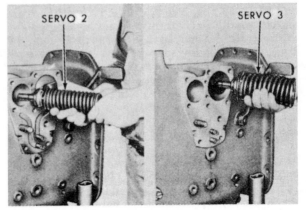

NOTE: On early production Model 5000 tractors, the bore in transmission housing for Servo 3 was 3¼ inches (8.255 cm) in diameter, whereas Servo 3 bore diameter in late production units is three inches (7.62 cm). Service parts affected are transmission housing, Servo 3 piston and sealing ring and Servo 3 timing valve.

To reassemble servos, follow procedure as outlined for Servo 1 in paragraph 231, making sure that sharp edge of snap ring (6 – Fig. 286 or 27 – Fig. 287) is placed away from piston. Install new sealing "O" rings on pistons and piston rod guides and in guide bores. Lubricate "O" rings with petroleum jelly and install guides over piston rods with small diameter of guides away from pistons. Insert assemblies in their bores. Push against outer end of Servo 3 while slowly backing out Band 3 adjusting screw immediately when servo contacts inner end of bore and stops moving inward. Repeat procedure was outlined to install Servo 2 assembly.

CAUTION: If either Band 2 or Band 3 adjusting screw is backed out farther than necessary, band struts may fall out of place.

Reinstall Servo 2 and 3 cover as outlined in paragraph 235.

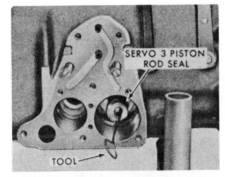

Fig. 285 – View showing method of removing servo guides using a hooked wire. Unit shown is not Model 5000, however procedure remains the same.

239. **R&R REAR SUPPORT, OUTPUT SHAFT AND "C" RING GEAR.** To remove rear support, first split tractor between transmission and rear axle center housing as outlined in paragraph 254, 255 or 256, drain transmission, then proceed as follows:

Remove cap screws which retain rear support, then thread two cap screws into tapped holes in rear support for use as jack screws (Fig. 288) to remove rear support from transmission housing. Then, remove rear support as shown in Fig. 289. If not removed with rear support, remove output shaft and "C" ring gear assembly. Refer to Fig. 290, for exploded view of rear support, output shaft and "C" ring gear; be careful not to lose thrust washer (2) located between output shaft and Clutch 2 and 3 assembly.

To remove pto shaft assembly from rear support, remove snap rings at rear side of bearing, press shaft from bearing and bearing from support. Pto shaft oil seal can be renewed at this time; install new seal with lip to inside (forward) using a suitable driver.

Remove output shaft and "C" ring gear from rear support. Bushings (13 – Fig. 290) are not serviced separately; renew rear support if bushings are excessively worn or scored.

Inspect output shaft bearing journals; renew output shaft if journals are excessively worn or rough. To remove output shaft, remove rear snap ring (5 – Fig. 290) and push gear or shaft forward out of "C" ring gear.

NOTE: Later production units will have ends of snap ring (5) brazed together; break weld with chisel, then remove and discard snap ring.

To reassemble, install new snap ring (1) in front groove in output shaft and insert shaft through "C" ring gear from front side. Install new late type snap ring (5) in rear groove and braze ends of snap ring together using a 1/16-inch (1.587 mm) bronze rod. Be careful not to

overheat ring gear and sun gear or output shaft and remove all brazing flash.

To reassemble transmission proceed as follows: Install new output shaft oil

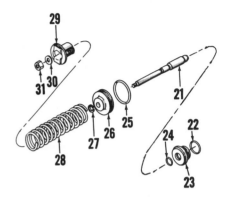

Fig. 287 – Exploded view of Servo 2 assembly and piston rod guide.

21. Piston rod	27. Snap ring
22. "O" ring	28. Spring
23. Guide	29. Retainer
24. "O" ring	30. Washer
25. "O" ring	31. Nut
26. Piston	

Fig. 288 – Removing transmission rear support with jackscrews.

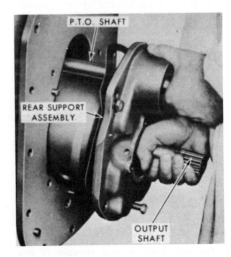

Fig. 289 – Removing rear support.

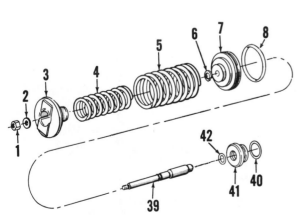

Fig. 286 – Exploded view of Servo 3 assembly and piston rod guide.

1. Nut
2. Washer
3. Retainer
4. Inner spring
5. Outer spring
6. Snap ring
7. Piston
8. "O" ring
39. Piston rod
40. "O" ring
41. Guide
42. "O" ring

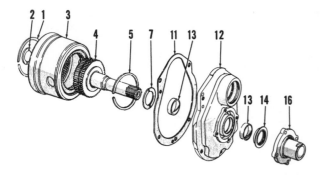

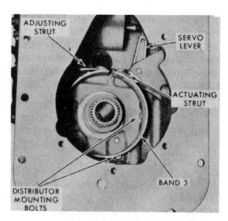

Fig. 290 — Exploded view of rear support, output shaft and "C" ring gear. Bushings (13) in rear support (12) are not serviced separately.

1. Snap ring
2. Thrust washer
3. "C" ring gear
4. Output shaft
5. Snap ring
7. Thrust washer
11. Gasket
12. Rear support plate
13. Bushings
14. Oil seal
16. Hydraulic pump idler gear support

Fig. 294 — Band 3 and struts installed in transmission.

seal in rear support using suitable driver so flange of seal (14 – Fig. 290) contacts rear support evenly. Use a light film of grease to stick thrust washer (2) over Clutch 2 and 3 retaining snap ring with counterbore of washer over snap ring. Carefully install assembled "C" ring gear and output shaft over Clutch 2 and 3 assembly so "C" ring gear engages gears on "C" carrier. Place selective thickness thrust washers (7) over end of output shaft as shown in Fig. 291.

Install rear support using a new gasket (11 – Fig. 290) taking care not to damage any seals passing over output shaft or pto shaft. Tighten retaining cap screws to a torque of 35-40 ft.-lbs. (48-54 N·m). Check transmission rear end play as outlined in paragraph 229 and if necessary renew selective thickness thrust washer with a new washer of suitable thickness to bring end play within correct limits.

240. **MAINSHAFT, CLUTCH 2 AND 3 ASSEMBLY, "C" CARRIER, "C" SUN GEAR AND BAND 3.** After removing rear support, output shaft and "C" ring gear as outlined in paragraph 239, mainshaft, "C" carrier and Clutch 2 and 3 assembly can be removed as a unit. Back out Band 3 adjusting screw until Band 3 is loose, then remove mainshaft, Clutch 2 and 3 assembly and "C" carrier as shown in Fig. 292. The "C"

sun gear and sealing rings can then be removed from distributor; refer to Fig. 293. Refer to Fig. 294, compress Band 3 and remove band, adjusting strut and actuating strut. Remove thrust washer (30 – Fig. 297) from rear face of distributor.

Remove the four cast iron sealing rings (14 – Fig. 298) from front end of mainshaft (13) and slide "C" carrier (37 – Fig. 297) and thrust washer (31) off shaft as shown in Fig. 295. Remove snap ring (16 – Fig. 298) from rear end of mainshaft and remove mainshaft from

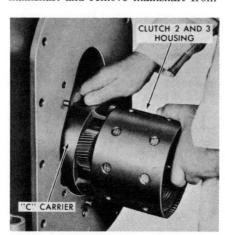

Fig. 292 — Removing transmission mainshaft. Clutch 2 and 3 assembly and "C" carrier from transmission.

Clutch 2 and 3 assembly as shown in Fig. 296. To overhaul removed Clutch 2 and 3 assembly, refer to exploded view in Fig. 298 and to paragraph 242.

The "C" carrier is serviced as a complete assembly only. Check to see that "C" carrier pinions are in good condition and that pinion shafts are tight in carrier. Pinion end play in carrier should be 0.010-0.028 inch (0.254-0.711 mm). Also check end thrust surfaces and bushing in carrier; renew carrier as an assembly if defect is noted.

Renew Band 3 if friction material inside band is worn, pitted or eroded or if

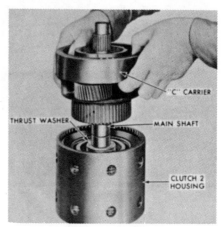

Fig. 295 — Removing "C" carrier from mainshaft.

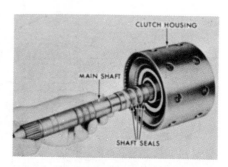

Fig. 296 — Removing mainshaft from Clutch 2 and 3 housing.

Fig. 291 — View showing selective fit thrust washer installed on output shaft.

Fig. 293 — Removing "C" sun gear and sealing rings from transmission.

band has been overheated. Also inspect strut sockets at band ends for cracks or other damage.

241. To reassemble transmission, proceed as follows: Install Band 3 and two struts as shown in Fig. 294, making sure that end of flat (actuating) strut having notch is towards band and end of band having an identifying notch is towards adjusting screw strut. Install thrust washer (30–Fig. 297) on rear of distributor. Install new sealing rings (28) on "C" sun gear, lubricate rings with petroleum jelly, align ring end gaps along top side of sun gear and carefully insert sun gear through distributor. Rotate sun gear slightly from side to side to align splines on front end of gear with splines in Clutch 1 housing.

Place thrust washer (1–Fig. 298) over front end of mainshaft (13), insert mainshaft through "C" carrier (37–Fig. 297), then place thrust washer (31) over front end of mainshaft and against "C" carrier. Carefully install four new sealing rings (14–Fig. 298) on front end of mainshaft, lubricate rings with petroleum jelly, align ring end gaps along top side of mainshaft and holding carrier and shaft assembly with ring end gaps up, carefully insert shaft through "C" sun gear. Work shaft slowly up and down and from side to side while pushing forward to allow rings to enter "C" sun gear. Rotate mainshaft slightly from side to side to engage splines on front end of shaft with splines in "B" carrier. When "C" carrier and mainshaft are in place and fully forward, tighten Band 3 adjusting screw.

Install three new sealing rings (15) on rear end of mainshaft, lubricate rings with petroleum jelly and align ring end gaps at top side of shaft. Carefully install Clutch 2 and 3 assembly over mainshaft to avoid breaking sealing rings. Partly support clutch housing and rotate housing slightly from side to side while lightly pushing forward on housing to align Clutch 2 discs with spline on "C" carrier. When housing is in position against thrust washer and "C" carrier, pull mainshaft rearward while lifting up on clutch housing so snap ring (16) can be installed in groove on rear end of mainshaft.

242. **OVERHAUL CLUTCH ASSEMBLIES (Except Direct Drive Clutch).** The following service procedures will apply to all transmission clutch assemblies except Direct Drive Clutch. The illustrations (Figs. 299 through 302) are of the 2-speed pto clutch used in other Select-O-Speed transmissions. However, other clutches are of similar construction and these pictures are used for illustrative purposes. Refer to Fig. 298 for exploded view of Clutch 2 and 3 assembly and to Fig. 305

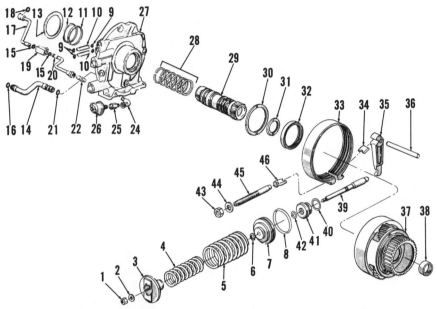

Fig. 297 – Exploded view of tubing, oil distributor, "C" carrier and sun gear, Band 3 and Servo 3 assemblies. Bushings (32 and 38) for "C" carrier (37) are not serviced. Switch assembly (items 24, 25 and 26) is no longer used.

1. Nut	13. Thrust washer	24. Elbow
2. Washer	14. Pressure tube	25. Coupling
3. Retainer	15. "O" rings	26. Oil pressure switch
4. Inner spring	16. "O" rings	27. Oil distributor
5. Outer spring	17. Direct drive clutch	28. Sealing rings
6. Snap ring	tube	29. "C" sun gear
7. Piston	18. "O" ring	30. Thrust washer
8. "O" ring	19. Servo 1 "T" connector	31. Thrust washer
9. "O" rings	20. Servo 1 – direct drive	32. Bushings
10. Servo tubes	clutch tube	33. Band 3
11. Sealing ring	21. "O" ring	34. Actuating strut
12. Sealing ring	22. Fitting	35. Lever arm

36. Pivot pin
37. "C" carrier
38. Bushing
39. Piston rod
40. "O" ring
41. Guide
42. "O" ring
43. Nut
44. Washer
45. Adjusting screw
46. Adjusting strut

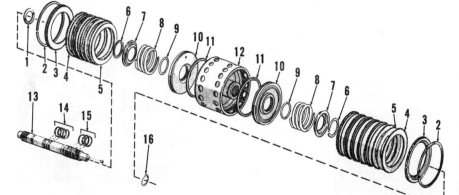

Fig. 298 – Exploded view of Clutch 2 and 3 assembly and transmission mainshaft.

1. Thrust washer	5. Internal spline plates	9. "O" rings	13. Mainshaft
2. Snap rings	6. Snap rings	10. Clutch pistons	14. Front sealing rings
3. Pressure plates	7. Retainers	11. "O" rings	15. Rear sealing rings
4. External spline plates	8. Piston return springs	12. Clutch housing	16. Snap ring

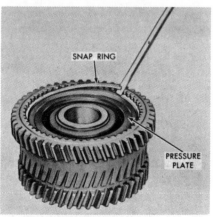

Fig. 299 – Removing clutch pressure plate retaining snap ring.

for Clutch 1 assembly. To disassemble clutch, proceed as follows:

Remove pressure plate retaining snap ring as shown in Fig. 299.

NOTE: Ends of snap ring in Clutch 1, 2 and 3 assemblies are welded together; break weld with sharp chisel, then remove and discard snap ring.

On Clutch 1 assembly, the "B" ring gear also functions as clutch pressure plate. After removing snap ring, remove pressure plate and bronze and steel discs; refer to Fig. 300.

With pressure plate and clutch discs removed, place clutch assembly in a press and compress piston return spring using Nuday tool N-775 as shown in Fig. 301. With spring compressed, remove retaining snap ring and slowly release spring to prevent cocking retaining washer on clutch housing. Remove return spring, retaining washer and snap ring, then remove piston using compressed air in port to piston as shown in Fig. 302. Remove "O" rings from inside and outside diameters of piston.

Clean all parts in solvent, air dry and inspect for wear or other damage. Renew clutch pressure plate if cracked, scored or showing signs of overheating. Fit piston into clutch housing to check for any binding condition; minor imperfections that cause binding or would cause damage to piston "O" rings can be removed with fine emery cloth. Check clutch discs for wear, signs of overheating or warping. All clutch discs should be flat. Renew any bronze discs that will snap over center. Check clutch housing, especially in hub area, and piston for cracks. Lubricate all parts with transmission fluid and reassemble as follows:

Install new "O" rings in inside diameter and on outside diameter of piston, lubricate with petroleum jelly and install piston, flat side in, in clutch housing. Be sure piston is seated in housing, then place housing in press. Place piston return spring, retaining washer and snap ring on piston, compress spring using Nuday tool N-775 as shown in Fig. 301 and install retaining snap ring in groove of clutch housing.

Place one of the bronze discs that is lined on one face only next to piston with unlined face down (against piston). Then, alternately place steel and double faced bronze discs. Install the other single faced bronze disc on top of last steel disc with unlined face up (against pressure plate). Install pressure plate with machined surface down (against unlined face of top bronze disc).

After installing pressure plate, install retaining snap ring. On all clutch assemblies be sure snap ring end gap straddles two teeth of housing, then weld ends of snap ring together using an AWS 312-16 Electrode.

CAUTION: Do not attempt to weld snap ring to pressure plate and be careful not to overheat plate or housing. Remove all weld flash.

243. **DISTRIBUTOR, CLUTCH 1 AND "B" CARRIER.** The oil distributor, Clutch 1 assembly and "B" carrier can be removed after removing control valve assembly as in paragraph 223, mainshaft "C" sun gear and Band 3 as outlined in paragraph 240 and Servo 2 and 3 cover as outlined in paragraph 234. Check transmission front end play as outlined in paragraph 228, then proceed as follows:

Loosen the four distributor retaining cap screws and with a pair of pliers, remove Servo 2 and 3 pressure tubes from distributor; do not completely remove tubes from housing. Disconnect fitting retaining Servo 1 – Direct Drive Clutch tube to distributor. Remove cap screws retaining distributor and insert long screwdriver through center of distributor against Clutch 1 housing. Hold Clutch 1 forward with screwdriver while removing distributor as in Fig. 303. Then, remove Clutch 1 assembly and "B" carrier as a unit as in Fig. 304.

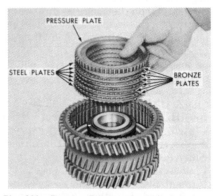

Fig. 300 — Bronze plates and steel plates are alternately placed between pressure plate and clutch piston.

Fig. 301 — Clutch piston return spring retaining snap ring being removed. Special tool (Nuday N-775) is used in press to compress piston return spring.

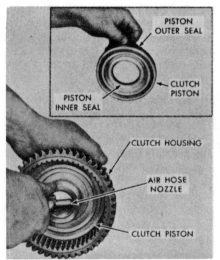

Fig. 302 — With piston return spring removed, remove clutch piston with air pressure as shown above. Inset shows piston inner and outer seals ("O" rings).

Fig. 303 — Removing oil distributor from Clutch 1 and "B" carrier assembly.

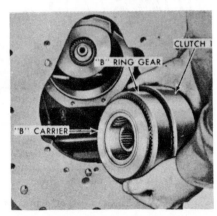

Fig. 304 — Removing Clutch 1 and "B" carrier as an assembly after removing oil distributor as shown in Fig. 303.

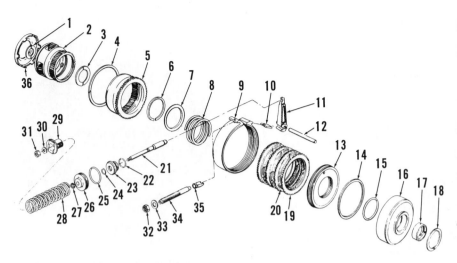

Fig. 305 — Exploded view of Clutch 1, "B" carrier and Servo 2 assemblies showing component parts and their relative positions.

1. Thrust washer
2. "B" carrier
3. Thrust washer
4. Snap ring
5. "B" ring gear – clutch 1 pressure plate
6. Snap ring
7. Retainer
8. Piston return spring
9. Band 2
10. Actuating strut
11. Lever arm
12. Pivot pin
13. Clutch piston
14. "O" ring
15. "O" ring
16. Clutch 1 housing
17. Bushing – not serviced
18. Thrust washer
19. External spline plates
20. Internal spline plates
21. Piston rod
22. "O" ring
23. Guide
24. "O" ring
25. "O" ring
26. Piston
27. Snap ring
28. Spring
29. Retainer
30. Washer
31. Nut
32. Nut
33. Washer
34. Adjusting screw
35. Adjusting strut
36. Oil collector

Remove "B" carrier from Clutch 1 assembly. Refer to paragraph 242 for Clutch 1 assembly overhaul instructions. Inspect pinion gears in "B" carrier for loose or rough bearings or for loose journal pins in carrier. Pinion end play should be 0.010-0.028 inch (0.254-0.711 mm). Check thrust and bearing surfaces of distributor for scoring or excessive wear. Renew parts as necessary and reassemble as follows:

Place "B" carrier on work bench, pinion side down, and stick thrust washer (3 – Fig. 305) in counterbore of carrier with a light film of grease. Place assembled Clutch 1 and "B" ring gear unit, ring gear side down, over "B" carrier. While partially supporting clutch assembly, rotate clutch assembly back and forth to align notches in clutch discs with splines on "B" carrier and to align teeth of "B" ring gear with pinion gear teeth on "B" carrier. Be sure thrust washer (1) is in place in hub of "B" sun gear in transmission. Then, holding Clutch 1 and "B" carrier together, install them on "B" sun gear.

If transmission front end play was incorrect when measured before disassembly and no new parts were installed which could change end play, measure old thrust washer (13 – Fig. 297) and renew with new washer of proper thickness to bring end play into limits of 0.005-0.015 inch (0.127-0.381 mm). Place washer on front side of distributor and install new seal rings on distributor with seal ring having locking ends in rear groove and ring with square ends in front groove.

NOTE: It is advisable to renew both seal rings with rings having locking ends (Ford part No. 313864) as this will make for easier assembly.

Install new "O" rings on main pressure tube. Lubricate sealing rings and "O" rings with petroleum jelly and align ring end gaps at top side of distributor through Clutch 1 hub and if square cut end snap ring is used in front groove, compress snap ring with two long screwdrivers and push distributor into place. It will be necessary to align main supply tube, Servo 1 – Direct Drive Clutch tube with holes fitting on distributor while distributor is being pushed into place. It may be necessary to lift distributor up slightly to engage pilot diameter with bore in transmission case. Start fitting onto threads of Servo 1 – Direct Drive Clutch connection. Loosely install the four distributor retaining cap screws. Insert Servo 2 and 3 pressure tubes through left side of transmission case, then install new "O" rings on each end of tubes. Lubricate "O" rings, push tubes into bores of distributor and turn tubes so beveled outer ends are aligned with oil passage slots in Servo 2 and 3 cover. Tighten distributor retaining cap screws to a torque of 20-25 ft.-lbs. (27-34 N·m). Recheck end play and if not within limits of 0.005-0.015 inch (0.127-0.381 mm), remove distributor and renew selective fit washer with one of correct thickness. Tighten Servo 1 – Direct Drive Clutch fitting.

Reassemble remainder of transmission as outlined in paragraphs 223, 235 and 236.

244. "A" RING GEAR – "B" SUN GEAR, "A" CARRIER AND DIRECT DRIVE CLUTCH. After removing distributor, Clutch 1 assembly and "B" carrier as outlined in paragraph 243, the "A" Ring Gear – "B" Sun Gear, "A" Carrier and Direct Drive Clutch can be removed as shown in Fig. 306.

NOTE: Two types of Direct Drive Clutch assemblies have been used. Refer to exploded view of Fig. 307 showing the Direct Drive Clutch (items 43 through 56). Early production clutch assemblies use only one piston return (Belleville) spring (49). Later production clutch assemblies have two piston return springs and internally splined bronze discs (45) have radial grooves in faces on disc instead of circumferential grooves as on early production discs. Clutch housings and pistons are not interchangeable between

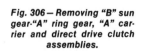

Fig. 306 — Removing "B" sun gear-"A" ring gear, "A" carrier and direct drive clutch assemblies.

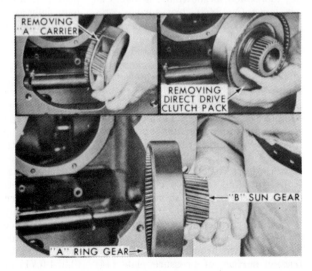

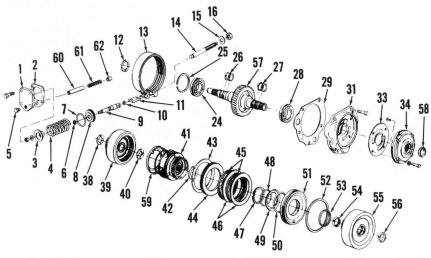

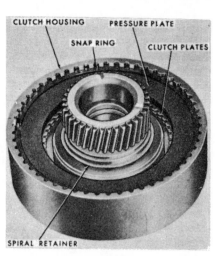

Fig. 307—Exploded view of front support plate, "A" planetary system and related parts.

Fig. 308—Direct Drive Clutch and "A" sun gear assembly.

1. Servo 1 cover	14. Adjusting screw	38. Thrust washer	50. Pressure ring
2. Gasket	15. Washer	39. "A" ring–"B" sun gear	51. Clutch piston
3. Retainer	16. Nut	40. Thrust washer	52. "O" ring
4. Spring	24. Ball bearing	41. "A" carrier	53. "O" ring
5. Plug	25. Snap ring	42. Thrust washer	54. Bushing
6. Snap ring	26. Rear sealing rings	43. Snap ring	55. Clutch housing
7. "O" ring	27. Front sealing rings	44. Pressure plate	56. Bushing
8. Servo piston	28. Ball bearing	45. Internal spline discs	57. Countershaft
9. Piston rod	29. Gasket	46. External spline discs	58. Plug
10. "O" ring	31. Front support plate	47. Spiral retainer	59. Oil collector
11. Actuating strut	33. Gasket	48. Pivot ring	60. Check valve pin
12. Thrust washer	34. Pump assy.	49. Spring	61. Valve spring
13. Band 1			62. Check valve

early and late production units and only new housings and pistons are available for service. If either a piston (51) or a housing (55) of early production assembly requires renewal, then a new type housing, piston and two return springs must be installed. The clutch discs are interchangeable; however, late type bronze discs with radial grooves should be installed for improved clutch performance.

To disassemble Direct Drive Clutch, refer to Fig. 308 and remove snap ring, pressure plate and clutch plates. Then, using snap ring removed from housing, install Belleville spring compressor (Nuday tool N-488) as shown in Fig. 309. Using lever (N-488-3), pry spring compressor down and turn a stepped stud under snap ring as shown in Fig. 310. Repeat this at the other two studs, then again at all three studs until highest step of studs is under snap ring. Then remove spiral retainer ring (Fig. 308) from hub of clutch housing and release compressor by prying it down with lever and turning studs from under snap ring.

Remove compressor tool, refer to Fig. 307 and lift out small pivot ring (48), piston return spring(s) (49) and large pivot ring (50). Remove clutch piston (51) with compressed air as shown in Fig. 302.

Carefully inspect all parts and renew as required. The two steel plates (46– Fig. 307), two bronze plates (45) and friction surface of pressure plate (44)

should be flat; renew plates if they are warped, scored or excessively worn. Free height of piston return (Belleville) spring(s) should be 0.115-0.125 inch (2.92-3.17 mm); renew if cracked or if free height measures less than 0.115 inch (2.92 mm). The bushings in clutch housing (55) are renewable. Front bushing (56) inside diameter (new) is 1.440-1.441 inches (3.657-3.660 cm); rear bushing (54) is 1.315-1.316 inches (3.340-3.342 cm). Bushings are pre-sized and should not require reaming if carefully installed.

To reassemble, install new inner "O" ring (53) and outer "O" ring (52) on piston, lubricate "O" rings with petroleum jelly and install piston in housing. Place large pivot ring (50) on piston, piston return spring(s) on large pivot ring and place small pivot ring on return (Belleville) spring(s). Compress piston return spring(s) with compressor tool as during disassembly, then install spiral retaining ring (47). Remove compressor tool and place steel disc on piston (either side down), install bronze disc (either side down), then second steel disc and second bronze disc. Place pressure plate, machined side down, on top bronze disc and install retaining snap ring (43).

The "A" carrier is renewable as a complete assembly only. Check pinions and thrust surfaces and renew carrier if defect is noted. Pinion end play in carrier should be 0.010-0.028 inch (0.254-0.711 mm).

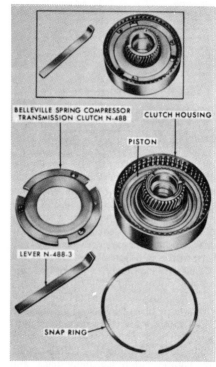

Fig. 309—Special tools used to disassemble Direct Drive Clutch assembly.

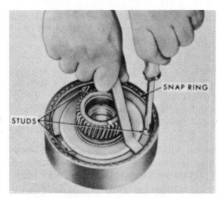

Fig. 310—Using special tools shown in Fig. 309 to compress direct drive clutch spring.

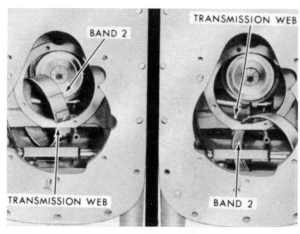

Fig. 311 — To remove or install Band 2, it must be rotated around transmission web as shown.

Fig. 314 — Model 5000 transmission main drive shaft (59), bearing (60), snap ring (61) and seal (62).

Check the two sealing rings (26) and thrust washer (12) on rear end of transmission shaft and renew if worn, scored or broken. Place new thrust washer against shaft rear bearing (24), then install the two lock-end rings in grooves on shaft. Lubricate sealing rings with petroleum jelly and align ring ends at top of shaft.

Place "A" carrier with splines up on work bench. Stick thrust washer (42) in recess of carrier hub with light film of grease. Set assembled Direct Drive Clutch on carrier and rotate clutch back and forth while partially supporting it to align splines of clutch discs with splines on carrier. Holding carrier and clutch together, install them as a unit, taking care to work clutch housing over sealing rings on transmission shaft. Place thrust washer (40) on shaft against "A" carrier, then install "A" ring gear – "B" sun gear taking care not to dislodge thrust washer (40). Install thrust washer (38) in hub of "B" sun gear.

245. BANDS 1 AND 2, STRUTS AND ACTUATING LINKAGE. After removing "A" carrier and Direct Drive Clutch as outlined in paragraph 244, Bands 1 and 2 can be removed as follows: Loosen Band 1 adjusting screw and remove band and strut. Turn adjusting screw in to remove from housing. Compress Band 2 and remove band struts. Remove Band 2 from housing by rotating it around cast web in bottom of transmission as shown in Fig. 311.

NOTE: To reinstall Band 2, steering gear housing must be removed to gain access to band struts; therefore, if not necessary and steering gear assembly is not removed, do not remove Band 2 or dislodge struts.

With Band 2 removed, actuating lever pivot pin for both Bands 2 and 3 can now be removed and Bands 2 and 3 actuating levers removed from housing. However, it is usually not necessary to remove pivot pin and actuating levers.

NOTE: Servo 3 actuating lever is the longest lever.

246. PUMP ADAPTER PLATE (TRANSMISSION FRONT COVER). To remove pump adapter plate, tractor must be split between engine and transmission as outlined in paragraph 164 or 165 then proceed as follows:

Unbolt transmission pump, tap pump body with a soft-faced hammer to break gasket seal and remove pump assembly.

Remove cap screws and nut securing adapter plate to transmission housing and thread two jack screws into tapped holes in pump adapter plate as shown in Fig. 312. Tighten jack screws evenly to pull adapter plate loose from gasket.

CAUTION: Do not allow countershaft to move forward while removing pump adapter plate, or thrust washer may drop from between "A" carrier and "A" ring gear requiring disassembly of transmission to reinstall thrust washer.

NOTE: During complete transmission overhaul, main supply pressure tube, Servo 1 and Direct Drive Clutch tubes should be removed and new sealing "O" rings installed on tube ends. The oil distributor must be removed as well as front plate to allow removal of all tubes.

Remove oil passage tubes as state of transmission disassembly will permit, renew sealing "O" rings on tube ends and reinstall in housing. Lubricate "O" rings with petroleum jelly before reinstalling tubes.

Install new sealing rings on front end of countershaft, lock ring ends together, lubricate with petroleum jelly and align ring ends at top side of shaft. Carefully install pump adapter plate with new gasket over countershaft to avoid damage to sealing rings and tighten cap screws to a torque of 25-30 ft.-lbs. (34-41 N·m). Install transmission pump as outlined in paragraph 226.

247. MAIN DRIVE (INPUT) SHAFT, BEARING AND SEAL. To renew main drive (input) shaft seal, first split tractor between engine and transmission as outlined in paragraph 164 or 165, refer to Figs. 313 and 314, then proceed as follows: Insert screwdriver or other suitable tool under outer metal lip of seal (62) and pry seal from housing. Be sure that sealing surface of shaft is clean and free of rust or burrs, then lubricate lip of new seal with petroleum jelly, slide seal over main drive shaft and drive into place with suitable driver.

To remove main drive shaft, transmission must be removed from tractor and all components removed as outlined in paragraph 222 through 244. Then, tap shaft rearward until bearing (60–Fig. 314) is free of bore in transmission housing, then remove shaft and bearing assembly from rear end of transmission.

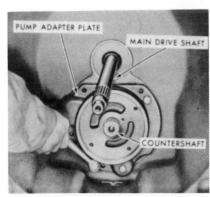

Fig. 312 — Using jackscrews to remove pump adapter (front support) plate.

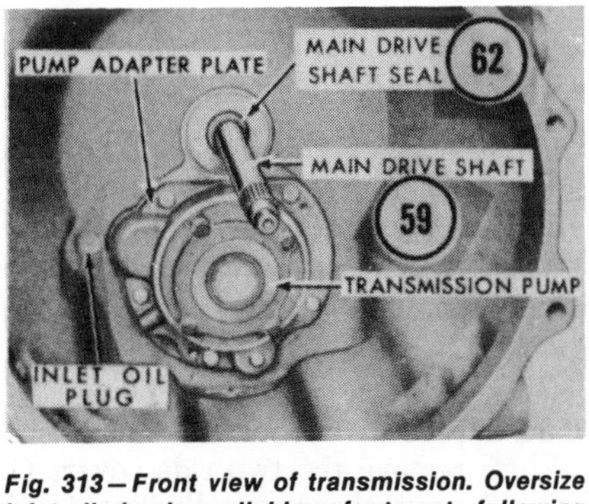

Fig. 313 — Front view of transmission. Oversize inlet oil plug is available; refer to note following paragraph 248.

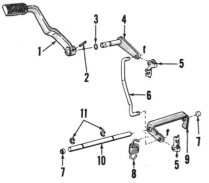

Fig. 315 — Exploded view of feathering pedal and linkage.

1. Foot pedal	6. Link
2. Roll pin	7. Plugs
3. "O" ring	8. Return spring
4. Shaft & arm	9. Rocker
5. Retainers	10. Pivot shaft

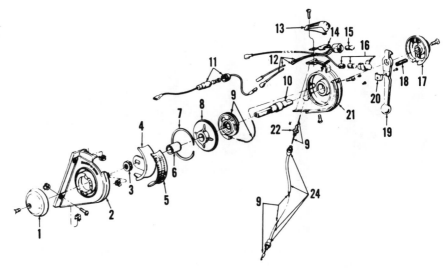

Fig. 316 — Exploded view of transmission speed (gear ratio) selector assembly.

1. Cover		8. Cover	12. Wiring harness	18. Spring
2. Housing, L.H.		9. Control wheel &	13. Warning light lens	19. Control wear
3. Nut		cable assy.	14. Clip	20. Wear plate
4. Dial		10. Shaft	15. Warning light	21. Housing, R.H.
5. Dial decal		11. Oil pressure warning	16. Dial light	22. Connector
6. Spacer		light wire	17. Cover	24. Conduit
7. Snap ring				

It is not usually necessary to remove snap ring (61) from groove in bore of housing. Drive seal (62) out forward, then install new seal as in preceding paragraph after transmission is reassembled.

248. COUNTERSHAFT AND BEARINGS. To remove countershaft assembly, transmission must be removed from tractor, all components removed as outlined in paragraph 222 through 244 and pump adapter (front cover) plate removed as outlined in paragraph 246. Then, remove countershaft and bearing assembly from front of transmission.

Remove cast iron sealing rings (26 and 27–Fig. 307) and thrust washer (12) if not already removed. If necessary to renew bearings or shaft, remove bearing from ends of shaft with bearing puller. Rear bearing (24) and front bearing (28) are alike; however, a snap ring is installed in outer race of rear bearing only. Install both front and rear bearings with snap ring groove forward, then install snap ring in groove on rear bearing only. Install countershaft and bearing assembly so snap ring on rear bearing contacts rear wall of front compartment in transmission, then reassemble transmission by referring to appropriate paragraphs and reversing disassembly procedure.

NOTE: In event of oil leakage at inlet oil plug (see Fig. 313) in front wall of transmission housing, hole can be reamed to 1.120-1.125 inch (2.845-2.857 cm) and a 1.135-1.137 inch (2.882-2.887 cm) oversize plug installed. Standard hole size is 0.9995-1.0015 inch (2.538-2.543 cm) and standard plug size is 1.010-1.012 inch (2.565-2.570 cm). Oversize plug may be installed during production of transmission at factory.

249. TRANSMISSION CONTROL LINKAGE. Usually control linkage components do not need to be disassembled when overhauling transmission. However, if necessary to renew linkage or sealing "O" rings at inching pedal shaft, proceed as follows:

249A. INCHING PEDAL AND FEATHERING VALVE LINKAGE. Refer to Fig. 315 for exploded view of inching pedal and feathering valve linkage. Inching pedal (1) can be renewed by driving out pin (2) and removing transmission bottom cover and holding pry bar against lever shaft (4) while installing new pedal. If necessary to renew "O" ring (3), shaft (4) or link (6), remove Clutch 2 and 3 assembly, mainshaft and "C" carrier as outlined in paragraph 240.

To remove feathering valve lever shaft (10) or components located on shaft, remove bottom transmission cover, disconnect inching pedal return spring (8), remove snap rings (11) from shaft and drive against either sealing plug (7) with a thin punch. Shaft need not be driven all the way out. After reinstalling shaft and components, insert sealing plugs and center feathering valve levers on valves by driving against either plug as required.

250. SPEED SELECTOR ASSEMBLY. To remove speed selector asembly, refer to paragraph 220. To disassemble, refer to exploded view in Fig. 316 and proceed as follows:

First, remove the two side covers (1 and 17) and the three Phillips head screws securing left-hand selector hous-

ing (2). Remove hex nut (3) from shaft (10) and remove snap ring (7) and retainer plate (8) to expose wheel and cable assembly.

NOTE: Impact torque is usually required to loosen nut (3) without twisting shaft (10).

Further disassembly procedure is obvious from inspection of unit and reference to exploded view in Fig. 316. To reassemble, reverse disassembly procedure. Adjust speed selector assembly as outlined in paragraph 214, after reinstalling assembly as in paragraph 220.

REDUCTION GEARBOX
Models 5610-6610-7610

251. Reduction gearbox (creeper gearbox) is available as an option on either non-synchromesh or synchromesh transmission. Reduction gearbox assembly is mounted at rear of transmission in place of output shaft gear used on standard transmissions. Reduction gearbox major components are: carrier, planetary gears, outer ring gear, intermediate ring gear and coupling. On non-synchromesh transmission, optional planetary gear assemblies are available to provide a creep range ratio below the low range of either 5.7:1 or 10:1. On synchromesh transmission, creep range ratio is 5.7:1 below the low range speed.

Coupling contained within reduction gear set slides by use of a selector fork and gear shift rail to engage low range or creep range position. In low range position, power is transmitted from sec-

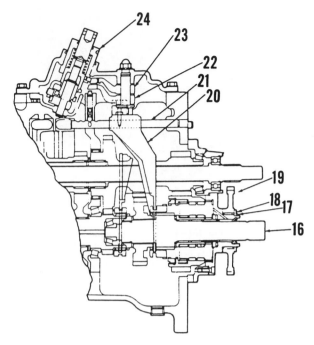

Fig. 316A — Cross-sectional view of gear reduction assembly used on Models 5610, 6610 and 7610 with synchromesh transmission. Gear reduction assembly used on models with non-synchromesh transmission is similar.

16. Output shaft & reduction gear set assy.
17. Snap ring
18. Thrust washer
19. Idler gear
20. Shift fork
21. Rail
22. Bellcrank
23. Bellcrank
24. Selector assy.

withdrawn. Remove secondary countershaft and withdraw shift coupling from selector fork as needed.

Remove screws securing output shaft retainer (5), then separate retainer (5) and outer ring gear (6) from rear support plate (4). Note shim(s) located between retainer (5) and outer ring gear (6). Separate output shaft from reduction gear set using a suitable soft-faced mallet.

NOTE: Rear bearing will fall from shaft end as shaft is extracted from reduction gear set. Intermediate ring gear needle roller bearing may remain on shaft as shaft is being extracted.

Lift thrust washer and as required, needle roller bearing from output shaft. Lay reduction gear set with intermediate ring gear facing up as shown in Fig. 318. Lift intermediate ring gear (7) from carrier (8) and remove thrust washer (9) and two needle roller bearings with spacer from carrier. Lift planetary gears (10) from carrier (8).

NOTE: Use care not to allow rollers to fall free from mounting lips.

Inspect all bearings for cracks, corrosion, roughness, excessive wear or any other damage. Inspect splines and gear teeth for missing teeth, chips, cracks, excessive wear or any other damage. Examine output shaft retainer and rear support plate for cracks, excessive wear or any other damage. Inspect thrust washers for cracks, excessive wear or any other damage. Inspect shift mechanism and coupling for excessive wear or any other damage. Renew all parts as needed.

NOTE: Complete reduction gear set should be renewed if several individual components must be serviced.

ondary countershaft on non-synchromesh transmissions or range cluster gear on synchromesh transmissions through carrier and coupling to output shaft. In creep range position, power is transmitted from secondary countershaft on non-synchromesh transmissions or range cluster gear on synchromesh transmissions to carrier. Rotation of carrier causes planetary gears to rotate. Rear teeth of planetary gears engage stationary outer ring gear and forward teeth engage and drive intermediate ring gear at a reduced speed. Coupling is engaged with intermediate ring gear, thereby driving output shaft at a reduced speed.

Non-Synchromesh Transmissions

252. **OVERHAUL.** Drain transmission oil. Separate transmission from rear axle as outlined in paragraph 254 or 256. Remove shift cover as outlined in paragraph 179. Remove snap ring (17–Fig. 316A), thrust washer (18) and hydraulic pump idler gear (19) located on output shaft retainer (5–Fig. 317). Remove cap screws securing rear support plate (4) to transmission housing. Using suitable tools, pry support plate (4) rearward from transmission. Pto countershaft (3) will be withdrawn with support plate. Partially withdraw secondary countershaft assembly (2), then tilt countershaft high enough to allow output shaft and reduction gear set (1) to be

Fig. 317 — View showing reduction gear set (1), secondary countershaft assembly (2), pto shaft (3), rear support plate (4), output shaft retainer (5) and outer ring gear (6) used on non-synchromesh transmissions.

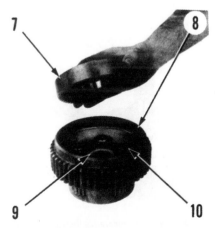

Fig. 318 — View showing removal of intermediate ring gear (7) from carrier (8). Note thrust washer (9) and planetary gears (10).

Fig. 319 — Position carrier (8) on outer ring gear (6) and install planetary gear (10) with rollers (11).

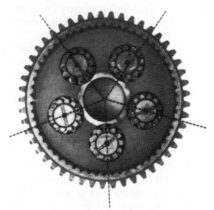

Fig. 320 — Punched tooth (master tooth) on all planetary gears should face toward center of carrier for correct installation.

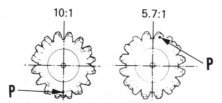

Fig. 322 — (P) identifies master tooth after determining alignment of front and rear planetary gears. Refer to text.

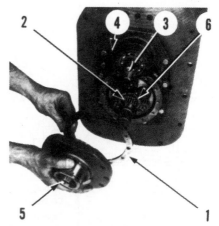

Fig. 324 — View showing output shaft retainer (5), needle roller bearing (2), rear support plate (4), thrust bearing (6), shim(s) (1) and pto shaft (3) used on synchromesh transmissions.

To reassemble, use Ford ITMIC-137A grease or a suitable equivalent to hold planetary gear rollers (11 – Fig. 319) in position. Set carrier (8) on outer ring gear (6). Outer ring gear acts as an alignment gage for planetary gears. Install planetary gears (10) and turn until punched tooth (master tooth) is facing toward center of carrier Fig. 320.

NOTE: Early production models are not equipped with an identified master tooth. To identify master tooth proceed as follows: Lay a rod (12 – Fig. 321) between gears (13 and 14) to identify which pair of teeth on gear (13) aligns with a pair of teeth on gear (14). On 5.7:1 (12-tooth gear) reduction ratio, four pair of teeth will align. Identify any one tooth on the right-hand side (Fig. 322) as the master tooth. On 10:1 (13-tooth gear) reduction ratio, only one pair of teeth will align. Identify opposite tooth (Fig. 322) as the master tooth.

Position thrust washer with chamfered side up in carrier hub, then install intermediate ring gear into carrier. Place coupling on output shaft and install thrust washer on shaft with chamfered side facing coupling. Lubricate and install narrow needle roller bearing in intermediate ring gear and two wide needle roller bearings with spacer between them in carrier. Position assembled output shaft vertically and install carrier assembly on output shaft. Install output shaft rear bearing on shaft until bearing seats against shaft shoulder. Use feeler gage (F – Fig. 323) to measure clearance between bearing (15) and carrier (8). Clearance should be 0.012-0.032 inch (0.3-0.8 mm).

Install assembled carrier and output shaft unit in transmission, engage selector fork in coupling groove. Withdraw outer ring from reduction gear set assembly. Install secondary countershaft assembly. Install rear support plate and tighten retaining cap screws to 32 ft.-lbs. (43 N·m) torque. Position outer ring gear, then install shim(s) and output shaft retainer. Install securing cap screws and tighten to 32 ft.-lbs. (43 N·m) torque.

Check output shaft preload as follows: Place transmission in neutral position. Use a pull scale and string wrapped around output shaft and measure amount of force required to turn output shaft. Correct preload is 13-22 pounds (6-10 kg). If preload is incorrect, shim(s) located between outer ring gear and output shaft retainer must be adjusted. Shims are available in 0.004, 0.006 and 0.019 inch (0.1, 0.15 and 0.5 mm). Adjust shim thickness as needed until correct preload is attained. Complete reassembly in reverse order of disassembly.

Synchromesh Transmission

253. OVERHAUL. Drain transmission oil. Separate transmission from rear axle as outlined in paragraph 254 or 256. Remove reverse/low/creep range rail and fork from transmission case as outlined in paragraph 191. Remove snap ring (17 – Fig. 316A), thrust washer (18), hydraulic pump idler gear (19) and

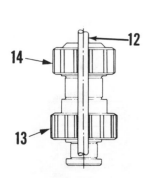

Fig. 321 — On early production models without an identified master tooth, use rod (12) to identify which pair of teeth on gear (13) aligns with a pair of teeth on gear (14).

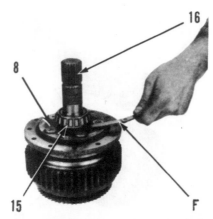

Fig. 323 — Measure clearance between bearing (15) and carrier (8) using feeler gage (F). Bearing should seat on shoulder of output shaft (16). Correct clearance is 0.012-0.032 inch (0.3-0.8 mm).

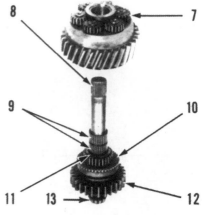

Fig. 325 — View of output shaft and reduction gear set assembly.

7. Reduction gear set
8. Output shaft
9. Needle roller bearings
10. Coupling
11. Thrust bearing
12. Reverse gear
13. Bearing

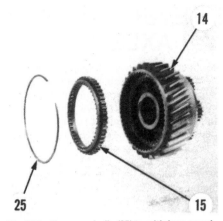

Fig. 326 — Remove circlip (25) to withdraw coupling (15) from carrier (14).

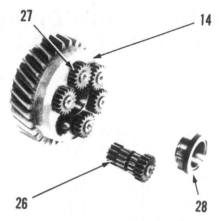

Fig. 327 — To remove planetary gear (27) from carrier (14), use special puller to withdraw retainer (28), then extract planetary gear assembly (26) from carrier bore.

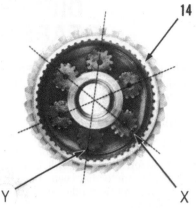

Fig. 328 — Planetary gears are correctly installed when "X" and "Y" marks on gear ends alternate and face away from center of carrier (14).

needle roller bearing located on output shaft retainer (5–Fig. 324). Remove screws securing output shaft retainer (5) to rear support plate (4), then remove retainer (5), shim(s) (1) needle roller bearing (2) and thrust bearing (6). Remove cap screws retaining rear support plate (4) to transmission housing. Using suitable tools, pry support plate (4) rearward from transmission. Pto countershaft (3) will be withdrawn with support plate. Withdraw output shaft and reduction gear set as an assembly, then remove high range coupling.

Lift reduction gear set (7–Fig. 325) from output shaft (8). Withdraw two needle roller bearings (9) and thrust bearing (11) from shaft. Slide coupling (10) off shaft. Use a suitable bearing knife and puller assembly to withdraw bearing (13) from front of output shaft. Remove snap ring, inner coupling, reverse gear and thrust washers from output shaft. Remove circlip (25–Fig. 326) and withdraw coupling ring (15) from carrier (14). Use special planetary gear retainer puller (Churchill tool 954C or Nuday tool 9527) and withdraw retainer (28–Fig. 327) from carrier (14). Extract planetary gear assemblies with rollers (26) from carrier bores.

NOTE: Use care not to allow rollers to fall free from mounting lips.

Lift intermediate ring gear from carrier.

Inspect all bearings for cracks, corrosion, roughness, excessive wear or any other damage. Inspect splines and gear teeth for missing teeth, chips, cracks, excessive wear or any other damage. Examine output shaft retainer and rear support plate for cracks, excessive wear or any other damage. Inspect thrust

washers and planetary gear retainer for cracks, excessive wear or any other damage. Inspect shift mechanism and coupling for excessive wear or any other damage. Renew all parts as needed.

NOTE: Complete reduction gear set should be renewed if several individual components must be serviced.

To reassemble, use Ford ITMIC-137A grease or a suitable equivalent to hold planetary gear rollers in position. Install planetary gears with rollers into carrier bores, alternating "X" marked gears with "Y" marked gears. Install planetary gear retainer (28–Fig. 327) to hold gears in position. Turn carrier assembly over so assembly rests on planetary gear retainer. Rotate planetary gears so "X" (Fig. 328) and "Y" marks face away from center of carrier (14). Install intermediate ring gear and coupling ring (15–Fig. 326) in carrier and secure with circlip (25). Install thrust washer, reverse gear, second thrust washer, inner coupling and snap ring on front of output shaft. Slide selector coupling (10–Fig. 325) on shaft with shift fork groove facing reverse gear (12). Lubricate and install thrust bearing (11) and two needle roller bearings (9) on shaft. Position assembled output shaft vertically and install carrier assembly on output shaft.

Install high range coupling on countershaft. Install assembled carrier and output shaft unit in transmission. Install reverse/low/creep range rail and fork in transmission case as outlined in paragraph 191. Install rear support plate with pto shaft and tighten retaining cap screws to 32 ft.-lbs. (43 N·m) torque. Lubricate and install thrust bearing and needle roller bearing on output shaft. Install shim(s) and output shaft retainer.

Fig. 329 — View showing correct mounting position of dial indicator (D) for checking end play between output shaft (S) and countershaft. Refer to text.

NOTE: Be sure ring gear assembly contained within output shaft retainer correctly engages planetary gears.

Tighten cap screws securing output shaft retainer to 32 ft.-lbs. (43 N·m) torque.

A dial indicator mounted as shown in Fig. 329 is used to measure end play between output shaft and countershaft. Move output shaft forward and rearward while observing indicator needle movement. End play should be 0-0.004 inch (0-0.102 mm). If end play is incorrect, adjust shim(s) thickness located between rear support plate and output shaft retainer until correct preload is attained. Shims are available in thicknesses of 0.004, 0.012, 0.019 and 0.027 inch (0.1, 0.3, 0.5 and 0.7 mm). Complete reassembly in reverse order of disassembly.

DIFFERENTIAL, MAIN DRIVE BEVEL GEARS, FINAL DRIVE AND REAR AXLE

SPLIT TRACTOR BETWEEN TRANSMISSION AND REAR AXLE CENTER HOUSING

All Models With Cab Except 6700-6710-7700-7710

254. To remove main drive bevel pinion and perform other jobs, it is necessary to split tractor between transmission housing and rear axle center housing as follows:

To split tractor between transmission housing and rear axle center housing on models equipped with a Ford cab, proceed as follows:

Remove interfering knobs and levers, unscrew retaining screws (S–Fig. 330) and remove control quadrant (2). Detach flow control knob (5), remove boot retainer (4) and push boot (3) lip down to cab underside. Disconnect hydraulic lift control rods, unscrew nuts securing bracket for hydraulic lift control levers and remove control assembly. On models without Load Monitor, detach position control link from selector shaft. On models with Load Monitor, detach load monitor link from selector shaft. Remove flow control cable guide. On models so equipped, detach lower end of Dual Power control rod from pivot assembly. On models so equipped, detach transmission hand brake lever and push control cable under cab. Extract pin connecting flow control cable to flow control valve. On models so equipped, disconnect rear wheel handbrake cable at lower end. Note position of adjusting nut on differential lock pedal, loosen adjusting nut and unscrew

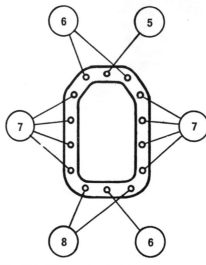

Fig. 331—When reconnecting rear axle center housing to transmission, tighten bolt at (5) to a torque of 55-75 ft.-lbs. (75-102 N·m), bolts at (6) to a torque of 110-135 ft.-lbs. (150-184 N·m), bolts at (7) to a torque of 65-86 ft.-lbs. (88-117 N·m) and bolts at (8) to a torque of 130-160 ft.-lbs. (177-218 N·m).

pedal. Detach lower end of pto rod from actuation lever. Unscrew retaining screws and remove engine hood rear support. Remove scuff plates beneath each cab door and pull back floor matting to uncover access covers. Remove access covers and unscrew front cab retaining nuts until nuts are flush with bolt ends. Unscrew four bolts securing rear cab struts to rear axle housings.

Drain rear axle lubricant and disconnect oil cooler lines at rear axle center housing. Disconnect brake pedal rods at pedal shanks. Using suitable jacks, raise rear of cab three inches and place supports under cab.

NOTE: Do not raise rear of cab more than 3½ inches (89 mm) as windshield may fracture.

Support front of rear axle center housing, unscrew rear axle to transmission bolts and roll rear axle away from tractor.

When reconnecting transmission and rear axle, refer to Fig. 331 for tightening torque. Refer to paragraph 286 when refilling center housing.

Models 6700-6710-7700-7710

255. To split tractor between transmission housing and rear axle center housing on Models 6700 and 7700, proceed as follows:

Disconnect all control rods leading from hydraulic lift control console. Dis-

connect Dual Power control rod. Disconnect any control cables which will prevent cab or platform from tilting three inches at rear. Remove front sheet metal which will prevent tilting of cab or platform. Remove scuff plates beneath each cab door, on models so equipped, and pull back floor matting to uncover access covers. Remove access covers on models with a cab. Unscrew front cab or platform retaining nuts until nuts are flush with bolt ends. Unscrew four bolts securing rear cab or platform struts to rear axle housings.

Drain rear axle lubricant and disconnect oil cooler lines at rear axle center housing. Disconnect brake rods. Using suitable jacks raise rear of cab or platform three inches (76 mm) and place supports under cab.

NOTE: Do not raise rear of cab more than 3½ inches (89 mm) as windshield may fracture.

Support front of rear axle center housing, unscrew rear axle to transmission bolts and roll rear axle away from tractor.

Refer to Fig. 331 when tightening rear axle-to-transmission bolts. Refill with lubricant as outlined in paragraph 286.

All Other Models

256. Drain lubricant from rear axle center housing. On models with hydraulic lift, move selector lever to draft control, push selector valve in on models so equipped, move control lever to bottom of quadrant and force lift arms to lowest position. Disconnect wiring to rear light, disconnect brake pedal return springs and brake rods and, on models with engine clutch, disconnect clutch release

Fig. 330—View of hydraulic control quadrant used on models equipped with a Ford cab except Models 6700, 6710, 7700 and 7710.

S. Screws
1. Control levers
2. Control quadrant
3. Boot
4. Boot retainer
5. Flow control knob

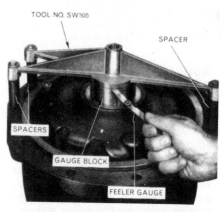

Fig. 332—Checking axle housing, final drive and brake assembly for proper differential carrier bearing spacer shim thickness. Refer to text.

rod. Remove brake lock rod and handle. Unbolt and remove step plates, or operator's platform on rowcrop models. On models with Select-O-Speed transmission disconnect traction coupling.

If tractor has been split between engine and transmission, proceed as follows: Attach a hoist to transmission, support rear axle center housing, then unbolt and remove transmission assembly.

If performing split on assembled tractor, proceed as follows: On models with

horizontal exhaust, disconnect pipe from muffler. On rowcrop models, adequately support tricycle front end to keep it from tipping. On models with wide front axle, drive wood wedges between front axle and front support and place a jack or work stand under rear end of transmission. Support rear axle with suitable moving hoist or floor jack, then unbolt rear axle center housing from transmission and move rear unit away.

To reconnect transmission and rear axle center housing, reverse disassembly procedure and on models with

Load Monitor, be sure oil tube mates with front portion of the lubrication tube. For transmission-to-center housing bolt tightening torque, refer to Fig. 331. Refill center housing with proper lubricant as outlined in paragraph 286.

DIFFERENTIAL AND BEVEL GEARS

257. **R&R DIFFERENTIAL ASSEMBLY.** To remove differential assembly, right axle and housing assem-

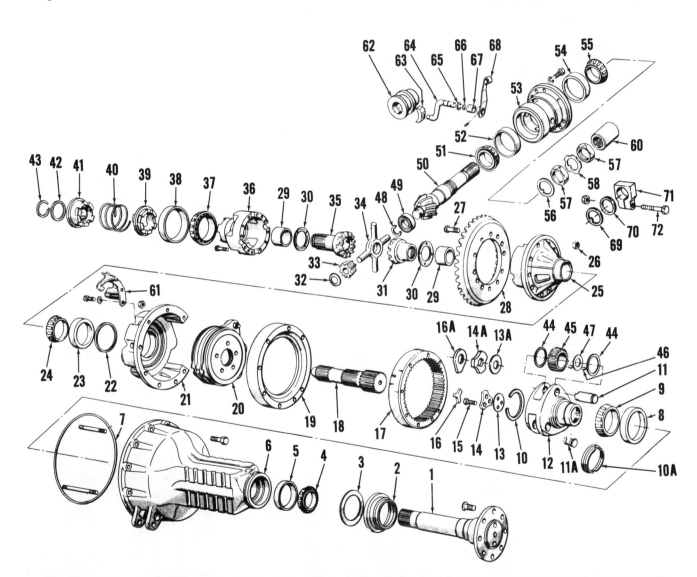

Fig. 333 — Exploded view of typical rear axle, final drive and differential assemblies. Models with Select-O-Speed transmission use sliding coupling (62) on transmission output shaft. Models with 8-speed transmission use solid coupling (60) except when equipped with load monitor. Model 7000, 7600, 7610, 7700 and 7710 tractors use only one cap screw (15) to retain axle shaft.

1. Axle shaft	22. Spacer shim	35. Side gear	48. Snap ring	61. Thrust block
2. Oil seal	23. Bearing cup	36. Diff. case half, L.H.	49. Pilot bearing	62. Sliding coupling
3. Gasket	24. Bearing cone	37. Bearing cone	50. Bevel pinion	63. Thrust block
4. Bearing cone	25. Diff. case half, R.H.	38. Bearing cup	51. Bearing cone	64. Shaft & arm
5. Bearing cup	26. Self locknut	39. Adapter	52. Bearing cup	65. Snap ring
6. Axle housing	27. Ring gear bolt	40. Spring	53. Bearing carrier	66. "O" ring
7. "O" ring	28. Bevel ring gear	41. Coupling	54. Bearing cup	67. Bushing
8. Bearing cup	29. Bushing	42. Washer	55. Bearing cone	68. Disconnect lever
9. Bearing cone	30. Thrust washer	43. Snap ring	56. Thrust washer	69. Thrust washer
10. Retainer (late)	31. Side gear	44. Thrust washers	57. Nut	70. Thrust washer
10A. Retainer (early)	32. Thrust washers (4)	45. Planetary pinions	58. Locking washer	71. Clamp nut
11. Pinion shafts (late)	33. Diff. pinions	46. Needle bearings	59. Nut	72. Clamp nut bolt
11A. Pinion shafts (early)	34. Spider	47. Spacer washer	60. Coupling	
12. Planetary carrier				
13. Spacer shims				
13A. Spacer shims				
14. Retainer				
14A. Retainer				
15. Cap screws				
16. Retainer				
16A. Retainer				
17. Planetary ring gear				
18. Planetary sun gear				
19. Brake outer housing				
20. Brake assy.				
21. Brake inner housing				

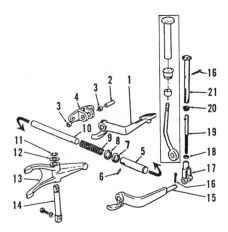

Fig. 334—Exploded view of differential lock linkage mounted in rear axle center housing. Models 5200, 7200, 6600C and 7600C use items 15 through 21. Pedal components shown in inset are used on some late model tractors.

1. Foot pedal (5100 & 7100)	11. Snap ring
2. Pivot pin	12. Washer
3. Bushings	13. Fork
4. Pedal bracket	14. Pivot shaft
5. Sleeve	15. Arm (5200-7200)
6. Pin	16. Cotter pin
7. Oil seal sleeve	17. Clevis
8. Oil seal	18. Jam nut
9. Spring	19. Operating rod
10. Actuator shaft	20. Bushing
	21. Pedal

bly must be removed. To remove and reinstall axle assembly, refer to paragraph 262.

With axle assembly removed, remove differential assembly from rear axle center housing. Refer to paragraph 259 for differential overhaul procedure.

258. DIFFERENTIAL CARRIER BEARING ADJUSTMENT. Differential carrier bearing preload is non-adjustable, however bearing preload can be considered correct when differential carrier bearing cup position in axle housing is properly established by installation of correct shim thickness behind bearing cup in bore of inner brake housing. Left carrier bearing cup is retained in a bore in rear axle center housing and does not require shimming. Once proper shim thickness is established for a particular assembly of axle housing, final drive ring gear, brake outer housing and brake inner housing, original shim should be retained for reassembly. However, if shim is lost or when renewing an axle housing, final drive ring gear, brake outer housing and/or brake inner housing, assembly must be gaged and proper shim thickness installed in new assembly. To gage axle housing, ring gear and brake housing assembly for proper shim thickness, proceed as follows:

Assemble axle housing, final drive ring gear, brake outer housing and brake inner housing without brake assembly and final drive sun gear and with carrier bearing cup and shim re-

moved from inner brake housing. Tighten brake housing retaining stud nuts to a torque of 72-90 ft.-lbs. (98-122 N·m) and install gage frame and block as shown in Fig. 332.

Measure resulting gap between gage frame and gage block with feeler gage taking care to obtain accurate measurement. For final assembly of axle housing, axle, final drive and brake assembly, select a shim of thickness nearest to measured gap between gage frame and block. Eight different shim thicknesses are available from 0.038-0.040 inch (0.965-1.016 mm) to 0.080-0.082 inch (2.032-2.082 mm) in steps of 0.004 inch (0.102 mm).

259. OVERHAUL DIFFERENTIAL. Remove differential as outlined in paragraph 257. Remove snap ring (43–Fig. 333), washer (42) coupling (41), coupling spring (40) a adapter (39). Place correlation marks on both halves (25 and 36) of differential case so they can be reassembled in the same relative position. Cut cap screw locking wire and remove cap screws evenly while lifting up left half of differential case (36).

Remove thrust washer (30), left-hand side gear (35), spider (34), pinion (33) and washer (32) assembly, right-hand side gear (31) and right thrust washer (30). Check carrier bearings (24 and 37) and renew if worn or damaged. To renew carrier bearing cups (23 and 38), left-hand axle housing must be removed from center housing and axle assemblies removed from both axle housings to provide clearance for removing cups.

Differential right half (25) and bevel ring gear (28) are riveted together during factory assembly. If renewing either case or ring gear separately, drill through rivet heads with a ½-inch (12.7 mm) drill, then remove rivets with a

punch. Assemble new ring gear and/or case with special bolts (27) and self-locking nuts (26) that are available for service. Be sure ring gear is not cocked on differential case and tighten nuts to a torque of 40-45 ft.-lbs. (54-61 N·m).

NOTE: Bevel ring gear and bevel pinion are available as a matched set only. When installing a new ring gear, refer to paragraph 260 and install matched bevel pinion.

Renew thrust washers (30 and 32) if worn or scored and other parts if excessively worn or damaged. Differential case bushings (29) may be renewed if case halves are otherwise serviceable. Bushings are presized and should not require reaming if carefully installed.

Reassemble differential by reversing the disassembly procedure and tighten cap screws retaining case to a torque of 67-75 ft.-lbs. (91-102 N·m) and secure with lock wire.

260. BEVEL PINION. To remove main drive bevel pinion, tractor must be split as outlined in paragraph 254, 255 or 256. Remove hydraulic lift cover as outlined in paragraph 354, 355 or 356, differential as outlined in paragraph 257, hydraulic pump as outlined in paragraph 379 or 381, and on models so equipped, Load Monitor as outlined in paragraph 370.

Remove cap screws and lockwashers retaining bevel pinion bearing carrier (53–Fig. 333) to rear axle center housing, then remove carrier and pinion assembly by using jack screws.

On models with no Load Monitor, straighten tabs on lockwasher (58) and remove nuts (57), then press shaft (50) out of bearing (55) and retainer (53). Remove bearing (51) from shaft.

Fig. 335—Removing brake assembly from axle housing.

Fig. 336—Pressing three planetary pinion shafts, retainer and bearing from early type planetary carrier. A flat plate and three pins of equal length are used to press shafts and retainer evenly. On late models, refer to Figs. 337 and 338.

On models with Load Monitor, loosen clamp bolt (72), remove clamp nut (71), then press shaft out of bearings and retainer.

Bevel pinion is available only in a matched set with bevel ring gear, and if necessary to renew bevel pinion, refer to paragraph 259 and install matched bevel ring gear.

Renew bearings if rough or excessively worn and adjust bearing preload as follows: Tighten nut (57–Fig. 333) or clamp nut (71), until pull required to rotate pinion in bearings is 16-21 pounds (7.26-9.52 kg) when checked with pull scale and cord wrapped around pinion shaft.

On models with no Load Monitor, install new tab washer (58) and second nut (57). Tighten nut and recheck bearing preload. Readjust if necessary and when adjustment is correct, bend tabs of washer against nuts.

On models with Load Monitor, tighten clamp bolt (72) when bearing preload adjustment is correct.

When reinstalling pinion, bearing and carrier assembly, tighten retaining cap screws to a torque of 80-100 ft.-lbs. (109-136 N·m).

261. OVERHAUL DIFFERENTIAL LOCK. Differential lock clutch can be removed from differential assembly after removing left axle and housing assembly. Refer to differential overhaul as outlined in paragraph 259.

Refer to Fig. 334 for exploded view of differential lock operating linkage. Operating fork (13) and shaft (14) can be removed from center housing after removing left axle and housing assembly. Actuating shaft (10), spring (9) and sleeve (5) can be removed after removing hydraulic lift cover and pedal (1) and bracket (4) assembly. With shaft and sleeve removed, pry out old seal shield (7) and oil seal (8). Install new seal with lip inward, then install shield with scraper edge out.

NOTE: Components which comprise pedal assembly for Models 5200, 7200, 6600C and 7600C are shown in items 15 through 21 in Fig. 234. Some late model tractors use pedal components shown in inset. Any service differences are apparent.

REAR AXLE AND FINAL DRIVE

262. R&R REAR AXLE ASSEMBLY. First, drain differential center housing and hydraulic system, then proceed as follows: Remove wheel weights if so equipped. On Models 6700, 6710, 7700 and 7710, refer to paragraph 255 and tilt cab or platform forward. Tilt cab forward on other models so equipped as

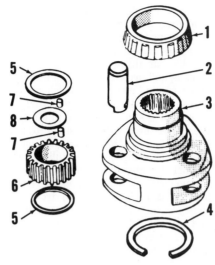

Fig. 337 – Exploded view of late type planetary unit. Refer to Fig. 338 for disassembly techniques.

1. Bearing	5. Thrust washer
2. Pinion shaft	6. Planet pinion
3. Planet carrier	7. Needle roller
4. "C" ring	8. Spacer washer

outlined in paragraph 254. On all other models, remove rear fender. Disconnect hydraulic lift lower link and brake linkage. Support tractor under center housing and attach hoist to rear wheel and axle unit. Remove cap screws retaining axle housing to center housing, then remove wheel and axle assembly from tractor. Lay wheel flat on floor with axle housing up to provide solid base for overhauling brakes, final drive and rear axle unit. When removing right axle assembly, differential assembly should also be removed from center housing to prevent assembly from accidentally falling out and becoming damaged.

When reinstalling, place new sealing "O" ring on axle housing. Tighten axle housing retaining cap screws to a torque of 110-135 ft.-lbs. (150-184 N·m).

263. OVERHAUL FINAL DRIVE AND REAR AXLE ASSEMBLY. With rear axle and housing assembly removed as outlined in paragraph 262, refer to exploded view in Fig. 333 and proceed as follows:

Unbolt and remove ring gear thrust block (61) from right axle housing. Remove stud nuts retaining brake inner housing (21) and remove inner housing, brake assembly (20) and brake outer housing (19). Refer to Fig. 348 for exploded view of brake assembly. Lift out final drive sun gear (18–Fig. 333) and locking plate (16 or 16A). On Models 7000, 7600, 7610, 7700 and 7710, it may be necessary to turn axle housing to align planetary (Fig. 343) so lockplate can be removed.

NOTE: It may be necessary to apply heat to loosen lock plate except on 7000, 7600, 7610, 7700 and 7710 models.

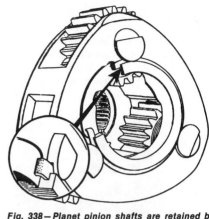

Fig. 338 – Planet pinion shafts are retained by "C" ring as shown. Bend ends of ring into depression in carrier (inset) to secure ring.

Remove cap screw (15–Fig. 333) retaining final drive planetary (12) to axle shaft on Model 7000, 7600, 7610, 7700 and 7710 or the three cap screws on all other models and remove retainer (14), spacer shim (13) and planetary assembly. Drive axle shaft seal (2) from outer end of axle housing and lift housing from shaft.

To remove bearing cone (9) and retainer (10) from planetary carrier (12) and to disassemble early type planetary unit, refer to Fig. 336 and press out all three pinion shafts at one time which will also press retainer and bearing cone from carrier. Each planetary pinion (45–Fig. 333) is fitted with two rows of needle rollers (46) which are separated by a washer (47). The bearings are serviced in a kit of 58 rollers. Renew planetary pinions, bearing rollers, thrust washers and/or bearing cone (9) as necessary. Carrier (12) is serviced only as a complete assembly including pinions and bearings. To reassemble, proceed as follows: Coat bore of pinion gear with heavy grease, stick row of bearings in one side of pinion, insert center thrust washer (47) and stick second row of bearing rollers in pinion. Place a thrust washer (44) on each side of pinion, place pinion, bearings and thrust washers in carrier and insert pinion shaft (11) with flat on end towards center of carrier. Press pinion shaft into carrier as shown in Fig. 340. Repeat procedure for remaining two pinions to complete assembly of carrier, then press retainer (10) and bearing cone (9) onto carrier housing.

In September 1969, a new type of planetary unit was introduced. Refer to Fig. 337. Planet pinion shafts (2) are a slip fit in carrier (3) and are retained by "C" ring (4) which rotates into a slot in carrier and shafts as shown in Fig. 338. To disassemble unit, straighten end of "C" ring and rotate ring until shafts are free. Withdraw shaft and slide out pinion and bearing, being careful not to lose any of the loose needle rollers (7 – Fig. 337). Assemble by reversing disassembly procedure. Bend ends of "C" ring into depression in carrier as shown in inset, Fig. 338. Planet pinions and bearings are interchangeable in old and new units.

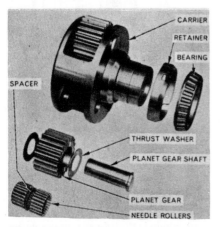

Fig. 339 – Exploded view of planetary carrier showing two pinions already reinstalled; refer to Fig. 340

Fig. 340 – Installing planetary gear unit in planetary carrier.

Fig. 341 – View showing special planetary ring gear tool (Nuday tool SW6-56) placed behind ring gear in position to press gear from axle housing. Refer also to Fig. 342.

On all models, if necessary to renew final drive ring gear, position removal tool (Nuday tool SW6-56) at outer side of gear as shown in Fig. 341, insert suitable bar through outer end of axle housing and press ring gear out of housing. To install new gear, attach removal tool to gear and position gear and tool on retaining studs as shown in Fig. 342, then press gear into housing. Remove tool and using a thin feeler gage, check to be sure ring gear is fully seated against shoulder of housing.

Using suitable bearing pullers, remove bearing cone (4 – Fig. 333) from axle shaft (1), then remove seal assembly (2) and gasket (3). Inspect sealing surface of shaft and remove all dirt, rust, paint and/or burrs which may damage new

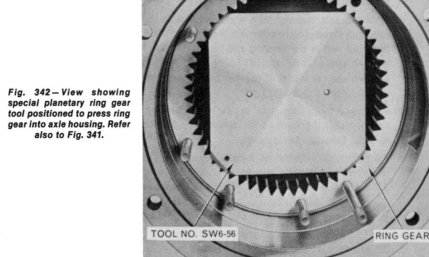

Fig. 342 – View showing special planetary ring gear tool positioned to press ring gear into axle housing. Refer also to Fig. 341.

Fig. 343 – View of planetary carrier installed in Model 7000, 7600, 7610, 7700 and 7710 rear axle housing.

seal. Coat shaft and lip of new seal with grease, then install new seal and gasket over shaft. Inspect bearing cups (5 and 8) in housing and renew if pitted, excessively worn or damaged in any other way. Lubricate bearing cone (4) and press cone tightly against shoulder on axle shaft. Lubricate bearing cups and lower axle housing over shaft, seal and bearing. Lubricate bearing (9) and install planetary carrier assembly into axle housing and over axle shaft.

Axle shaft end play/bearing preload must be adjusted at this time; proceed as follows: Obtain thickest spacer shim (13) available and install this shim with retainer (14) on inner end of axle shaft. On Models 7000, 7600, 7610, 7700 and 7710, install axle retaining cap screw and

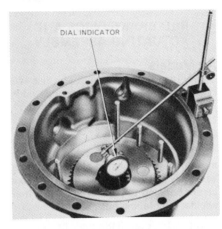

Fig. 344 — View showing method of checking rear axle shaft bearing adjustment (end play) with dial indicator. Refer to text.

tighten to a torque of 300-350 ft.-lbs. (408-476 N·m). On other models, install and tighten the three axle retaining cap screws to a torque of 95-120 ft.-lbs. (129-163 N·m). Mount a dial indicator with extended plunger against head of axle retaining cap screw as shown in Fig. 344 and zero indicator dial. Lift axle housing upward and note dial indicator reading. Remove cap screw(s), retainer and spacer shim, measure spacer shim thickness using a micrometer, then subtract dial indicator reading from measured shim thickness. For proper axle bearing adjustment (axle end play of 0.001 inch [0.025 mm] to bearing preload of 0.003 inch (0.076 mm), select a shim of thickness not more than 0.001 inch (0.025 mm) larger or 0.003 inch (0.076 mm) smaller than the figure resulting from subtracting dial indicator reading from thick shim measured thickness. Shim thickness may vary 0.001

inch (0.025 mm), so measure selected shim to be sure it is of appropriate thickness. Shims are available in 11 different thicknesses of 0.049-0.050 inch (1.245-1.270 mm) to 0.089-0.090 inch (2.260-2.286 mm) in steps of 0.004 inch (0.102 mm).

On Models 7000, 7600, 7610, 7700 and 7710, install selected shim, retainer and cap screw, tighten cap screw to a torque of 300-350 ft.-lbs. (408-476 N·m) and install locking plate (16A – Fig. 333). It may be necessary to tighten or loosen cap screw slightly in order to install locking plate.

On other models, install selected shim, retainer and three cap screws and tighten cap screws to a torque of 95-120 ft.-lbs. (129-163 N·m). Align a point of cap screw hex head with a mark on retainer (14 – Fig. 333). This will allow locking plate (16) to be installed between cap screw heads.

Be sure axle seal (2 – Fig. 333) is fully over end of axle housing, then stake rim of seal down into groove around axle housing in at least four equally spaced points. Reinstall final drive sun gear, outer brake housing, brake assembly and inner brake housing. If renewing differential carrier bearing cup (23) in brake inner housing, be careful not to lose or damage shim (22) located between bearing cup and brake housing. If shim is lost, or if brake inner housing, brake outer housing, final drive ring gear and/or axle housing have been renewed, a new shim (22) should be selected as outlined in paragraph 258. Reinstall ring gear thrust block (61), where removed, and differential assembly, if removed. Then, reinstall axle and wheel assembly as outlined in paragraph 262.

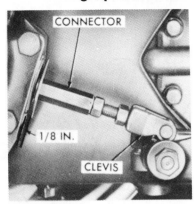

Fig. 345 — To adjust brake linkage on early Model 5000, loosen locknut next to connector and turn connector to obtain ⅛ to ¼-inch (3.18-6.35 mm) clearance between disc and seal as shown; then adjust pedal height as outlined in text.

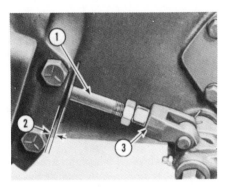

Fig. 346 — View of brake linkage used on late models except 6700, 6710, 7700 and 7710. Refer to text to adjust pedal height.

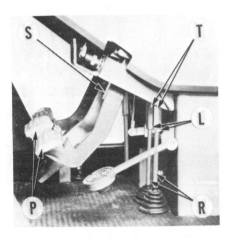

Fig. 347 — View of brake linkage used on Models 6700, 6710, 7700 and 7710. Refer to text for adjustment.

BRAKES

ADJUSTMENT

All Models Except 6700-6710-7700-7710

264. Refer to Fig. 345 or 346 and proceed as follows: With right-hand pedal in the up (released) position, loosen locknut next to connector on early 5000 models, then turn connector until there is ⅛-inch (3.2 mm) clearance between connector disc and seal in rear axle housing. Tigthen locknut and repeat this procedure on left-hand brake linkage. Note linkage difference for late 5000 and other models, Fig. 346.

Disconnect brake return spring on right brake pedal, allowing pedal to drop, and push pedal down to take up any slack in linkage. Loosen locknut

next to clevis and turn connector (early models) or pull rod nut (late models) until right pedal is 1½ to 1¾ inches (37.8-44.5 mm) below left pedal; then tighten locknut. Disconnect return spring on left brake pedal and push pedal downward to take up any slack in linkage. Loosen locknut next to clevis on left linkage and turn connector (early models) or pull rod nut (late models) until left pedal is level with right pedal. Check adjustment by engaging brake pedal lock; lock should engage easily if pedals are aligned. Tighten locknut and reinstall both pedal return springs.

NOTE: On Rowcrop models make adjustment at brake pull rods, not at the vertical control rods (19 – Fig. 349).

Models 6700-6710-7700-7710

265. To adjust brakes, depress pedal and place a ⅛-inch (3 mm) spacer between pedal shank and stop as shown in Fig. 347. Loosen locknut (L) and rotate turnbuckle (T) to remove free play in linkage. Retighten locknut and adjust remaining brake pedal. Check operation to be sure brakes are balanced.

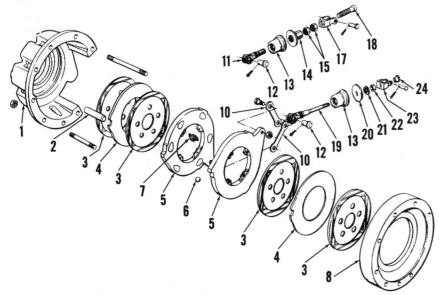

Fig. 348 — Exploded view of brake assembly. Note that pull rod assembly (items 11 through 18) used on tractors prior to production date 6-68 differs from that used on later tractors (items 19 through 24). Late Model 5000 tractors use three brake discs (3) whereas Model 7000 tractors use four. Models 7600 and 7700 changed from 4-disc to 3-disc brake in 1977 and all current models use 3-disc brakes.

1. Brake inner housing
2. Torque pin
3. Brake discs
4. Secondary discs
5. Actuating discs
6. Steel balls
7. Return springs
8. Brake outer housing
10. Actuating links
11. Pull rod (early)
12. Clevis
13. Seal
14. Connector
15. Locknuts
17. Clevis
18. Clevis bolt
19. Pull rod (late)
20. Washer
21. Fastener
22. Locknut
23. Clevis
24. Clevis nut

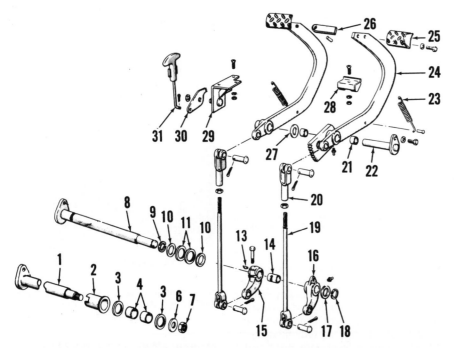

Fig. 349 — Exploded view showing brake pedal arrangement for Model 5200, 7200, 6600C and 7600C tractors. Brake pedals on "All Purpose" model tractors operate directly on brake cross shaft (8). Note brake cross shaft (1) used on tractors prior to production date 6-68.

1. Pedal shaft (early)
2. Spacer
3. Seal
4. Bushings
6. Washer
7. Nut
8. Pedal shaft (late)
9. Lock ring
10. Thrust washer
11. Seal assy.
13. Woodruff key
14. Bushing
15. Actuating lever, L.H.
16. Actuating lever, R.H.
17. Thrust washer
18. Snap ring
19. Control rod
20. Clevis
21. Bushing
22. Pivot pin
23. Return spring
24. Pedal shank
25. Pedal
26. Interlock bar
27. Washer
28. Return stop
29. Pawl bracket
30. Pawl
31. Pawl control rod

R&R BRAKE DISCS AND ACTUATING ASSEMBLY

All Models

266. Brakes are of multiple disc wet type and are located in rear axle housings. To gain access to brakes, remove rear axle housing as outlined in paragraph 262.

NOTE: When right axle housing is removed, it is suggested that differential assembly be removed from rear axle center housing to prevent assembly from accidentally falling out.

Refer to Fig. 348, and proceed as follows: Unbolt and remove inner disc brake housing assembly from inner end of axle housing. Remove brake control rod connector (early) or fastener (late) and if damaged, remove brake rod seal. To remove seal, place a sharp tool between seal flange and housing and pry seal out. Remove brake disc assemblies, intermediate discs, actuating disc assembly and outer brake housing.

NOTE: Where differential carrier bearing cup is installed in inner brake housing, differential carrier bearing adjustment is made by adding or removing shims from between bearing cup and brake housing. Therefore, if either inner or outer brake housing are renewed, differential carrier bearing adjustment may be affected. Refer to paragraph 258.

The brake actuating assembly can be disassembled if necessary as follows: Remove clevis pin to disconnect control rod from actuating links. Remove the four actuating disc return springs, separate actuating discs and remove the six steel balls. Inspect all parts and renew as necessary. To reassemble, lay one disc on bench with inner side up and lay the six balls in ramped seats. Lay other disc on top of balls so lugs for torque pin are about one inch apart and install the four return springs. If self-locking nuts have been removed from links, install new nuts. Reconnect control rod to links.

NOTE: In early 1969, Model 5000 tractors changed from a 4-disc brake of the type shown in Fig. 348 to a 3-disc brake. In 1977, Models 7600 and 7700 also changed from a 4-disc to a 3-disc brake. In current production, all models use a 3-disc design. The brake differs in that ball ramps of actuating discs (5) were reduced from a 25-degree angle to 20-degree angle, permitting greater disc pressure with lighter pedal pressure. Longer links (10) are also used. New brake discs (3) have greater strength in spline area and should be used with new actuating discs.

To reinstall brake components in axle housing, proceed as follows: Install outer brake housing on the eight studs in

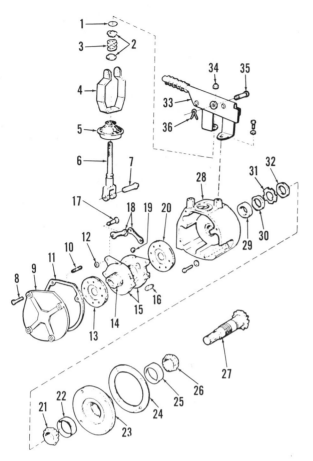

Fig. 350 – Exploded view of transmission handbrake. Brake shaft (27) is engaged by a bevel gear splined to bevel pinion of differential.

1. Flat washer
2. Rod collar (2)
3. Spring
4. Clevis
5. Rod seal
6. Control rod
7. Clevis pin
8. Cover bolt (4)
9. Cover
10. Stud
11. Gasket
12. Jam nut (2)
13. Brake disc
14. Brake spring
15. Actuator plate (2)
16. Return spring (2)
17. Link bolt (2)
18. Link (2)
19. Ball (4)
20. Brake disc
21. Bearing cone
22. Bearing cup
23. Retainer
24. Shims*
25. Bearing cup
26. Bearing cone
27. Brake shaft
28. Housing
29. Oil seal
30. Bearing locknut
31. Tab washer
32. Bearing locknut
33. Lever assy.
34. Rod locknut
35. Clevis pin
36. Cotter pin

and Load Monitor fork on models so equipped. Be particularly careful when inserting shaft through right seal. Complete reassembly of tractor by reversing disassembly procedure. Refill center housing with proper lubricant.

TRANSMISSION HAND BRAKE

Models So Equipped

Some late model tractors may be equipped with a transmission hand brake. Brake shaft (27–Fig. 350) is meshed with a bevel gear which is splined to forward end of bevel pinion (50–Fig. 333) within rear axle center housing. With unit so equipped, gears must be renewed in matched sets.

268. **ADJUSTMENT.** Release lever (33–Fig. 350) does not contain a locking mechansim to keep brake applied. It is so designed to prevent use as a parking brake. Locknut (34) on control rod (6) should be set so spring (3) is free to turn with lever (33) in relaxed position. Pull of 60-70 pounds (27-32 kg) should apply brake firmly. If brake application proves unsatisfactory, it will be necessary to dismantle housing (Fig. 351) to check condition of brake disc facings.

269. **R&R BRAKE ASSEMBLY.** Steps for removal of transmission brake assembly are as follows: Drain rear axle center housing and hydraulic system and unbolt and remove left side step plate. Remove cotter pin (36–Fig. 350), pull out clevis pin (35), then unbolt and remove lever assembly (33) from housing (28). Back off rod locknut (34) from rod (6) and slip spring (3) with keepers and clevis from rod. Carefully pry out seal (5) from top opening in housing, then unbolt and remove cover (9) and gasket (11). Lift out outer brake disc (13) as shown in Fig. 351, then remove clevis

axle housing and insert torque pin in brake housing. Then, refer to Fig. 348 for reassembly of brakes. Brake disc (3) and secondary disc (4) nearest brake outer housing (8) are not installed on 3-disc units. After inner brake housing is installed, tighten retaining nuts to a torque of 72-90 ft.-lbs. (98-122 N·m).

Reinstall axle housing to center housing using a new "O" ring on axle housing. Tighten axle housing retaining cap screws to a torque of 110-135 ft.-lbs. (150-170 N·m). Refill center housing to proper level with correct lubricant; refer to paragraph 286.

BRAKE CROSS SHAFT AND SEALS

All Models

267. All models are equipped with a brake cross shaft to transmit braking motion from right tractor side to left brake. Brake pedals may attach directly to cross shaft or pivot arms may attach to cross shaft with linkage between pivot arms and brake pedals.

To renew cross shaft, shaft bushings, pedal bushings, pivot arm bushing and/or cross shaft seals on models without Load Monitor, proceed as follows: Drain center housing. Disconnect brake pedal return springs and brake pull rods. On Rowcrop models, disconnect control

rods from actuating levers (Fig. 349).

Remove snap ring (or hex nut) and washer from right end of cross shaft and remove right brake pedal, or actuating lever. Loosen clamp bolt in left brake pedal or actuating lever, remove pedal or lever from shaft, then extract Woodruff key from cross shaft and remove thrust washer. Remove all burrs from right end of shaft, then withdraw shaft from center housing.

On models equipped with Load Monitor, the same basic procedure can be used to remove cross shaft, however, remove hydraulic pump so Load Monitor fork assembly is accessible. The Load Monitor fork assembly is mounted on brake cross shaft and must be installed as cross shaft is being installed.

Pry oil seals from each side of center housing. Inspect bushings in right brake pedal and in bores of center housing. Renew any bushing that is excessively worn. To remove bushings from center housing, drive bushings inward and remove through inspection plate opening or hydraulic pump opening. Bushings are presized and should not require reaming if carefully installed. Install new shaft oil seals with lip inward. Before inserting shaft, remove all burrs and lubricate shaft and seals. Place spacer (washer) on cross shaft, then insert cross shaft through center housing

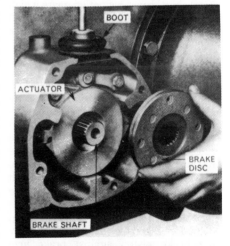

Fig. 351 – View to show handbrake cover removed and partial disassembly for inspection. See text.

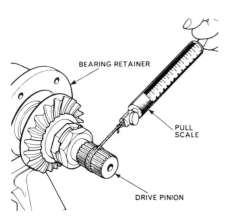

Fig. 352—Technique for checking brake shaft bearing preload by use of a spring scale. Refer to text.

pin (7 – Fig. 350) and slide complete actuator assembly out of housing when control rod (6) is lifted from housing. Inner brake disc (20) can now be slipped from brake shaft (27) for inspection. If only brake discs (13 and 20) require renewal, further disassembly is not needed. However, if brake shaft (27) or its bearings require attention, remove

four cap screws which retain housing (28) and bearing retainer (23) to center housing to completely remove unit. Handle shim pack (24) with care. It is used to set brake shaft backlash. If brake shaft (27) or its bearings (21 and 26) are renewed, backlash between brake shaft pinion and its drive gear will require adjustment. Use this procedure:

Check for required shim thickness by measuring with a feeler gage between bolt flange of retainer (23) and center housing surface, using no shims, with assembled brake shaft and retainer set in mesh with its drive gear on differential pinion shaft. Thickness of shim pack (24) should be greater by 0.009 inch (0.0228 mm) than this measured gap. Make up needed shim pack from shims which are available in thicknesses of 0.005 inch (0.127 mm) and 0.020 inch (0.508 mm).

Reinstall brake assembly by reversing removal procedure and adjust lever operation as in paragraph 268.

270. OVERHAUL. Remove brake assembly as covered in preceding paragraph. When housing (28) is separated from bearing retainer (23), be sure to check condition of brake shaft seal (29).

If renewal is required, install new seal with its lip toward shaft bearings. To disassemble actuating unit, remove return springs (16) and separate actuator plates (15) to determine condition of ball ramps and four steel balls (19). Note that actuator, parts 15 through 19, is serviced as an assembly. In reassembly, nuts (12) must be staked to bolts (17).

To disassemble bearings from brake shaft, clamp flange of bearing retainer in a padded jaw vise and flatten tabs of washer (31) so bearing locknuts (30 and 32) can be removed, then bump shaft (27) from bore of retainer (23). Press out bearing cups (22 and 25) from retainer and pull bearing cones (21 and 26) from brake shaft if renewal is necessary. Reverse this order to reassemble. When new bearing cups and cones are in place, tighten inner locknut (32) on brake shaft until rolling torque, measured by spring scale from a cord wrapped around splined portion of brake shaft as shown in Fig. 352 registers a steady 20-26 pounds (9.07-11.79 kg). Install tab washer (31 – Fig. 350) and outer locknut (30) and recheck bearing preload. When correct, lock tab washer against flats of both inner and outer nuts.

POWER TAKE-OFF

OPERATING PRINCIPLES

271. All models use an independent type power take-off with a hydraulically operated multiple disc clutch that can be engaged or disengaged at any time tractor engine is running. Pto input shaft is driven by a splined coupling attached to engine flywheel and passes through the hollow transmission upper shaft to drive the pto clutch hub. Clutch output (pto rear) shaft drives reduction gears which transfer power to output shaft at 540 rpm on single-speed models and 540 or 1000 rpm on two-speed models.

On all models, the pto clutch is engaged by the control valve directing hydraulic pressure to the clutch piston which gradually applies pressure to the clutch discs through the feathering spring. To release the pto clutch, the control valve blocks the oil pressure passage to the clutch piston and allows oil in the piston circuit to bleed off into the sump. The return spring then forces the clutch piston to release pressure against the clutch discs.

Brake piston on early models (Fig. 353), is engaged by spring pressure forcing brake piston against brake arm which contacts clutch housing causing housing to stop turning. To release

brake arm, control valve directs hydraulic pressure to brake piston which compresses return spring and releases brake arm.

Brake piston on late models (Fig. 354 or Fig. 355), is engaged by control valve directing hydraulic pressure to brake piston which slides to compress return spring. Brake piston rod end is forced against brake arm which tightens band around clutch housing to prevent rotation of clutch housing. To release brake band, control valve blocks oil pressure passage to brake piston and allows oil in piston circuit to bleed off into sump. Return spring then forces brake piston rod end away from brake arm to release tension on brake band.

TROUBLESHOOTING

272. OPERATING CHECKS. When troubleshooting problems are encountered with the independent pto, refer to the following:

PTO CLUTCH DOES NOT ENGAGE OR WILL NOT FULLY ENGAGE (SLIPS). Trouble could be caused by:
  A. Low rear axle oil level.
  B. Failure of hydraulic pump.
  C. Failure of connecting pipe.

D. Pressure control valve stuck open.
  E. Pressure control valve spring broken.
  F. Cast iron sealing rings on clutch housing broken.
  G. Clutch piston sealing rings leaking.
  H. Brake piston sealing rings leaking (early models).
  I. Brake piston return spring weak or broken (late models).

PTO CLUTCH WILL NOT DISENGAGE. Trouble could be caused by:
  A. Control valve stuck.
  B. Control valve return spring broken.
  C. Clutch piston return spring broken.
  D. Control to valve linkage disconnected or broken.

CLUTCH DISENGAGES, BUT PTO STILL TURNS. Trouble could be caused by:
  A. Cold oil.
  B. Brake piston stuck.
  C. Clutch piston return spring weak or broken.
  D. Worn brake pad or brake band lining.
  E. Clutch plates distorted or seized.
  F. Brake band out of adjustment.

G. Broken brake band adjustment screw.

H. Brake piston return spring weak or broken (early models).

I. Brake piston sealing rings leaking (late models).

PTO HYDRAULIC PRESSURE CHECK

Models-5000-5600-6600-6700-7000-7600-7700

273. To check pto clutch system hydraulic pressure, first operate tractor until rear axle and hydraulic oil is at normal operating temperature, then stop engine and connect a 0-400 psi (0-2760 kPa) hydraulic gage at plug opening in hydraulic pump as shown in Fig. 356. Start engine and operate at low idle speed, then engage and disengage pto clutch several times, noting pressure gage reading when clutch is engaged. The gage reading should be constant and not less than 140 psi (966 kPa) nor more than 155 psi (1069.5 kPa). A high or fluctuating gage reading would indicate a sticking pressure control valve (24 – Fig. 353 or 27 – Fig. 354). A low pressure reading would indicate a worn pump, leak in hydraulic circuit, low oil level or a sticking pressure control valve.

Models 5610-6610-6710-7610-7710

274. **PRESSURE REGULATING VALVE.** Operate tractor until rear axle and hydraulic oil is at normal operating temperature, then stop engine and connect a suitable gage (1 – Fig. 357) with adapter tube (4) to outlet/test port (2) on hydraulic pump (3).

Start engine and place pto control lever in disengaged position. Set engine speed to 850 rpm and observe pressure gage reading. Move pto control lever to engaged position and observe reading. Pressure readings should be 160-180 psi (1104-1242 kPa). If pressure gage reading is below minimum with lever in disengaged position, control valve assembly must be removed and pressure regulating valve shimming increased or damaged brake piston seal renewed.

If pressure gage reading is normal with lever in disengaged position but falls below minimum in engaged position, clutch assembly must be removed and clutch piston and sealing rings between clutch housing and support inspected for excessive wear or any other damage.

If pressure gage readings are high with lever in both positions, clutch and control valve assembly must be removed and inspected for restricted oil passages or a sticking pressure regulating valve.

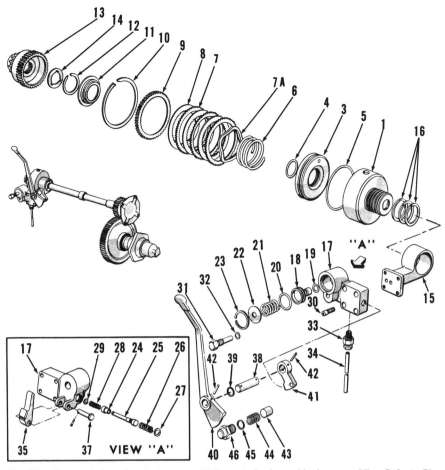

Fig. 353 — Exploded view of early type pto clutch, control valve and brake assemblies. Refer to Fig. 359 for exploded view of reduction gears and output shaft.

1. Clutch housing	13. Drive hub	24. Pressure regulating valve
3. Clutch piston	14. Thrust washer	25. Control valve
4. "O" ring	15. Brake support	26. Valve return spring
5. "O" ring	16. Sealing rings	27. Snap ring
6. Piston return spring	17. Brake & control valve housing	28. Pressure regulating spring
7A. Feathering spring	18. Brake piston	29. Snap ring
7. External spline plates	19. "O" ring	30. Socket head screws
8. Internal spline plates	20. "O" ring	31. Locating bolt
9. Pressure plate	21. Spring	32. "O" ring
10. Snap ring	22. Washer	33. Connector
11. Retainer	23. Snap ring	34. Nylon tube
12. Snap ring		35. Brake lever
		37. Pin
		38. Control shaft
		39. "O" ring
		40. Control lever
		41. Control arm
		42. Pin
		43. Detent
		44. Spring
		45. "O" ring
		46. Control lever stop

After making necessary adjustments or repairs, recheck pressure readings, then remove pressure gage and adapter tubes.

275. **COOLER/LUBRICATION CIRCUIT RELIEF VALVE.** Disconnect banjo bolt (1 – Fig. 358) from cooler tube.

NOTE: Be sure plastic transfer tube remains correctly positioned in control valve housing.

Install gage (2), adapter (3) and hose (4) to outlet port using banjo bolt (1). With hydraulic fluid at normal operating temperature and engine set at 1250 rpm, note pressure gage reading. Correct pressure reading is 45-65 psi (310.5-448.5 kPa). If reading obtained is above or below recommended range, control valve assembly must be removed and

valve shimming increased or decreased as required.

After making necessary adjustments, recheck pressure reading, then remove pressure gage assembly.

OVERHAUL

276. **R&R AND OVERHAUL PTO PUMP.** The pto hydraulic pump is an integral part of the hydraulic lift pump assembly; refer to paragraph 379 or 381 in HYDRAULIC SYSTEM section for pump service information.

277. **R&R PTO REAR (CLUTCH OUTPUT) SHAFT.** To remove power take-off rear (clutch output) shaft (63 – Fig. 359), first remove hydraulic lift top link rocker. Then, unbolt bearing retainer (59) from rear face of rear axle

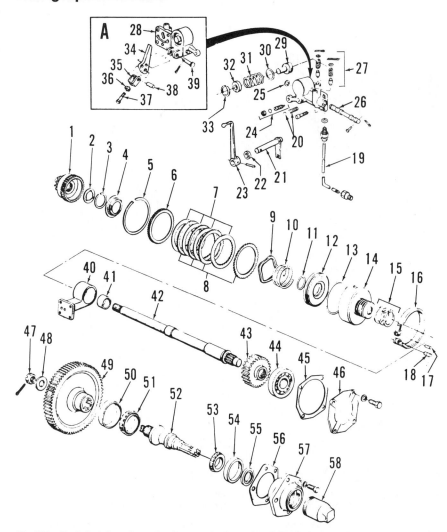

1. Clutch hub
2. Thrust washer
3. Spring retainer
4. Spring seat
5. Pressure plate retainer
6. Pressure plate
7. Clutch plate (int.) (4)
8. Clutch plate (ext.) (4)
9. Feathering spring
10. Piston spring
11. Inner "O" ring
12. Drive clutch piston
13. Outer seal
14. Clutch housing
15. Support seal (3)
16. Brake band
17. Band-to-arm pin
18. Adjuster anchor
19. Pressure tube
20. Detent ball & spring
21. Clutch arm
22. Seal
23. Pto lever
24. Locator pin, seal & nut
25. Valve retainer
26. Control valve
27. Regulating valve (2)
28. Control valve housing
29. Brake piston
30. "O" ring
31. Spring
32. Guide
33. Retainer
34. Brake lever
35. Adjustment clevis
36. Locknut (10-32)
37. Adjusting screw (10-32)
38. Clevis pin
39. Brake lever pin
40. Control valve support
41. Shaft bushing
42. Shaft
43. Drive gear
44. Upper shaft bearing
45. Gasket
46. Bearing retainer
47. Shaft nut (7/8-20)
48. Washer
49. Driven gear
50. Bearing cup
51. Bearing cone
52. Output shaft
53. Bearing cone
54. Bearing cup
55. Seal
56. Shims
57. Bearing retainer
58. Cap

Fig. 354 — Exploded view of pto clutch, control valve and brake with shafts and reduction gears which became available for Models 5000 and 7000 in June, 1973. Same design is used on late model tractors. Models 5610, 6610, 6710, 7610 and 7710 use five internal and external clutch plates (7 and 8) and use control valve assembly in Fig. 355. Reduction gears for two-speed (540 and 1000 rpm) version are shown in Fig. 360

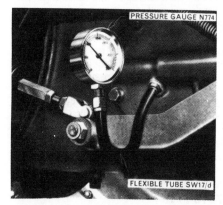

Fig. 356 — View showing pressure gage connected to hydraulic pump for checking pto system relief pressure on Models 5000, 5600, 6600, 6700, 7000, 7600 and 7700.

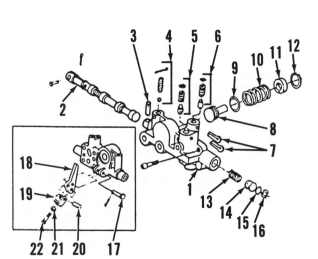

Fig. 355 — Exploded view of control valve assembly used on Models 5610, 6610, 6710, 7610 and 7710.

1. Control valve housing
2. Control valve
3. Pin
4. Detent ball, spring & cotter pin
5. Pressure regulating valve
6. Cooler/lubrication circuit relief valve
7. Cotter pins
8. Brake piston
9. "O" ring
10. Spring
11. Guide
12. Circlip
13. Spring
14. Retainer
15. "O" ring
16. Circlip
17. Pin
18. Brake lever
19. Adjustment clevis
20. Clevis pin
21. Locknut
22. Adjusting screw

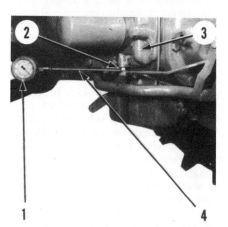

Fig. 357 — View showing gage (1) with adapter tube (4) connected to outlet/test port (2) on hydraulic pump (3). Use gage to test pressure regulating valve on Models 5610, 6610, 6710, 7610 and 7710.

center housing and withdraw retainer (59), bearing (61), gear (62) and shaft as an assembly. Remove shaft and bearing from retainer, then pull bearing from shaft and remove gear. Bushing (64) is located in bore of rear axle center housing and can be renewed after removing hydraulic lift cover assembly as outlined in paragraph 354, 355 or 356.

Proceed in this same manner to remove upper rear shaft from later design pto, shown in Fig. 354, or from optional two-speed pto shown in Fig. 360.

Reassemble and reinstall by reversing removal procedure. Install assembly

Fig. 358 — View showing gage (2), adapter (3) and hose (4) connected at outlet port with banjo bolt (1). Use gage to test cooler/lubrication circuit relief valve on Models 5610, 6610, 6710, 7610 and 7710.

Fig. 360 — View of reduction gear set and output shaft for two-speed (540 and 1000 rpm) pto. Models 5610, 6610, 6710, 7610 and 7710 use helical cut gears, an "O" ring and shim(s) in place of gasket (7) and a thrust needle bearing between two steel thrust washers in place of thrust washer (17) located between components 16 and 18.

1. Shaft
2. Snap ring
3. Shims*
4. Front bearing
5. Drive gear
6. Rear bearing
7. Gasket
8. Bearing retainer
9. Output shaft
10. Snap ring
11. "O" ring
12. Bearing retainer
13. Rear bearing
14. Shaft sleeve
15. Driven gear
16. Driven gear
17. Thrust washer (3)
18. Front bearing
19. Shaft seal
20. Gasket
21. Bearing retainer
22. Cap
*As required

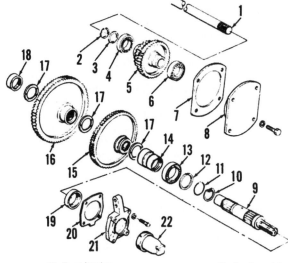

with new gasket or "O" ring and tighten retaining cap screws to a torque of 40-50 ft.-lbs. (54-68 N·m).

278. R&R SINGLE-SPEED PTO OUTPUT SHAFT AND GEAR. To remove power take-off output shaft (53 – Fig. 359) and gear (56), first refer to paragraph 262 and remove right rear axle and differential assembly. Then, proceed as follows:

Remove cotter pin from nut (58) and shaft. Hold shaft from turning with a pto yoke installed on rear end of shaft and with suitable bar inserted through

yoke, then remove nut and washer (57) from front end of shaft. Remove cap screws holding bearing retainer (48) to rear axle center housing and remove output shaft and retainer as a unit taking care not to lose or damage shims (49). Remove bearing cone (54) from gear (56), then remove gear from housing. Withdraw shaft from bearing retainer, then remove bearing cup (51) and oil seal (50). Install new oil seal with lip forward, then reinstall bearing cup or install new cup if necessary. Remove bearing cone (52) from shaft and bearing cup

(55) from center housing if worn or damaged.

Reassemble and reinstall by reversing disassembly and removal procedures. Tighten bearing retainer cap screws to a torque of 40-50 ft.-lbs. (54-68 N·m). Tighten gear retaining nut to a torque of 200-250 ft.-lbs. (272-340 N·m). Rotate nut so cotter pin can be installed through nut and shaft.

NOTE: If parts were installed that would affect shaft end play, or if shims (49) were lost, proceed as follows:

Install bearing retainer without shims (49) and tighten retaining cap screws finger tight. Rotate shaft to be sure bearings are seated, retighten cap screws finger tight and measure gap between bearing retainer and rear axle center housing with a feeler gage as shown in Fig. 362. Remove bearing retainer and reinstall with shim thickness

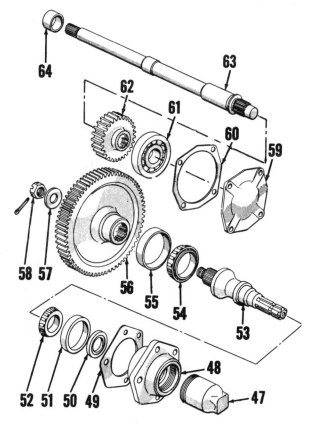

Fig. 359 — Exploded view of single-speed reduction gears and output shaft.

47. Pto cover
48. Bearing retainer
49. Shims
50. Oil seal
51. Bearing cup
52. Bearing cone & roller
53. Output shaft
54. Bearing cone & roller
55. Bearing cup
56. Driven gear
57. Washer
58. Slotted nut
59. Bearing retainer
60. Gasket
61. Ball bearing
62. Drive gear
63. Pto rear shaft
64. Bushing

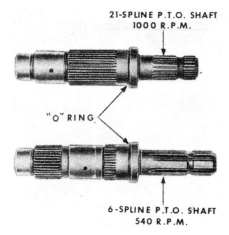

Fig. 361 — Views of 1000 rpm (upper) and 540 rpm (lower) pto shafts. Note placement of "O" ring (11 — Fig. 360) on shafts.

of 0.005 inch (0.127 mm) more than measured gap and tighten retaining cap screws to a torque of 40-50 ft.-lbs. (54-68 N·m).

Same general procedures apply for removal and reinstallation of output shaft (52 – Fig. 354) of newer design 540 rpm pto used in later 5000 and 7000 models (after June, 1973) and in later model tractors. Note that these steps do not apply for models equipped with two-speed pto. For these, refer to paragraph 279.

279. R&R TWO-SPEED PTO OUTPUT SHAFT AND GEARS.

If tractor is equipped with a sump cover directly beneath pto driven gears (15 and 16 – Fig. 360), remove sump cover and drain oil from pto compartment of center housing. If tractor is not equipped with a sump cover, drain lubricant from center housing. Remove rear clutch output shaft (1) as covered in paragraph 277. Remove cap (22 – Fig. 360), then release snap ring (10) and pull output shaft (9) from splines of driven gear (15 or 16). When drained, remove eight cap screws from bottom cover of pto sump to remove cover, then unbolt and remove bearing retainer (21) with gasket (20) and output shaft seal (19).

Support driven gears (15 and 16) and pull rear bearing (13) with shaft sleeve (14) from bearing bore of center housing, then remove gears through sump opening. Note that gears pilot on one another and that three thrust washers (17) are used. One is inserted between gear (16) and front bearing (18), another between gears (15 and 16) and one between hub of gear (15) and shaft sleeve (14).

NOTE: On Models 5610, 6610, 6710, 7610 and 7710, a thrust needle bearing between two steel thrust washers is used in place

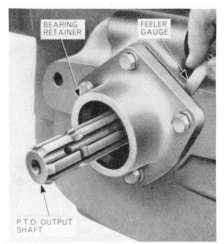

Fig. 362 – Checking for shim pack thickness to be installed between bearing retainer and rear axle center housing. Refer to text. Procedure does not apply to two-speed pto.

of thrust washer (17) located between gear (16) and front bearing (18).

Renew all damaged or excessively worn parts, paying special attention to thrust washers (17) and bearings (13 and 18). To remove bearing (13) from shaft sleeve (14) remove bearing retainer (12). Front bearing (18) is sealed on its forward side. If seal is broken, renew bearing. To reset bearing (18) in its bore, use a driver sleeve of 1⅞ inch (47.6 mm) outside diameter. When fitting a new shaft seal (19) into retainer (21), set with seal lip inward and press into retainer using a 3¾-inch (95.25 mm) OD driver.

Reinstall parts in reverse of removal order. Torque upper shaft cover (8) cap screws to 40-50 ft.-lbs. (54-68 N·m) and bearing retainer (21) cap screws to 60-70 ft.-lbs. (82-95 N·m). When reinstalling sump cover, use a new gasket and torque retaining cap screws evenly to 20-26 ft.-lbs. (27-35 N·m).

If only pto compartment was drained, refill capacity is one U.S. gallon (3.8 liters).

280. R&R PTO CLUTCH AND VALVE ASSEMBLY – EARLY MODELS.

To remove pto clutch and valve assembly, first remove hydraulic lift cover as outlined in paragraph 354 or 355 and pto rear (clutch output) shaft as in paragraph 277, then proceed as follows:

NOTE: If tractor is equipped with Load Monitor draft sensing system (including Models 5000 and 7000), pto clutch and valve assemblies cannot be removed unless tractor is split between transmission and rear axle center housing. See paragraph 256.

Disconnect hydraulic pressure line from clutch valve housing and remove locating bolt (31 – Fig. 353) from outer left side of rear axle center housing. The pto clutch and valve assembly can then be removed from clutch hub and center housing. With clutch assembly removed, clutch hub can now be removed from transmission pto (clutch input) shaft.

To reinstall assembly, place pto clutch hub on splines of transmission shaft and stick thrust washer (14) onto clutch housing (1) hub with heavy grease so the three prongs on washer enter holes in housing hub. Place clutch and valve assembly in center housing and install on input shaft and clutch hub. Turn assembly from side to side to align splines in clutch discs with splines on clutch drive hub. When clutch housing is fully forward, reinstall locating bolt through outer left side of rear axle center housing and into valve housing, using a new seal ring (32) on locating bolt. Reconnect hydraulic pressure line to clutch valve housing. Reinstall pto rear shaft as in paragraph 277 and hy-

draulic lift cover assembly as outlined in paragraph 354 or 355.

281. R&R PTO CLUTCH AND VALVE ASSEMBLY – LATER (After May, 1973) MODELS.

As with earlier models, first step in access to pto clutch is removal of hydraulic lift cover. See paragraph 355 or 356.

NOTE: Tractors equipped with Load Monitor must be split between transmission and rear axle center housing for removal of pto clutch.

Remove pto rear (clutch output) shaft as outlined in paragraph 277, then follow these steps: Disconnect hydraulic pressure tube (19 – Fig. 354) from control valve housing (28). Remove outer locknuts, seals and pins (24). Two of each are used to locate and align valve housing (28). Remove clutch valve and brake assembly (parts 3 to 18) as a unit which will include support (40) and all parts of control valve housing (28) after separation of clutch arm (21) from control valve plunger (26). Clutch hub (1) and thrust washer (2) can now be removed from transmission pto (clutch input) shaft. If tractor is Load Monitor equipped, removal of parts will be forward through transmission opening of center housing rather than from top. Order for removal will be apparent.

Reconnect and reinstall in reverse of removal order, taking special care to install thrust washer (2) into clutch assembly (14) so its prongs are engaged in housing hub. Use heavy grease to hold thrust washer in place during installation. When clutch housing (14) is fully forward against clutch hub (1), reinstall locator pins (24) with new seals. When rear shaft (42) is reinstalled and in alignment, tighten locator pins (24) and locknuts to 30-35 ft.-lbs. (41-48 N·m) torque. Adjustment procedure for brake band (16) is outlined in paragraph 283 or 284.

Reinstall hydraulic lift cover as in paragraph 355 or 356 and rejoin tractor if split was made per paragraph 254, 255 or 256. See paragraph 177 or 186 for lubricant fluid capacities.

282. OVERHAUL CLUTCH AND VALVE ASSEMBLY – EARLY MODELS.

With pto clutch and valve asembly removed as outlined in paragraph 280, proceed as follows:

Unbolt valve housing (17 – Fig. 353) from valve support (15) and remove support from clutch housing. Place valve housing in a press and compress brake piston spring by pushing washer (22) inward and remove retaining snap ring (23). Release spring and remove spring and brake piston (18), then remove "O" rings (19 and 20) from piston. Remove snap ring (27 – View "A") and withdraw spring (26) and control valve (25), then remove snap ring (29) and remove pressure control valve (24) and spring (28) from valve spool. Remove pin (37) if necessary to renew brake arm (35).

Remove snap ring (10) from clutch housing (1), then remove pressure plate (9), clutch discs (7 and 8) and feathering spring (7A) from housing. Remove cast iron sealing rings (16) from rear hub of housing. Place housing in a puller as shown in Fig. 363 and with Select-O-Speed transmission clutch spring compressor (Nuday tool N-775), compress spring (6—Fig. 353) far enough to remove snap ring (12). Gradually release spring pressure, then remove retainer (11) and spring. Use air pressure through clutch piston port in rear hub of housing to remove piston (3) from housing, then remove sealing rings (4 and 5) from piston.

Check valve, brake piston and clutch piston springs against the following values:

NOTE: At production date 4-68, pressure control valve was changed.

Clutch piston return spring:
Free length . . . . 2.29 inches
(58.16 mm)
Lbs. at 1.378 inches 305-335
Kg at 35.0 mm 138.3-151.9

Brake piston spring:
Free length . . . . 1.58 inches
(40.1 mm)
Lbs. at 0.78 inches 76-84
Kg at 19.8 mm 34.5-38.1

Pressure control valve spring
(prior 4-68):
Free length . . . . 1.62 inches
(41.44 mm)
Lbs. at 1.10 inches 13-14
Kg at 27.9 mm 5.89-6.35

Pressure control valve spring
(after 3-68):
Free length . . . . 1.48 inches
(37.6 mm)
Lbs. at 1.10 inches 21.5-23.5
Kg at 27.9 mm 9.75-10.66

Control valve return spring:
Free length . . . . 2.04 inches
(51.8 mm)
Lbs. at 1.56 inches 2.8-3.2
Kg at 39.6 mm 1.27-1.45

Carefully clean and inspect all other parts and renew any that are excessively worn or otherwise damaged. Reassemble using all new seals and "O" rings as follows:

Install new sealing rings on outer and inner perimeters of clutch piston (3), lubricate sealing rings and install piston in housing (1). Position piston return spring (6) and retainer (11) over piston, compress spring as during disassembly

and install retaining snap ring (12). Remove assembly from press and place feathering spring on piston. Alternately install external and internal splined clutch discs. Install external splined disc against piston, then install clutch pressure plate and retaining snap ring. Install new cast iron sealing rings (16) on rear hub of housing, lubricate sealing rings and install valve support over hub and rings.

Install pressure control valve and spring and retaining snap ring on control valve spool, then insert valve assembly in housing, install return spring and retain with snap ring. Install new "O" rings on brake piston (18), lubricate "O" rings and insert piston in valve housing. Install brake spring and washer, compress spring and install retaining snap ring. Be sure mating surfaces of valve housing and valve support are clean, then bolt valve housing to support. Reinstall clutch and valve assembly as outlined in paragraph 280.

283. OVERHAUL PTO CLUTCH, VALVE AND BRAKE ASSEMBLIES —MODELS 5000 AND 7000 (After May, 1973) AND MODELS 5600-6600-6700-7600-7700. When pto clutch valve and brake are removed as an assembly, as covered in paragraph 281, proceed in this order:

Place unit on bench and slide valve housing (28—Fig. 354), support (40) and brake (16) from clutch housing (14). Remove pins (17 and 18) and separate brake band (16). Back out four 5/16-18 socket head screws to remove valve housing (28) from support (40). Remove retainer (25) and pull out control valve (26), watching for spring and ball detent (20). Pull out cotter pins and remove both regulating valves (27) taking care not to drop parts. Remove retainer (33), guide (32), spring (31) and brake piston (29). Remove "O" ring (30) from piston. Note that it may be necessary to use compressed air if piston is tight in valve body bore.

Carefully clean and inspect all parts making necessary renewals. All parts are serviced. Parts, especially seals and "O" rings, should be coated with hydraulic oil and assembled in reverse of disassembly order. Brake band (16) should be refitted to housing (28) and housing should be bolted to support (40). Tighten four socket head screws evenly to 10-14 ft.-lbs. (14-19 N·m) torque. Set assembly aside.

Proceed with overhaul of pto clutch assembly as follows:

Remove retaining ring (5), pressure plate (6) and all clutch plates (7 and 8) with feathering (wave) spring (9) from clutch housing (14). Set housing up in a shop press or in a puller as shown in Fig.

363 and use tool N-775 as illustrated to relieve pressure so spring retainer (3) can be removed, followed by spring seat (4) and spring (10). Use compressed air to remove piston (12), then remove inner "O" ring (11) and outer seal (13) from piston.

Carefully clean and inspect all parts and renew those which are not serviceable, then reassemble housing parts in reverse of removal order. Use tool N-775 to reset spring retainer (3) in hub of housing (14).

Set brake band around clutch housing when clutch overhaul is complete, with control valve support (40) collar fitted over rear hub of clutch housing (14). Seals (15), especially if renewed, should be coated with hydraulic fluid. Adjust pto brake band as follows:

With locknut (36) loose, use a torque wrench of proper range to tighten adjusting screw (37) to 9-11 inch-pounds (1.017-1.243 N·m), then back off by one and one-half turns on Models 5000 and 7000 and two turns on all other models and tighten locknut (36). Note that adjusting screw is 10-gage, 32 thread. Adjust with care.

Reinstall pto clutch brake and valve assembly in center housing as covered in paragraph 281.

284. OVERHAUL PTO CLUTCH, VALVE AND BRAKE ASSEMBLIES —MODELS 5610-6610-7610-7710. When pto clutch assembly, brake and control valve are removed as a unit, as covered in paragraph 281, proceed as follows:

Place unit on a clean work bench and separate control valve and brake band from clutch assembly. Remove clevis pin (20—Fig. 355) and remove brake band and brake lever (18) from control valve

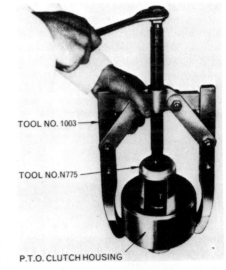

TOOL NO. 1003

TOOL NO. N775

P.T.O. CLUTCH HOUSING

Fig. 363—View of typical tool set up for removal of spring retainer from pto clutch housing. Refer to text.

housing. Back out four 5/16-18 socket head screws to remove valve housing from support. Remove cotter pin and withdraw spring and detent ball (4). Remove pin (3) and pull out control valve (2). Remove circlip (16), retainer (14) and spring (13). Pull out cotter pins (7) and remove pressure regulating valve (5) and cooler/lubrication circuit relief valve (6).

NOTE: Lay parts out in order keeping components of each valve assembly together. Exercise care when handling valves.

Remove circlip (12), guide (11), spring (10) and brake piston (8). Remove "O" ring (9) from piston. Note that it may be necessary to use compressed air if piston is tight in valve body bore.

Carefully clean and inspect all parts. Reassembly is reverse order of disas-

sembly. Coat "O" rings with hydraulic fluid. Attach support to valve housing and secure with four socket head screws. Tighten screws to 15 ft.-lbs. (20 N·m) torque. Refit brake lever and brake band to housing and secure with clevis pin. Set assembly aside.

Proceed with overhaul of pto clutch assembly as follows:

Remove retaining ring (5 – Fig. 354), pressure plate (6) and all clutch plates (7 and 8) with feathering (wave) spring (9) from clutch housing (14). Set housing up in a shop press or in a puller as shown in Fig. 363 and use tool N-775 as illustrated to relieve pressure so spring retainer (3) can be removed, followed by spring seat (4) and spring (10). Use compressed air to remove piston (12), then remove inner "O" ring (11) and outer seal (13) from piston.

Carefully clean and inspect all parts

and renew those which are excessively worn or damaged in any other way. Reassemble housing parts in reverse order of removal. Coat "O" ring, seal and clutch discs with hydraulic oil during reassembly. Use tool N-775 to reset spring retainer (3) in hub of housing (14).

Lubricate seals (15) with hydraulic fluid and slip brake band around clutch housing, then slide rear hub of clutch housing into support (40). Adjust pto brake band as follows:

With locknut (21 – Fig. 355) loose, use a torque wrench of proper range to tighten adjusting screw (22) to 9-11 inch-pounds (1.017-1.243 N·m), then back off by two and one-half turns and tighten locknut (21).

Reinstall pto clutch brake and valve assembly in center housing as covered in paragraph 281.

BELT PULLEY

OVERHAUL BELT PULLEY ASSEMBLY

All Models

285. Refer to exploded view of unit in Fig. 364 and proceed as follows: Drain lubricant and unbolt pulley from hub (3). Unbolt pulley shaft bearing retainer (8) from housing (12) and remove assembly. Unscrew nut (1) from pulley shaft and remove hub from shaft and bearing (10) from retainer. Remove seals (4 and 5) and bearing (6) from outer end of retainer. Remove bearing cups (7 and 9) if worn or scored.

Remove cover (14) and drive gear (17) and bearing assembly. Remove bearing cup (19) and oil seal (20) from housing. Remove bearing cup (15) from cover if cup is worn or scored.

Reassemble by reversing disassembly procedure. Install new seals (4 and 20) with lip to inside. Select a gasket shim (21) that will give 0.002 inch (0.05 mm) end play of drive gear (17). Tighten self-locking nut (1) to provide 0.002 inch (0.05 mm) end play of pulley shaft in bearings, then install bearing retainer (8) to housing with proper thickness of shim gasket (11) so there is some back-

lash between pulley shaft pinion and drive gear. Shim gaskets (11 and 21) are available in thicknesses of 0.005, 0.010

and 0.020 inch (0.127, 0.254 and 0.508 mm). Refill housing to oil level plug with SAE 80 EP gear lubricant.

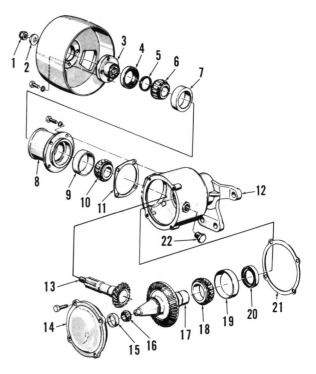

Fig. 364 – Exploded view of typical belt pulley assembly available for all tractors.

1. Hex nut
2. Washer
3. Pulley hub
4. Oil seal
5. Seal ring
6. Bearing cone & roller
7. Bearing cup
8. Bearing retainer
9. Bearing cup
10. Bearing cone & roller
11. Gasket shim
12. Housing
13. Drive gear
14. Cover
15. Bearing cup
16. Bearing cone & roller
17. Drive gear
18. Bearing cone & roller
19. Bearing cup
20. Oil seal
21. Gasket shim
22. Plug

HYDRAULIC LIFT SYSTEM

The hydraulic lift system for early Model 5000 tractors, without Load Monitor, incorporates automatic draft control, automatic position control, pump flow (rate of lift) control and a selector valve with remote outlets for

operating a single acting remote cylinder either independently or in conjunction with the 3-point hitch lift cylinder. Provision is also made for installation of optional single or dual spool valves for control of single or double ac-

ting remote cylinders. A system for remote cylinder operation only is also available.

The hydraulic lift system for later Model 5000 tractors (1971 production) and Model 7000 tractors, is very similar

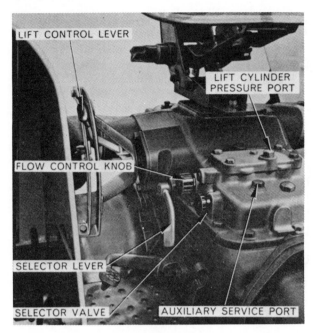

Fig. 365 — View of Model 5000 tractor without Load Monitor showing hydraulic controls, auxiliary service plug, lift cylinder pressure port and filler plug. Breather is incorporated into filler plug.

Newer Ford specification numbers for hydraulic fluid are M2C134-A, M2C86-A or Ford 134 which correspond to a general specification of SAE 80 EP and SAE 20W/30. System capacity is 33 quarts (31.2 liters) for Model 5000, or 55 quarts (52.0 liters) for Model 7000. Model 5600 without Dual Power calls for 43 quarts (40.7 liters). Model 5600 with Dual Power requires 56 quarts (53 liters) which is the same for all other models. If tractor is equipped with two-speed (540 and 1000 rpm) pto, capacity is increased by 3.8 U.S. quarts (3.6 liters). On Models 5600, 6600, 6700, 7600 and 7700 equipped with front-wheel drive, install an additional 6.4 U.S. pints. On Models 5610, 6610, 6710, 7610 and 7710 equipped with front-wheel drive, install an additional 3.8 U.S. pints. Hydraulic fluid should be drained and new fluid installed after each 1200 hours of service, or yearly, whichever occurs first. Remove transmission drain plug and rear axle center housing drain plug when draining Model 7000. Be sure the 3-point hitch lift arms are in lowered position and any remote cylinders are retracted before draining. See Fig. 209 for drain plug locations.

On Model 5000, maintain fluid level at check opening in right side of rear axle center housing at rear side of hydraulic pump. Model 5600 has a combination filler plug and dipstick in left rear corner of transmission cover and a dipstick in rear axle center housing. On Model 7000 and on Model 5600 with Dual Power and on all other models maintain fluid level to mark on dipstick which is located on left side of rear axle center housing.

NOTE: Although transmission and rear axle center housing of these models function as a common reservoir, fluid level in transmission housing is higher than in rear axle center housing. Therefore, it is important that tractor be level when checking fluid level.

to the early 5000 model tractors except that a Load Monitor unit has been added along with internal linkage to accomodate operation of Load Monitor. As with Model 7000, Load Monitor is standard on Models 7600, 7610, 7700 and 7710 along with dual-sensing upper link, but is optional on Models 5600, 5610, 6600, 6610, 6700 and 6710.

The Load Monitor is a torque sensing device interposed in tractor drive line and connects transmission output shaft and bevel pinion shaft. The Load Monitor senses (reacts to) the torque in tractor drive line and thus furnishes draft control based on draft imposed by implement as well as resistance of tractor movement. This operation provides a total load control.

Fluid for Model 5000 and 5600 (unless equipped with Dual Power option) hydraulic system is common with differential and final drive but is separated from transmission by oil seals. Fluid for hydraulic system of all other models is common with differential, final drive and transmission. Oil passes from transmission housing to rear center housing through oil transfer holes located in transmission rear bearing support plate.

Hydraulic power is supplied by a gear type pump mounted in right side of rear axle center housing. Pump is driven by a shaft splined into a hub attached to engine flywheel. Shaft passes through the hollow transmission shafts to gears at rear end of transmission.

See Fig. 367 for a schematic view of Load Monitor hydraulic lift system.

HYDRAULIC FLUID

All Models

286. Recommended hydraulic fluid is Ford part No. M2C53-B, or M2C53-A.

Fig. 366 — View of Model 7000 tractor with Load Monitor showing hydraulic controls. Filler plug is located in transmission top cover.

S. Adjustable stop
23. Selector valve
26. Selector lever
31. Flow control
34. Lift control lever

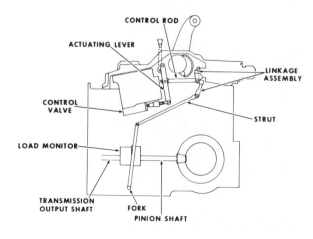

Fig. 367 — Schematic view showing the basic arrangement of components which comprise the Load Monitor hydraulic system.

HYDRAULIC SYSTEM FILTERS

Models 5000-5600-6600-6700-7000-7600-7700

287. Models without Load Monitor have a mesh strainer on hydraulic pump intake tube and a paper filter element on sump return tube. Screen should be cleaned and filter element renewed each 2400 hours of service, or every two years, to coincide with hydraulic system drain and refill.

Models with Load Monitor have a mesh strainer on hydraulic pump intake tube and a disposable filter element located on right side of rear center frame (Fig. 366). Filter element should be renewed every 300 hours and new filter should be hand tightened only. Do not use filter removal tool to tighten filter. The mesh strainer on hydraulic pump intake tube should be cleaned every 1200 hours to coincide with hydraulic drain and refill.

The hydraulic pump intake tube and mesh strainer can be removed from pump after pump is removed as outlined in paragraph 379. See Fig. 438 for a view of intake tubes and strainers.

Models 5610-6610-6710-7610-7710

288. Hydraulic pump intake tube passes oil across a removable plastic strainer (1–Fig. 368) to feed low pressure side of pump and through a disposable filter element (2) located on face of pump housing to feed high pressure side of pump. Tractors equipped with optional engine-mounted gear pump have an additional disposable filter element located on pump. Plastic strainer (1) should be cleaned periodically and disposable filter elements should be re-

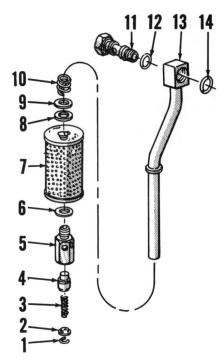

Fig. 369 — Exploded view of hydraulic fluid return line, filter and back pressure valve used on early models without Load Monitor.

1. Snap ring	8. Gasket
2. Plate	9. Washer
3. Spring	10. Spring
4. Back pressure valve	11. Hollow bolt
5. Valve seat	12. "O" ring
6. Gasket	13. Return tube
7. Filter element	14. "O" ring

newed every 300 hours. Coat new filter seal with oil and hand tighten only. Do not use filter removal tool to tighten filter. Hydraulic fluid should be changed every 1200 hours or once every year.

OIL COOLER BY-PASS VALVE

Models 5600-6600-6700-7600-7700

With development of Dual Power transmission option for these tractors, a

Fig. 368 — View showing location of plastic strainer (1) and disposable filter element (2) used on 5610, 6610, 6710, 7610 and 7710 models.

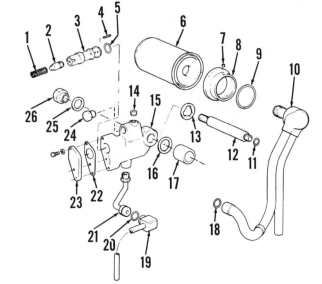

Fig. 371 — View of oil cooler by-pass valve and hydraulic oil filter installed on Models 5600, 6600 and 7600. Also see Fig. 370 and refer to text.

1. Oil cooler feed pipe
2. Valve cover plate
3. Dump pipe

hydraulic oil cooler was installed forward of tractor radiator core. Cooler is standard on 6600, 6700, 7600 and 7700 models but is installed only on those 5600 models which have Dual Power.

Return oil is directed to the two-stage cooler valve which replaces oil filter manifold assembly (Model 7000) shown in Fig. 410, and is routed by this spring-loaded valve to flow forward to cooler assembly. Oil returning from cooler is channeled to Dual Power housing on tractors so equipped, or if without Dual Power, directly to main countershaft bearing in transmission. On Model 5600 without Dual Power, oil flows directly from cooler valve to sump in rear axle center housing.

Basic function of this cooler by-pass valve is protection of oil cooler assembly and lines from excess pressure build-up. When system back pressure exceeds

Fig. 370 — Exploded view of hydraulic oil filter and cooler by-pass valve used on Models 5600, 6600 and 7600.

1. Valve spring
2. Valve spool
3. By-pass sleeve
4. Pin
5. "O" ring
6. Filter assy.
7. Retaining screw
8. Spacer-support
9. "O" ring
10. Support & tubes
11. "O" ring
12. Sleeve
13. Valve gasket
14. Plug (¼-in. NPT)
15. By-pass valve body
16. Gasket
17. Spacer
18. "O" ring
19. Return tube
20. "O" ring
21. Dump pipe & nut
22. Gasket
23. Cover plate
24. Plug (¾-18)
25. Nut seal
26. Cap nut

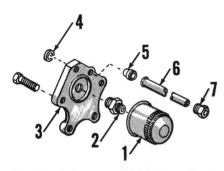

Fig. 372 – On tractors with hydraulic system for remote controls only (without 3-point hitch) plate (3) is mounted on lift cover instead of lift cylinder and return filter (1) is attached to plate.

1. Filter
2. Adapter
3. Manifold plate
4. Seal rings
5. Seal
6. Return tube
7. Nut

35-45 psi (241.5-310.5 kPa), second stage of by-pass valve will open to return surplus oil directly to center housing sump.

289. R&R AND OVERHAUL COOLER BY-PASS VALVE. Carefully clean external area around filter and by-pass valve shown in Fig. 371. Disconnect and remove feed pipe (1) which connects to cooler, then back off cap nut (26 – Fig. 370). Pull outward on valve body (15) and when assembly is clear, remove dump pipe (21) and discard "O" ring (20). Unbolt and remove cover plate (23) and discard gasket (22). Carefully check

spring (1), spool (2) and valve sleeve (3) for undue wear, burrs or other damage. Make necessary renewals.

NOTE: For test purpose, to determine if valve is functioning, when plug (14) is removed from valve body (15) a pressure gage can be installed for a check reading. If by-pass valve does not open at 35-45 psi (241.5-310.5 kPa), remove and renew valve spring (1), and if found defective, spool (2) and sleeve (3).

Reassemble by-pass valve parts when clean and with all passages clear, using all new gaskets and "O" rings. Tighten transfer tube cap nut (26) to 20-25 ft.-lbs. (27-34 N·m) torque, and dump pipe (21) nut to 40-50 ft.-lbs. (54-68 N·m). Cap screws for cover plate (23) should be tightened to 7-11 ft.-lbs. (9.5-15 N·m).

290. OIL FILTER, SUPPORT AND ADAPTER. Should it become necessary to service oil filter mount and spacer (10) due to oil leak or accidental damage, proceed as follows:

Drain rear axle center housing. Remove lift cover assembly by first disconnecting draft control yoke plunger from rocker and swing rocker away from yoke. Remove tractor seat assembly. Clean area of cooler by-pass valve and remove cap nut (26), then fit a hex (Allen) wrench and back out sleeve (12). Also see Fig. 371. Further removal of parts from cooler by-pass valve is un-

necessary. Clean around gasket joint of lift cover and remove fifteen cover bolts. Use of lifting bracket tool SW2 will help in removal of lift cover. If tractor is NOT equipped with Load Monitor, lift cover can now be removed from center housing. If tractor does have a Load Monitor system, raise lift cover about 3½ inches (89 mm) and reach through to detach strut from selector linkage, then complete removal of lift cover. Remove hydraulic pump as outlined in paragraph 379. If tractor is equipped with Load Monitor, it will also be necessary to partially withdraw brake pedal cross shaft as covered in paragraph 267 so Load Monitor unit can be shifted for work space.

To remove oil filter support (10 – Fig. 370), unscrew and remove filter (6), then back out set screw (7). Remove spacer (8) and discard "O" ring (9). Now, carefully tap support (10) with its tubes from side wall of housing and remove. Thoroughly clean assembly and inspect for damage or possible hydraulic oil leaks. Be sure to discard "O" ring (18) at pump end of adapter. Reverse removal order to install removed and renewed parts and assemblies.

NOTE: If support (10) and tubes are renewed, use of proprietary Ford sealing compound EST-M4G148-A is required for sealing tubes to support.

Install new "O" ring (9) and tighten filter (6) in place against spacer (8) by

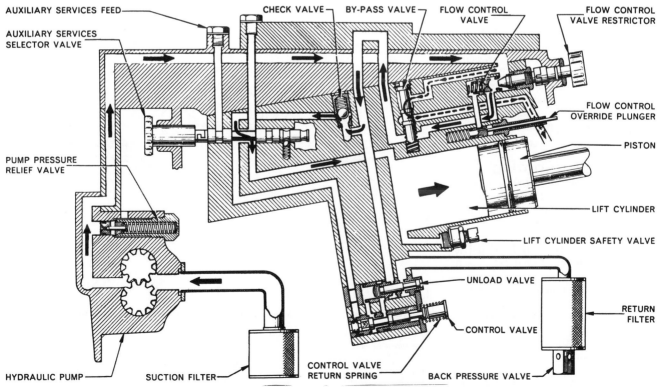

Fig. 373 – Schematic diagram of early hydraulic system without Load Monitor.

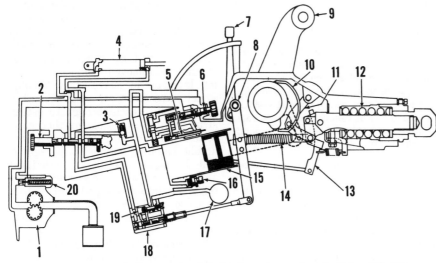

Fig. 374—Schematic diagram of early hydraulic system equipped with Load Monitor.

1. Hydraulic pump	6. Flow control valve	11. Position control link	16. Relief valve (safety)
2. Selector valve	7. Lift lever	12. Draft control spring	17. Filter
3. Check valve	8. Actuating lever	13. Selector linkage	18. Control valve
4. Load Monitor remote cylinder	9. Lift arm	14. Selector rod & roller	19. Unload valve
5. Override valve	10. Cam	15. Lift piston	20. System relief valve

hand to ensure that spacer is in full sealing contact with side wall of center housing, then tighten retaining set screw (7) to 4.6 ft.-lbs. (6.25 N·m). When removed assemblies are reinstalled, refill rear axle reservoir with hydraulic oil and check for external leaks in area of filter and cooler by-pass valve.

Models 5610-6610-6710-7610-7710

291. A cooler/lubrication circuit relief valve is located in independent power take-off valve housing. Basic function of relief valve is to protect oil cooler assembly and lines from excess pressure build-up. When system back pressure exceeds 45-65 psi (310.5-448.5 kPa), relief valve will open allowing surplus oil to return directly to center housing sump. Refer to paragraph 275 for testing relief valve pressure.

TROUBLESHOOTING

292. When troubleshooting problems are encountered with the hydraulic lift system, refer to the following malfunctions and possible causes:

A. FAILURE TO LIFT UNDER ALL CONDITIONS. Could be caused by:
1. Low oil level in rear axle center housing.
2. Flow control valve stuck in open position.
3. Unload valve stuck in open position.
4. Back pressure valve stuck in open position.
5. Control linkage damaged or disconnected.

6. Pump assembly faulty.
7. Pump drive shaft or gears broken.
8. Pump intake filter screen clogged.

B. FAILURE TO LIFT UNDER LOAD. Could be caused by:
1. Hydraulic pump worn.
2. Pressure relief valve setting low or faulty valve.
3. Cylinder safety valve faulty.

4. Internal leakage due to damaged "O" ring, faulty lift piston seals, cracked or porous lift cylinder casting or damaged hydraulic pressure lines.

C. EXCESSIVE CORRECTIONS (BOBBING OR "HICCUPS") IN RAISED OR TRANSPORT POSITION. Could be caused by:
1. Worn or damaged check valve ball or seal.
2. Selector valve worn or damaged.
3. Unload valve plug worn.
4. Lift cylinder safety valve damaged.
5. Faulty lift piston seals.
6. Control valve and/or bushing worn.
7. Internal linkage due to damaged "O" rings, etc.
8. Cracked or porous lift cylinder or lift cover castings.

D. OCCASIONALLY FAILS TO LIFT NOT DUE TO LOAD. Could be caused by:
1. Worn or loose selector valve.
2. Unload valve sticking.
3. Faulty back pressure valve.
4. Control valve incorrectly adjusted.

HYDRAULIC PRESSURE CHECK

All Models

293. On models manufactured before April, 1968, cracking pressure for hydraulic system relief valve should be 2100 psi (14.49 MPa) and operating

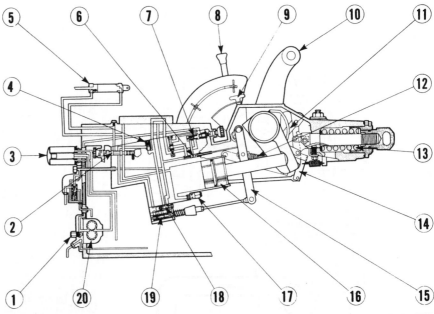

Fig. 375—Schematic diagram of hydraulic system with Load Monitor used on Models 5600, 6600, 6700, 7600 and 7700.

1. Pressure relief valve	8. Lift control lever	15. Actuating lever
2. Selector valve	9. System selector lever	16. Lift piston
3. Disposable filter	10. Lift arm	17. Lift cylinder shift valve
4. Check valve	11. Position control link	18. Unload valve
5. Load Monitor remote cylinder	12. Control rod and roller assy.	19. Control valve
6. Override valve	13. Draft control mainspring	20. Hydraulic pump
7. Flow control valve	14. Selector linkage assy.	

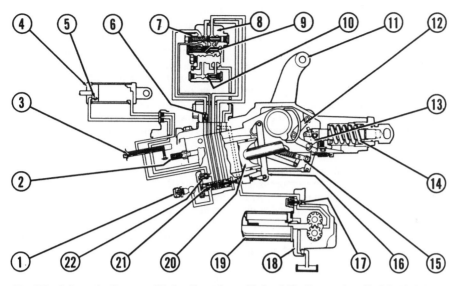

Fig. 376 — Schematic diagram of hydraulic system with Load Monitor used on Models 5610, 6610, 6710, 7610 and 7710.

1. Exhaust valve
2. Lift cylinder safety valve
3. Auxiliary services selector valve
4. Load Monitor remote cylinder
5. By-pass valve
6. Check valve
7. Flow control valve
8. Priority valve pack
9. Unload valve
10. Check valve
11. Lift arm
12. Position control link
13. Selector link
14. Draft control mainspring
15. Control rod and roller assy.
16. Actuating lever
17. Pressure relief valve
18. Hydraulic pump
19. Disposable filter
20. Lift piston
21. Control valve
22. Drop poppet valve

NOTE: Relief valve on late model tractors is non-adjustable, complete valve assembly must be renewed for servicing.

Proceed as follows to check system pressure:

On Models 5000, 5600, 6600, 6700, 7000, 7600 and 7700, lower lift arms and remove filler plug and auxiliary service port plug. Connect a "tee" as shown in Fig. 377. Connect a hose to shut-off valve and insert other end of hose in filler plug opening. Pull selector valve to out position and open shut-off valve. Start engine, operate at 1650 rpm and place lift control lever in fully raised position. Gradually close shut-off valve while observing pressure gage. The gage reading should gradually increase to 2500-2650 psi (17.25-18.28 MPa), or 2400-2600 psi (16.56-17.94 MPa), depending upon model, then remain constant with shut-off valve fully closed. On models with serviceable pressure relief valve, if pressure is higher than stated, shims should be removed from pressure relief valve to reduce relief pressure to specified value. If pressure gage reading is below specified value, insufficient shim thickness, a weak pressure relief valve spring and/or worn hydraulic pump is indicated. Renew relief valve spring if cracked or distorted or if free height is less than that of new spring. If spring appears satisfactory, add one shim and recheck pressure. Each additional 0.010 inch (0.254 mm) thick shim should increase pressure approximately 100 psi (690 kPa) on early type relief valves, or 63-70 psi (434.7-483 kPa) on late type relief valves.

pressure with relief valve wide open should be 2500-2650 psi (17.25-18.28 MPa). Beginning with April, 1968 models, relief valve cracking pressure was specified at 2150-2450 psi (14.83-16.90 MPa) and system operating pressure with relief valve open was specified at 2400-2600 psi (16.56-17.94 MPa). Mininum cracking pressure for relief valve on late model tractors is 2400 psi (16.56 MPa) and operating pressures range from 2550 to 2650 psi (17.59-18.28 MPa). Increased performance of newer tractors is a result of increased volume of flow, larger cylinder cross-section and redesigned valves for flow control.

Relief valve is located in external side of main hydraulic pump body. Refer to Fig. 438 for identification of two earlier types of relief valve. Adjusting shims for earlier (before April, 1968) valves are offered in thicknesses of 0.010 and 0.025 inch (0.254 and 0.635 mm), and for later years (since April, 1968) in thicknesses of 0.010-0.015 inch (0.254-0.318 mm). Maximum allowable thickness of shim pack is 0.080 inch (2.03 mm).

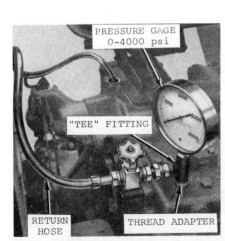

Fig. 377 — View showing pressure gage, shut-off valve and return tube connected to Model 5000 tractor for checking hydraulic pump relief pressure. Procedure for Model 7000 tractor is similar. Refer to text for procedure and specifications.

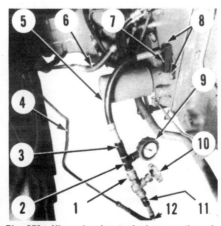

Fig. 378 — View showing typical connection of testing equipment for hydraulic pump pressure test on Models 5610, 6610, 6710 and 7710 with a single pump or main pump pressure on models with dual pumps.

1. Adapter
2. "T" adapter
3. Adapter
4. Tube assy.
5. Hose assy.
6. Hose assy.
7. Pressure tube adapter
8. Cap screws
9. Pressure gage
10. Shut-off valve
11. Adapter
12. Hose assy.

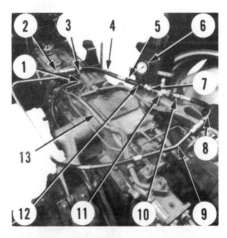

Fig. 379 — View showing typical connection of testing equipment for hydraulic pump pressure test on Models 5610, 6610, 6710, 7610 and 7710 with an engine-mounted auxiliary gear type pump.

1. Priority valve pack
2. Pump pressure pipe
3. Elbow
4. Hose assy.
5. Adapter
6. Pressure gage
7. Adapter
8. Hose assy.
9. Tube assy.
10. Adapter
11. Shut-off valve
12. "T" adapter
13. Remote control valve supply line

NOTE: On models with non-adjustable pressure relief valve, complete valve assembly must be renewed if pressure gage reading is not to recommended specification.

NOTE: On Models 5600, 6600, 6700, 7600 and 7700 there is a pressure test port in same horizontal passage as pressure relief valve.

Relief valve can be removed without removing hydraulic pump. If renewing pressure relief valve or relief valve spring, or adding shims does not increase system pressure, overhaul pump assembly as outlined in paragraph 380.

To test Models 5610, 6610, 6710, 7610 and 7710 with a single pump or main pump pressure on models with dual pumps, connect pressure testing equipment as shown in Fig. 378. A 0-4000 psi (0-27.6 MPa) hydraulic pressure gage must be used. After testing equipment is connected, fully open shut-off valve (10). Start engine and set engine speed at 1650 rpm. Push selector valve knob fully in and place lift control lever in full down position. Gradually close shut-off valve while observing gage pressure. Hydraulic pressure should be 2550-2650 psi (17.59-18.28 MPa).

Open shut-off valve, stop engine and disconnect pressure testing equipment. If pressure reading is below specified level, repair or renew faulty components and recheck pump pressure.

To test auxiliary engine-mounted gear type pump used on Models 5610, 6610, 6710, 7610 and 7710, connect pressure testing equipment as shown in Fig. 379. A 0-4000 psi (0-27.6 MPa) hydraulic pressure gage must be used. After testing equipment is connected, fully open shut-off valve (11). Start engine and set engine speed at 1650 rpm. Gradually close shut-off valve while observing gage pressure. Hydraulic pressure should be 2550-2650 psi (17.59-18.28 MPa).

NOTE: Do not close shut-off valve beyond 2650 psi (18.28 MPa) reading. If pressure reading is allowed to go above 2650 psi (18.28 MPa), pump damage could occur.

Open shut-off valve, stop engine and disconnect pressure testing equipment. If pressure reading is below specified level, repair or renew faulty components and recheck pump pressure.

ADJUSTMENTS

Models 5000-5600-6600-7000-7600 Without Load Monitor

294. **ADJUST DRAFT CONTROL MAIN SPRING.** Refer to Fig. 380 and proceed as follows:

Lower lift arms to fully lowered posi-

tion. Disconnect top link rocker from yoke and pivot rocker back out of way. Loosen set screw in retainer nut and turn nut in until tension of main draft control spring is felt. Turn yoke onto draft control plunger until all free play of yoke and plunger is eliminated, then turn yoke to nearest horizontal rocker pin attaching hole position and reconnect rocker arm. Tighten set screw in retainer nut.

295. **ADJUST DRAFT CONTROL LINKAGE.** With lift cover and cylinder assembly removed from tractor as outlined in paragraph 354, proceed as follows:

Adjust draft control main spring as outlined in paragraph 294. Position selector lever in draft control and move lift arms to fully raised position. Move lift control lever to a position 3¼ inches (82.55 mm) from top stop on control lever quadrant. Remove control valve plate (32 – Fig. 418) and spacer (33), loosen jam nut and turn turnbuckle (Fig. 381) so front end of control valve spool is exactly 0.200 inch (5.08 mm) below flush with front end of control valve bushing. A special adjustment gage (Nuday tool SW 508) is available which measures 0.200 inch (5.08 mm) when plunger of gage is flush with step cut in end of gage body.

NOTE: On Models 5600 and 6600 only, adjust turnbuckle to provide 0.030 inch (0.76 mm) from end of control valve to front face of control valve bushing. Use tool SW28 (or 6210) to check adjustment.

Recheck adjustment after tightening jam nut. With draft control linkage properly adjusted, adjust position control linkage as outlined in paragraph 296.

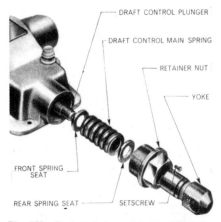

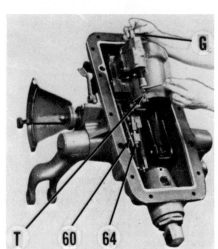

Fig. 380 – Exploded view of hydraulic system draft control main spring installation.

Fig. 381 – Adjusting draft control linkage on Models 5000, 5600, 6600, 7000 and 7600 without Load Monitor. Refer to text for procedure and specifications.

Fig. 382 – Adjusting position control linkage on Models 5000, 5600, 6600, 7000 and 7600 without Load Monitor. Refer to text for procedure and specifications.

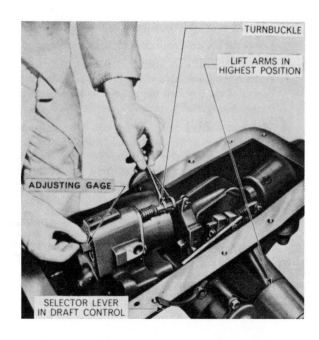

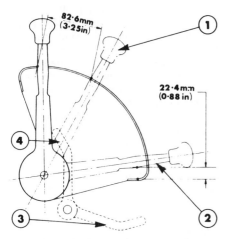

Fig. 383 — View of lift control lever and selector lever placements for setting adjustments of draft and position control on Models 5600 and 6600 without Load Monitor. Note measurements. Refer to text.

1. Draft control setting
2. Position control setting
3. Selector lever in draft control
4. Selector lever in position control

296. ADJUST POSITION CONTROL LINKAGE. Refer to paragraph 295 and adjust draft control linkage, then proceed as follows:

Place selector lever in position control and move lift arms to fully lowered position so cylinder arm is resting against stop in lift cover. Move lift control lever and adjustable stop so stop is against lower end of slot in control lever quadrant and lever is against stop as shown in inset in Fig. 382. With control valve plate (32 – Fig. 418) and spacer (33) removed from front end of lift cylinder casting, loosen locknut and turn position adjusting screw (64 – Fig. 381) so front end of control valve is exactly 0.200 inch (5.08 mm) below flush with front end of control valve bushing. Procedure varies slightly for Models 5600 and 6600. Refer to Fig. 383 and position lower edge of lift control lever exactly 0.88 inch (22.4 mm) from bottom edge of quadrant as shown. Now, turn position control adjusting screw to provide 0.030 inch (0.76 mm) from end of control valve to front face of control valve bushing by use of setting gage SW28 (or 6210). Recheck adjustment after tightening locknut to 15-20 ft.-lbs. (20-27 N·m). When adjustment is correct, reinstall control valve baffle and plate assembly on front end of lift cylinder casting, then carefully reinstall lift cover and cylinder assembly as outlined in paragraph 354 to avoid any damage to control linkage.

Models 5000-5600-6600-7000-7600 With Load Monitor

297. ADJUST DRAFT CONTROL MAIN SPRING. Refer to Fig. 384 and

Fig. 384 — View showing quadrants, control levers and adjustment positions of Models 5000, 5600, 6600, 7000 and 7600 equipped with Load Monitor.

place selector lever in position control setting. Place lift control lever at bottom of quadrant and fully lower lift arms. Disconnect top link rocker from yoke and pivot rocker back out of the way. Loosen set screw in retainer nut and turn nut in until tension of main draft control spring is felt. Turn yoke onto draft control plunger until all free play of yoke and plunger is eliminated, then turn yoke to the nearest horizontal rocker pin attaching hole position. Reconnect rocker arm and tighten set screw in retainer nut.

298. LIFT COVER LINKAGE. Prior to making any lift cover adjustments, lift cover must be removed as outlined in paragraph 355 and the following preparatory adjustments made.

Adjust strut (15 – Fig. 385), if necessary, until the distance between centerlines of attaching holes is 14.18-14.42 inches (36.02-36.63 cm). Loosen nut on draft control adjustment screw (12 – Fig. 436) and position offset in its highest position in respect to normal operating position of lift cover, then

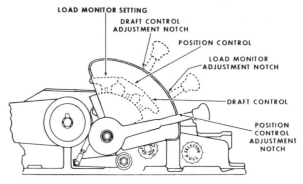

tighten nut. Install Adjustment Fixture SW25 (Fig. 385) on lift cover as shown, clamp tool in position so it will stay fixed for all remaining adjustments. Install strut on tool.

NOTE: Adjustment Fixture SW862-A can be used in same manner and all adjustments must be made in the following sequence.

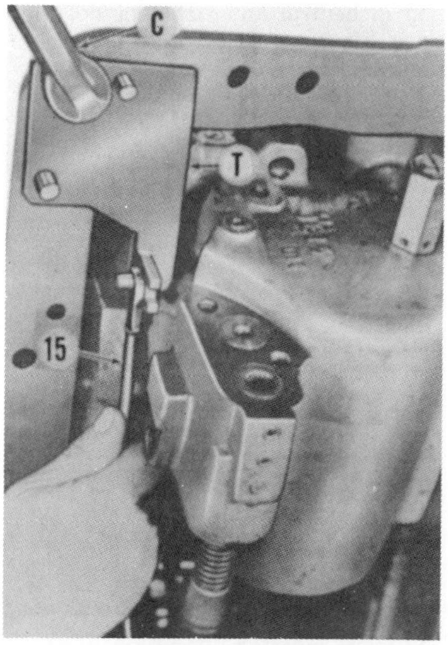

299. SELECTOR LEVER INDEX ADJUSTMENT. On models with Ford cab, set selector lever in draft control position and adjust length of attached rod so center-to-center distance is 6.68-6.72 in. (169.7-170.7 mm). On all models, manually place lift arms in fully raised position and set system selector lever for position control. Position lift control lever so upper edge (lower edge on Rowcrop tractors) is aligned with position control adjustment notch on quadrant. See Fig. 384. Remove cotter pin and washer from strut attaching pin and install tool SW862-B (Fig. 386), or tool SW26 (Fig. 387) between control rod roller and strut attaching pin as shown. Adjust selector lever clevis until foot of tool is in contact with control rod roller, then tighten clevis locknut.

NOTE: On Rowcrop tractors, loosen quadrant attaching bolts and realign

Fig. 385 — View showing tool SW-25 (T) installed and strut (15) attached. Item (C) is clamp. Tool SW862-A is similar.

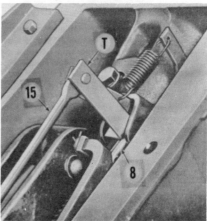

Fig. 386 — View showing tool SW862-B (T) installed and strut (15) attached. Item (8) is control rod and roller assembly. Also see Fig. 387.

145

quadrant until foot is in contact with control rod roller and tighten quadrant attaching bolts.

If tool SW862-B was used to make index adjustment, remove tool. If tool SW26 was used, leave tool in position while making following position control adjustment.

300. **POSITION CONTROL ADJUSTMENT.** With selector lever in position control and lift lever aligned with position control notch in quadrant, manually reposition lift arms in fully lowered position. Remove plate (32 – Fig. 418) and spacer (33) from front face of lift cylinder and control valve assembly. Adjust control valve spool turnbuckle (24 – Fig. 436) so front end of control valve spool is 0.030 in. (0.76 mm) from end of control valve to front face of control valve bushing. Control valve setting tool SW28 (or 6210) can be used to check this setting. Tighten turnbuckle locknut when adjustment is completed.

Remove tool SW26, if previously installed.

301. **LOAD MONITOR ADJUSTMENT.** With lift arms in fully lowered position, set system selector lever in full Load Monitor position. Set lift control lever so upper edge of lever aligns with Load Monitor notch in quadrant. Adjust length of strut so front end of valve spool is 0.030 inch (0.76 mm) from end of control valve to front face of control valve bushing. Use control valve setting gage SW28 (or 6210) to check this setting. Tighten locknut on strut when adjustment is complete.

302. **TOP LINK DRAFT CONTROL ADJUSTMENT.** With lift arms in fully lowered position, set system selector lever in draft control position. Set lift control lever so upper edge of lever aligns with draft control adjustment notch in quadrant. See Fig. 384. Adjust position of draft control adjustment screw (12 – Fig. 436) so front end of valve spool is 0.030 inch (0.76 mm) from end of control valve to front face of control valve bushing. Use control valve setting gage SW28 (or 6210) to check this setting. Tighten adjustment screw retaining nut to 24-31 ft.-lbs. (33-42 N·m) when adjustment is complete.

Install plate and spacer on lift cylinder and control valve assembly, then remove Adjustment Fixture SW862-A or SW25 from lift cover.

303. **LOAD MONITOR FORK ADJUSTMENT.** Install Adjustment Fixture SW862-A or SW25 on rear axle center housing as shown in Fig. 388 or 389. Place Load Monitor fork and lever assembly firmly in contact with Load Monitor, then pull firmly rearward on fork to remove any slack in components, including Load Monitor. The strut attaching pin of fork lever should locate freely in notch of tool. If strut pin does not locate in notch of tool, turn fork adjusting screw (Fig. 390) as required.

NOTE: Fig. 390 shows fork adjusting screw viewed with dipstick plate removed from center housing. Plate need not be removed if lift cover is off.

Remove adjustment tool and install lift cover as outlined in paragraph 355.

304. **POST ADJUSTMENT CHECK (MODELS WITHOUT CAB).** After tractor has been reassembled, the adjustments in paragraphs 299 through 303 can be verified as follows:

Block tractor wheels, place transmission in neutral and start engine, then place lift control lever at bottom of quadrant.

SPECIAL NOTE: When this after-adjustment check is being made, it is important to manually push system selector lever downward to firmly engage position control. This action is not required in normal operation as system selector lever will lock in position control in the first cycle of the hydraulic system. Be sure lever is fully engaged in position control so initial response of hydraulic system for POST-ADJUSTMENT CHECK will be accurate.

305. Put system selector lever in top link draft control position, then slowly raise lift control lever until the first raise signal is sensed. At this point, upper edge of lift control lever should be within ½-inch (12.7 mm) of draft control notch in quadrant.

306. Move lift control lever back to bottom of quadrant and place system

Fig. 389 – Tool SW25 (T) is used in same manner as tool SW862-A shown in Fig. 388. Note pin of fork lever (15) is located in notch of tool when properly adjusted.

Fig. 387 – View showing tool SW26 (T) installed and strut (15) attached. Item (8) is control rod and roller assembly. Also see Fig. 386.

Fig. 388 – With fork properly adjusted, pin on fork lever (15) will be located in notch of tool SW862-A (T). Also see Fig. 389.

Fig. 390 – Load Monitor fork adjustment screw (A) is accessible after removing dipstick plate. Adjustment screw is also accessible when hydraulic lift cover is removed.

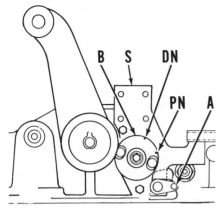

Fig. 391 – Diagram of lift cover on Models 6700 and 7700 not equipped with Load Monitor. Refer to text for adjustments.

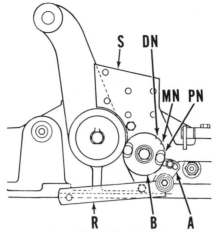

Fig. 392 – Diagram of right lift cover side on Models 6700 and 7700 with Load Monitor. Refer to text for adjustments.

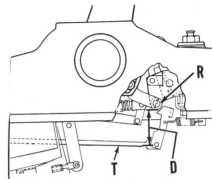

Fig. 393 – Diagram of lift cover linkage on Models 6700 and 7700 with Load Monitor. Refer to text for adjustments.

selector lever in position control. Slowly raise lift control lever until the first raise signal is sensed. At this point, upper edge (lower edge on Rowcrop tractors) of lift control lever should be within ½-inch (12.7 mm) of position control notch in quadrant.

307. Move lift control lever back to bottom of quadrant and place system selector lever in full Load Monitor position. Slowly raise lift control lever until the first raise signal is sensed. At this point, upper edge of lift control lever should be within ¾-inch (19.0 mm) of Load Monitor notch in quadrant.

308. If draft control and position control checks are not as outlined in paragraphs 305 and 306, it will be necessary to remove hydraulic lift cover and recheck adjustments and/or linkage. If Load Monitor check is not as outlined in paragraph 307, proceed to LOAD MONITOR FIELD ADJUSTMENT.

309. **LOAD MONITOR FIELD ADJUSTMENT (MODELS WITHOUT CAB).** Drain enough fluid from rear center housing so oil will not run out when dipstick plate is removed, then remove left step plate and dipstick plate. Block tractor wheels, place transmission in neutral and start engine. Place system selector lever in full Load Monitor position, then align upper edge of lift control lever with Load Monitor notch in quadrant. If lift arms have not raised, turn Load Monitor fork adjusting screw to move fork lever forward until the first raise signal is sensed. If lift arms have raised, lower lift control lever, then turn fork adjusting screw so fork lever moves slightly rearward. Repeat this operation until upper edge of lift control lever will align with Load Monitor notch in quadrant when the first raise signal is sensed.

Stop engine, reinstall dipstick plate and left step plate and refill hydraulic system.

Models 6700-7700
Without Load Monitor

310. **ADJUSTMENT DRAFT CONTROL MAIN SPRING.** Refer to paragraph 294 and adjust draft control main spring as outlined.

311. **ADJUST DRAFT CONTROL LINKAGE.** Remove lift cover as outlined in paragraph 354 and proceed as follows:

Set actuator (A–Fig. 391) in draft control position (down) and rotate body (B) so draft control notch (DN) is aligned with notch on support (S). Remove plate (32–Fig. 418) and spacer (33) from front face of lift cylinder and control valve assembly. Adjust turnbuckle (T–Fig. 381) so there is 0.030 inch (0.76 mm) from end of control valve to front face of control valve bushing. Use tool SW28 (or 6210) to check adjustment. Recheck adjustment after tightening turnbuckle jam nut.

312. **ADJUST POSITION CONTROL LINKAGE.** Lift cover must be removed as outlined in paragraph 354 prior to adjustment.

Move actuator (A–Fig. 391) to position control setting (up) and remove lift cylinder safety valve. Lift arms must be in full down position. Rotate body (B) so position control notch (PN) is aligned with notch on support (S). Remove plate (32–Fig. 418) and spacer (33) from front face of lift cylinder and control valve assembly. Turn position adjusting screw (64–Fig. 381) so there is 0.030 inch (0.76 mm) from end of control valve to front face of control valve bushing. Use tool SW28 (or 6210) to check adjustment. Recheck adjustment after tightening jam nut on adjusting screw.

313. **ADJUST SELECTOR LEVER CONTROL ROD.** Rotate turnbuckle on selector lever control so selector lever travel matches control panel markings.

314. **ADJUST LIFT CONTROL LEVER CONTROL ROD.** Place selector lever in position control setting. Place lift control lever in fully lowered setting and rotate turnbuckle on lift control lever rod so there is 0.38-0.50 inch (9.7-12.7 mm) lower stop in quadrant and lift control lever.

Models 6700-7700
With Load Monitor

315. **ADJUST DRAFT CONTROL MAIN SPRING.** Refer to paragraph 294 and adjust draft control main spring as outlined.

316. **LIFT COVER LINKAGE.** Follow procedure outlined in paragraph 298.

317. **ADJUST POSITION CONTROL LINKAGE.** Align position control notch (PN–Fig. 392) on body (B) with notch on support (S). Place lift arms in fully raised position. Detach strut (T–Fig. 393) and install gage tool SW862-B or SW26. With control rod roller (R) contacting linkage, rotate actuator assembly (A–Fig. 392) until roller contacts foot of gage. Tighten actuator locknut to retain setting. If used, remove tool SW862-B; tool SW26 may remain in position. Manually reposition lift arms to fully lowered position. Make sure control rod roller (R–Fig. 393) contacts linkage. Remove plate (32–Fig. 418) and spacer (33) from front face of lift cylinder and control valve assembly. Adjust control valve spool turnbuckle (24–Fig. 436) so front end of control valve spool is 0.030 inch (0.76 mm) from end of control valve to front face of control valve bushing. Control valve setting tool SW28 (or 6210) can be used to check

setting. Tighten turnbuckle locknut when adjustment is completed.

Remove tool SW26, if previously installed.

318. ADJUST LOAD MONITOR LINKAGE. With lift arms in fully lowered position, align Load Monitor notch (MN–Fig. 392) on body (B) with notch on support (S). Rotate actuator assembly (A) so roller (R–Fig. 393) is 1.44 inch (36.6 mm) from link hole (distance D). Remove plate (32–Fig. 418) and spacer (33) from front face of lift cylinder and control valve assembly. Adjust length of Load Monitor strut (T–Fig. 393) so front end of control valve spool is 0.030 in (0.76 mm) from end of control valve to front face of control valve bushing. Control valve setting tool SW28 (or 6210) can be used to check setting. Tighten strut locknut.

319. DRAFT CONTROL ADJUST-MENT. With lift arms in fully lowered position, align draft control notch (DN–Fig. 392) on body (B) with notch on support (S). Rotate actuator assembly (A) so roller (R–Fig. 393) is 4.05 inch (102.9 mm) from link hole (distance D). Remove plate (32–Fig. 283) and spacer (33) from front face of lift cylinder and control valve assembly. Adjust position of draft control adjustment screw (12–Fig. 436) so front end of valve spool is 0.030 in. (0.76 mm) from end of control valve to front face of control valve bushing. Use control valve setting gage SW28 (or 6210) to check setting. Tighten adjustment screw retaining nut to 24-31 ft.-lbs. (33-43 N·m) when adjustment is complete.

320. VERIFY POSITION CONTROL MARKS. Place lift arms in fully

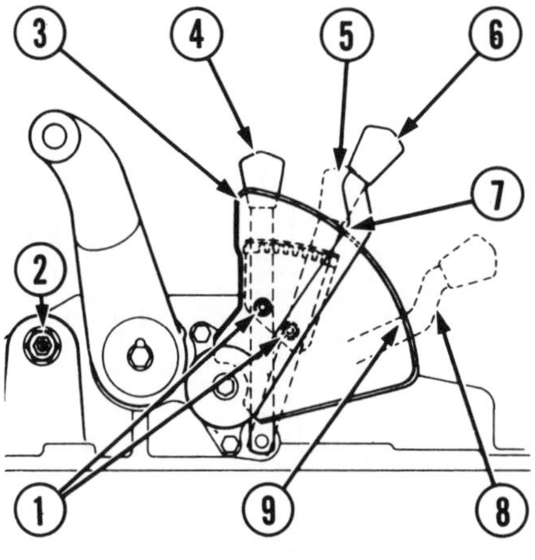

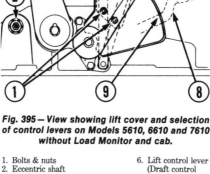

Fig. 395 – View showing lift cover and selection of control levers on Models 5610, 6610 and 7610 without Load Monitor and cab.

1. Bolts & nuts	6. Lift control lever (Draft control setting)
2. Eccentric shaft	
3. Quadrant	7. Notch
4. Selector lever (Position control setting)	8. Lift control lever (Position control setting)
5. Selector lever (Draft control setting)	9. Notch

raised position. Rotate actuator assembly (A–Fig. 392) so roller (R–Fig. 393) is 2.52 inch (64 mm) from link hole (distance D). Tighten actuator locknut to 11-15 ft.-lbs. (15-20 N·m). Position control notch (PN–Fig. 392) on body (B) should be aligned with notch on support (S), if not, stamp new marks for future reference.

321. ADJUST SELECTOR LEVER. To adjust selector lever, detach selector lever control rod from actuator arm (R–Fig. 392).

NOTE: If actuator has not been adjusted prior to installation of lift cover, hold actuator arm (R) to prevent internal

damage due to actuator return spring pressure.

Rotate actuator assembly (A) so position control notch (PN) on body (B) is aligned with notch on support (S) or new alignment marks noted in paragraph 320 are aligned. Set selector lever in position control notch of console and adjust control rod clevis so rod can be reattached to actuator arm.

322. LOAD MONITOR FORK ADJUSTMENT. Refer to paragraph 303 for Load Monitor fork adjustment procedure.

Models 5610-6610-7610 Without Load Monitor And Cab

323. ADJUST DRAFT CONTROL MAIN SPRING. Locate selector lever in position control. Position lift control lever at bottom of quadrant. Ensure lift arms are fully down, then turn yoke (2–Fig. 394) out to remove preload tension on main spring. Loosen set screw (1) and turn retainer nut (3) in until draft control main spring pressure is felt. Rotate yoke (2) on plunger (6) until all free play is eliminated, then continue to turn yoke (2) until hole in yoke is horizontally positioned. Tighten set screw (1).

324. ADJUST DRAFT CONTROL LINKAGE. With lift cover and cylinder assembly removed from tractor as outlined in paragraph 356, proceed as follows:

Lower lift arms to full down position. Align upper edge of lift control lever (6–Fig. 395) with draft control setting notch (7) on quadrant. Position selector lever in draft (seventh) control setting (5) in selector quadrant. Loosen selector quadrant retaining nuts (1), then move

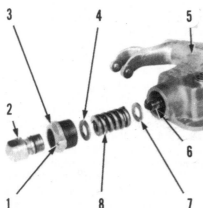

Fig. 394 – View showing draft control main spring components used on all "10" series tractors.

1. Set screw	5. Lift cover
2. Yoke	6. Plunger
3. Retainer nut	7. Front spring seat
4. Rear spring seat	8. Main spring

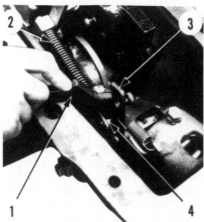

Fig. 396 – View showing control rod roller setting for draft control adjustment on Models 5610, 6610, 6710, 7610 and 7710 without Load Monitor. Refer to text.

1. Gage tool	3. Selector link
2. Control rod	4. Control rod roller

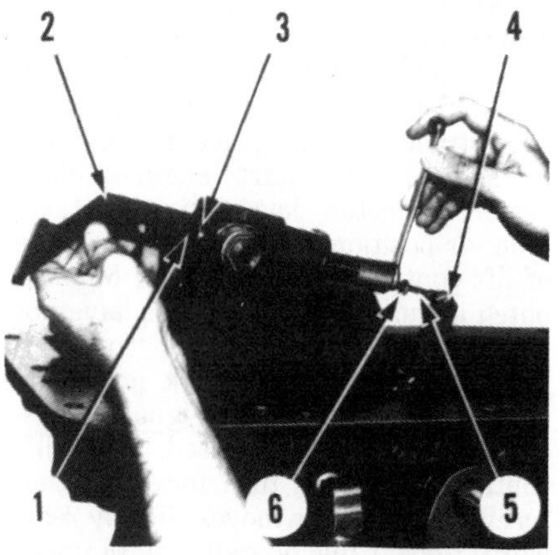

Fig. 397 – View showing procedure for checking distance control valve pin extends beyond control valve bushing on all "10" series tractors. Refer to text.

1. Control valve pin	4. Actuating lever
2. Gage tool	5. Turnbuckle
3. Control valve bushing	6. Locknut

quadrant until bottom of control rod roller (4–Fig. 396) is 1.95 inch (49.5 mm) above lift cover surface. Check by using Churchill gage tool FT.8527 or Nuday tool 4654 (1). Tighten selector quadrant retaining nuts to 18 ft.-lbs. (24 N·m) torque. Remove control valve front plate (30–Fig. 426).

NOTE: Do not allow drop poppet valve (26) and spring (27) to fall free from lift cylinder housing.

Loosen locknut (6–Fig. 397) and adjust turnbuckle (5) until control valve pin (1) extends 0.19 inch (4.83 mm) beyond control valve bushing (3). Use Churchill gage tool FT.8527 or Nuday tool 4654 (2) to check dimension, then tighten locknut (6) to 10 ft.-lbs. (14 N·m) torque. Unless position control linkage is to be adjusted, reassemble components. Tighten cap screws retaining control valve front plate to 34 ft.-lbs. (46 N·m) torque.

325. ADJUST POSITION CONTROL LINKAGE.

NOTE: Be sure draft control linkage is adjusted prior to adjustment of position control linkage.

With lift cover and cylinder assembly removed from tractor as outlined in paragraph 356, proceed as follows:

Lower lift arms to full down position. Align upper edge of lift control lever (8–Fig. 395) with position control setting notch (9) on quadrant. Locate selector lever in position (first) control setting (4) in selector quadrant. Top of control rod roller (3–Fig. 398) should be 0.03 inch (0.76 mm) below lift cover surface. Use Churchill gage tool FT.8527 or Nuday tool 4654 (1) to check setting.

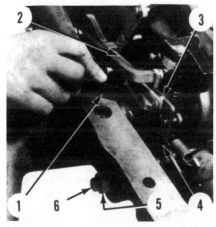

Fig. 398 – View showing control rod roller setting for position control adjustment on Models 5610, 6610, 6710, 7610 and 7710 without Load Monitor. Refer to text.

1. Gage tool
2. Control rod
3. Control rod roller
4. Selector link
5. Locknut
6. Eccentric shaft

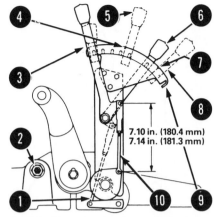

Fig. 399 – View showing lift cover and selection of control levers on Models 5610, 6610 and 7610 without Load Monitor and with cab.

1. Selector lever rod
2. Eccentric shaft
3. Selector lever (Draft control setting)
4. Notch
5. Lift control lever (Draft control setting)
6. Selector lever (Position control setting)
7. Notch
8. Lift control lever (Position control setting)
9. Quadrant
10. Lift control lever rod

Loosen locknut (5) and rotate eccentric shaft (6) until control valve pin (1–Fig. 397) extends 0.19 inch (4.83 mm) beyond control valve bushing (3). Use Churchill gage tool FT.8527 or Nuday tool 4654 (2) to check dimension, then tighten locknut to 18 ft.-lbs. (24 N·m) torque. Reinstall control valve front plate and tighten retaining cap screws to 34 ft.-lbs. (46 N·m) torque.

Models 5610-6610-7610 Without Load Monitor And With Cab

326. ADJUST DRAFT CONTROL MAIN SPRING. Refer to paragraph 323 and adjust draft control main spring as outlined.

327. LIFT COVER LINKAGE. Lift control lever rod should be adjusted to a length of 7.10-7.14 inch (180.4-181.3 mm) as shown in Fig. 399.

328. ADJUST DRAFT CONTROL LINKAGE. With lift cover and cylinder assembly removed from tractor as outlined in paragraph 356, proceed as follows:

Lower lift arms to full down position. Align upper edge of lift control lever (5–Fig. 399) with draft control setting notch (4) on quadrant. Position selector lever in draft (first) control setting (3) in selector quadrant. Loosen selector quadrant retaining nuts, then move quadrant until bottom of control rod roller (4–Fig. 396) is 1.95 inch (49.5 mm) above lift cover surface. Check by using Churchill gage tool FT.8527 or Nuday tool 4654 (1). Tighten selector quadrant retaining nuts to 18 ft.-lbs. (24

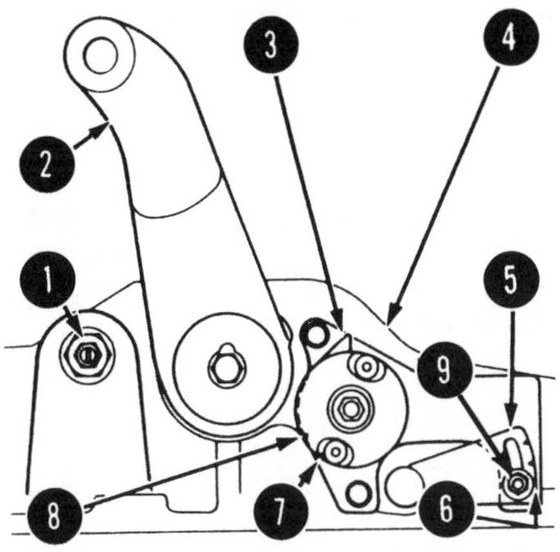

Fig. 400 – View of lift cover and hydraulic linkage on Models 6710 and 7710 with and without Load Monitor.

1. Eccentric shaft
2. Lift arm
3. Body
4. Lift cover
5. Actuator
6. Position control identification mark
7. Draft control notch on support
8. Position control notch on support
9. Cap screw

N·m) torque. Remove control valve front plate (30–Fig. 426).

NOTE: Do not allow drop poppet valve (26) and spring (27) to fall free from lift cylinder housing.

Loosen locknut (6–Fig. 397) and adjust turnbuckle (5) until control valve pin (1) extends 0.19 inch (4.83 mm) beyond control valve bushing (3). Use Churchill gage tool FT.8527 or Nuday tool 4654 (2) to check dimension, then tighten locknut to 10 ft.-lbs. (14 N·m) torque. Unless position control linkage is to be adjusted, reassemble components. Tighten cap screws retaining control valve front plate to 34 ft.-lbs. (46 N·m) torque.

329. ADJUST POSITION CONTROL LINKAGE. With lift cover and cylinder assembly removed from tractor as outlined in paragraph 356, proceed as follows:

Lower lift arms to full down position. Align upper edge of lift control lever (8–Fig. 399) with position control setting notch (7) on quadrant. Locate selector lever in position (seventh) control setting (6) in selector quadrant. Top of control rod roller (3–Fig. 398) should be 0.03 inch (0.76 mm) below lift cover surface. Use Churchill gage tool FT.8527 or Nuday tool 4654 (1) to check setting. Loosen locknut (5) and rotate eccentric shaft (6) until control valve pin (1–Fig. 397) extends 0.19 inch (4.83 mm) beyond control valve bushing (3). Use Churchill gage tool FT.8527 or Nuday tool 4654 (2) to check dimension, then tighten locknut to 18 ft.-lbs. (24 N·m) torque. Reinstall control valve front plate and tighten retaining cap screws to 34 ft.-lbs. (46 N·m) torque.

Models 6710-7710
Without Load Monitor

330. ADJUST DRAFT CONTROL MAIN SPRING. Refer to paragraph 323 and adjust draft control main spring as outlined.

331. ADJUST DRAFT CONTROL INTERNAL LINKAGE. With lift cover and cylinder assembly removed from tractor as outlined in paragraph 356, proceed as follows:

Lower lift arms to full down position. Align notch on body (3–Fig. 400) with draft control notch on support assembly (7). Loosen securing cap screw (9) and rotate actuator (5) until bottom of control rod roller (4–Fig. 396) is 1.95 inch (49.5 mm) above lift cover surface. Check by using Churchill gage tool FT.8527 or Nuday tool 4654 (1). Tighten actuator securing cap screw to 13 ft.-lbs. (18 N·m) torque. Remove cap screws

Fig. 401 — Refer to text and adjust draft control linkage on models with Load Monitor.

1. Turnbuckle
2. Locknut
3. Eccentric shaft
4. Strut

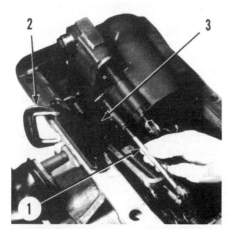

Fig. 402 — Position strut (1) on adjustment tool (3). Tool (3) is retained to lift cover surface by clamp (2).

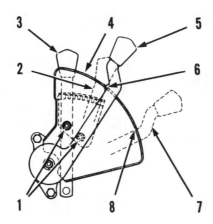

Fig. 403 — View showing quadrant and selection of control levers on Models 5610, 6610 and 7610 with Load Monitor and without cab.

1. Bolts & nuts
2. Selector lever (Draft control setting)
3. Selector lever (Load Monitor setting)
4. Quadrant
5. Lift control lever

   (Draft control & Load Monitor setting)
6. Notch
7. Lift control lever (Position control setting)
8. Notch

securing control valve front plate (30– Fig. 426).

NOTE: Do not allow drop poppet valve (26) and spring (27) to fall free from lift cylinder housing.

Loosen locknut (6–Fig. 397) and adjust turnbuckle (5) until control valve pin (1) extends 0.19 inch (4.83 mm) beyond control valve bushing (3). Use Churchill gage tool FT.8527 or Nuday tool 4654 (2) to check dimension, then tighten locknut (6) to 10 ft.-lbs. (14 N·m) torque. Unless position control linkage is to be adjusted, reassemble components. Tighten cap screws retaining control valve front plate to 34 ft.-lbs. (46 N·m) torque.

332. ADJUST POSITION CONTROL INTERNAL LINKAGE. With lift cover and cylinder assembly removed from tractor as outlined in paragraph 356, proceed as follows:

Lower lift arms to full down position. Align notch on body (3–Fig. 400) with position control notch on support assembly (8). Loosen securing cap screw (9) and rotate actuator (5) until top of control rod roller (3–Fig. 398) is 0.03 inch (0.76 mm) below lift cover surface. Use Churchill gage tool FT.8527 or Nuday tool 4654 (1) to check setting. Tighten actuator securing cap screw to 13 ft.-lbs. (18 N·m) torque. Loosen locknut (5) and rotate eccentric shaft (6) until control valve pin (1–Fig. 397) extends 0.19 inch (4.83 mm) beyond control valve bushing (3). Use Churchill gage tool FT.8527 or Nuday tool 4654 (2) to check dimension, then tighten locknut to 18 ft.-lbs. (24 N·m) torque. Reinstall control valve front plate and tighten retaining cap screws to 34

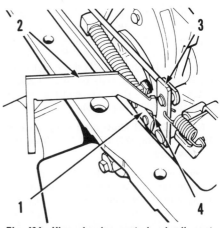

Fig. 404 — View showing control rod roller setting for position control adjustment on Models 5610, 6610, 6710, 7610 and 7710 with Load Monitor.

1. Control rod
2. Gage tool
3. Selector linkage assy.
4. Control rod roller

ft.-lbs. (46 N·m) torque. Scribe an identification mark on actuator (5–Fig. 400) and support assembly to identify setting.

Models 5610-6610-7610 With Load Monitor And Without Cab

333. ADJUST DRAFT CONTROL MAIN SPRING. Refer to paragraph 323 and adjust draft control main spring as outlined.

334. LIFT COVER LINKAGE. With lift cover and cylinder assembly removed from tractor as outlined in paragraph 356, proceed as follows:

Measure length of strut (4–Fig. 401) between center of mounting holes, recommended length is 8.92 inch (227 mm). Adjust length as needed. Loosen locknut (2) and rotate eccentric shaft (3) upward until shaft is closest to roof of lift cover, then tighten nut. Position and clamp Churchill tool FT.8529 or Nuday tool 4656 (3–Fig. 402) on lift cover and place strut (1) on tool.

NOTE: Strut must not move from position while the following adjustments are being performed. Adjustments must be performed in sequence outlined as follows:

335. ADJUST POSITION CONTROL LINKAGE. After performing steps outlined in paragraph 334, proceed as follows:

Place lift arms in fully raised position. Align upper edge of lift control lever (7–Fig. 403) with position control setting notch (8) on quadrant. Locate selector lever in position (middle) control setting in selector quadrant. Loosen selector quadrant retaining nuts. Ensure control rod roller is in contact with selector linkage assembly, then adjust quadrant until bottom of control rod

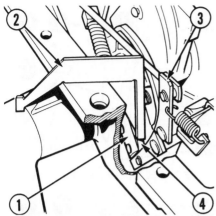

Fig. 405 — View showing control rod roller setting for draft control adjustment on Models 5610, 6610, 6710, 7610 and 7710 with Load Monitor.

| 1. Control rod | 3. Selector linkage assy. |
| 2. Gage tool | 4. Control rod roller |

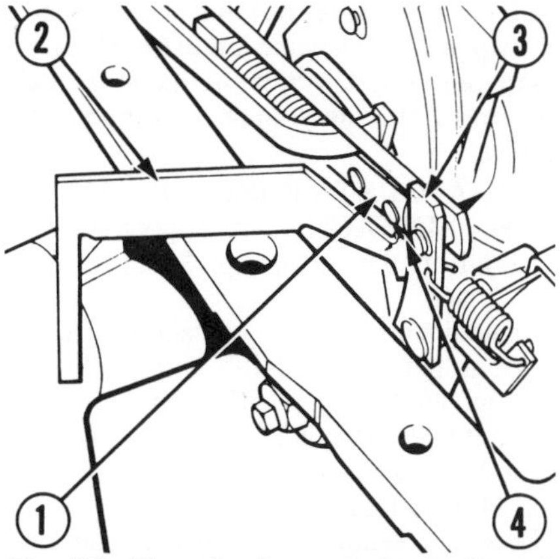

Fig. 406 — View showing control rod roller setting for Load Monitor adjustment on Models 5610, 6610, 6710, 7610 and 7710.

| 1. Control rod | 3. Selector linkage assy. |
| 2. Gage tool | 4. Control rod roller |

roller (4 – Fig. 404) is 0.72 inch (18.3 mm) above lift cover surface. Use Churchill gage tool FT.8528 or Nuday tool 4655 (2) to check setting. Tighten selector quadrant retaining nuts to 18 ft.-lbs. (24 N·m) torque. Lower lift arms to full down position. Remove cap screws securing control valve front plate (30 – Fig. 426).

NOTE: Do not allow drop poppet valve (26) and spring (27) to fall free from lift cylinder housing.

Loosen locknut (6 – Fig. 397) and adjust turnbuckle (5) until control valve pin (1) extends 0.19 inch (4.83 mm) beyond control valve bushing (3). Use Churchill gage tool FT.8527 or Nuday tool 4654 (2) to check dimension, then tighten locknut (6) to 10 ft.-lbs. (14 N·m) torque. Unless draft control linkage is to be adjusted, reassemble components. Tighten cap screws retaining control valve front plate to 34 ft.-lbs. (46 N·m) torque.

336. **ADJUST DRAFT CONTROL LINKAGE.** After performing steps outlined in paragraphs 334 and 335, proceed as follows:

Lower lift arms to full down position. Align upper edge of lift control lever (5 – Fig. 403) with draft control setting notch (6) on quadrant. Position selector lever in draft (seventh) control setting (2) in selector quadrant. Bottom of control rod roller (4 – Fig. 405) should be 2.10 inch (53.3 mm) above lift cover surface. Use Churchill gage tool FT.8528 or Nuday tool 4655 (2) to check setting. Loosen locknut (2 – Fig. 401) and rotate eccentric shaft (3) until control valve pin (1 – Fig. 397) extends 0.19 inch (4.83 mm) beyond control valve bushing (3).

Use Churchill gage tool FT.8527 or Nuday tool 4654 (2) to check dimension, then tighten locknut to 28 ft.-lbs. (38 N·m) torque.

337. **ADJUST LOAD MONITOR LINKAGE.** After performing steps outlined in paragraphs 334, 335 and 336, proceed as follows:

Lower lift arms to full down position. Align upper edge of lift control lever (5 – Fig. 403) with Load Monitor setting notch (6) on quadrant. Position selector lever in Load Monitor (first) control setting (3) in selector quadrant. Top of control rod roller (4 – Fig. 406) should be 0.11 inch (2.79 mm) above lift cover surface. Use Churchill gage tool FT.8528 or Nuday tool 4655 (2) to check setting.

Loosen locknut (6 – Fig. 397) and adjust turnbuckle (5) until control valve pin (1) extends 0.19 inch (4.83 mm) beyond control valve bushing (3). Use Churchill gage tool FT.8527 or Nuday tool 4654 (2) to check dimension, then tighten locknut (6) to 10 ft.-lbs. (14 N·m) torque.

Reinstall control valve front plate and tighten retaining cap screws to 34 ft.-lbs. (46 N·m) torque. Remove clamp and withdraw Churchill tool FT.8529 or Nuday tool 4656 (3 – Fig. 402) from lift cover.

338. **ADJUST LOAD MONITOR FORK.** After performing steps outlined in paragraphs 334, 335, 336 and 337, proceed as follows:

Install Churchill tool FT.8529 or Nuday tool 4656 (T – Fig. 407) on center housing surface as shown in Fig. 407. Position Load Monitor fork and lever assembly in contact with Load Monitor.

NOTE: Thrust blocks must be flat against flange of Load Monitor which must be pushed to the rear.

Rotate fork adjusting screw (S – Fig. 408) until pin (P – Fig. 407) of lever (L) slides smoothly in slot of tool (T).

Remove tool (T) and install hydraulic lift cover as outlined in paragraph 356.

339. **POST ADJUSTMENT CHECK.** After tractor has been reassembled, the adjustments in paragraphs 335 through 338 can be verified as follows:

Block tractor wheels, place transmission in neutral and start engine, set engine speed at 1000 rpm. Place lift control lever at bottom of quadrant. Put system selector lever in top link draft control position, then slowly raise lift control lever until first raise signal is sensed. At this point, upper edge of lift control lever should be within ½-inch (12.7 mm) of draft control notch in quadrant.

Move lift control lever back to bottom of quadrant and place system selector lever in position control. Slowly raise lift control lever until first raise signal is sensed. At this point, upper edge (lower edge on Rowcrop tractors) of lift control lever should be within ½-inch (12.7 mm) of position control notch in quadrant.

Move lift control lever back to bottom of quadrant and place system selector lever in full Load Monitor position. Slowly raise lift control lever until the first raise signal is sensed. At this point, upper edge of lift control lever should be within ¾-inch (19.0 mm) of Load Monitor notch in quadrant.

If draft control and position control checks are not as outlined, it will be necessary to remove hydraulic lift cover and recheck adjustments and/or linkage. If Load Monitor check is not as outlined, proceed to LOAD MONITOR FIELD ADJUSTMENT.

Fig. 407 — View showing procedure for checking Load Monitor fork adjustment on Models 5610, 6610, 6710, 7610 and 7710. Refer to text.

L. Fork lever
P. Pin
T. Tool

340. LOAD MONITOR FIELD ADJUSTMENT. Drain enough fluid from center housing so oil will not run out when dipstick plate is removed, then remove left step plate and dipstick plate. Block tractor wheels, place transmission in neutral, start engine and set engine speed at 1000 rpm. Place system selector lever in full Load Monitor position, then align upper edge of lift control lever with Load Monitor notch in quadrant. If lift arms have not raised, turn Load Monitor fork adjusting screw (W – Fig. 408A) to move fork lever forward until first raise signal is sensed. If lift arms have raised, lower lift control lever, then turn fork adjusting screw so fork lever moves slightly rearward. Repeat this operation until upper edge of lift control lever will align with Load Monitor notch in quadrant when first raise signal is sensed.

Stop engine, reinstall dipstick plate and left step plate and refill hydraulic system.

Models 5610-6610-7610 With Load Monitor And With Cab

341. ADJUST DRAFT CONTROL MAIN SPRING. Refer to paragraph 323 and adjust draft control main spring as outlined.

342. LIFT COVER LINKAGE. With lift cover and cylinder assembly removed from tractor as outlined in paragraph 356, proceed as follows:

Lift control lever rod (10 – Fig. 409) should be adjusted to a length of 7.10-7.14 inch (180.4-181.3 mm). Measure length of strut (4 – Fig. 401) between center of mounting holes, recommended length is 8.92 inch (227 mm). Adjust length as needed. Loosen locknut (2) and rotate eccentric shaft (3) upward until shaft is closest to roof of lift cover, then tighten nut. Position and clamp Churchill tool FT.8529 or Nuday tool 4656 (3 – Fig. 402) on lift cover and place strut (1) on tool.

NOTE: Strut must not move from position while the following adjustments are being performed. Adjustments must be performed in sequence outlined as follows:

343. ADJUST POSITION CONTROL LINKAGE. After performing steps outlined in paragraph 342, proceed as follows:

Place lift arms in fully raised position. Align upper edge of lift control lever (8 – Fig. 409) with position control setting notch (7) on quadrant. Locate selector lever in position (middle) control setting (4) in selector quadrant.

Loosen locknut (11), ensure control rod roller is in contact with selector linkage assembly, then adjust length of selector lever rod assembly (1) until bottom of control rod roller (4 – Fig. 404) is 0.72 inch (18.3 mm) above lift cover surface. Use Churchill gage tool FT.8528 or Nuday tool 4655 (2) to check setting. Tighten locknut (11) to 9 ft.-lbs. (12 N·m) torque. Lower lift arms to full down position. Remove cap screws securing control valve front plate (30 – Fig. 426).

NOTE: Do not allow drop poppet valve (26) and spring (27) to fall free from lift cylinder housing.

Loosen locknut (6 – Fig. 397) and adjust turnbuckle (5) until control valve pin (1) extends 0.19 inch (4.83 mm) beyond control valve bushing (3). Use Churchill gage tool FT.8527 or Nuday tool 4654 (2) to check dimension, then tighten locknut (6) to 10 ft.-lbs. (14 N·m) torque. Unless draft control linkage is to be adjusted, reassemble components. Tighten cap screws retaining control valve front plate to 34 ft.-lbs. (46 N·m) torque.

344. ADJUST DRAFT CONTROL LINKAGE. After performing steps

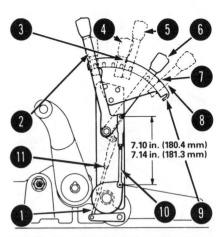

Fig. 409 – View showing lift cover and control levers on Models 5610, 6610 and 7610 with Load Monitor and cab.

1. Selector lever rod assy.
2. Selector lever (Draft control setting)
3. Notch
4. Selector lever (Position control setting)
5. Lift control lever (Draft control & Load Monitor setting)
6. Selector lever (Load Monitor setting)
7. Notch
8. Lift control lever (Position control setting)
9. Quadrant
10. Control lever rod assy.
11. Locknut

outlined in paragraph 342 and 343, proceed as follows:

Lower lift arms to full down position. Align upper edge of lift control lever (5 – Fig. 409) with draft control setting notch (3) on quadrant. Position selector lever in draft (first) control setting (2) in selector quadrant. Bottom of control rod roller (4 – Fig. 405) should be 2.10 inch (53.3 mm) above lift cover surface. Use Churchill gage tool FT.8528 or Nuday tool 4655 (2) to check setting. Loosen locknut (2 – Fig. 401) and rotate eccentric shaft (3) until control valve pin (1 – Fig. 397) extends 0.19 inch (4.83 mm) beyond control valve bushing (3). Use Chruchill gage tool FT.8527 or Nuday tool 4654 (2) to check dimension, then tighten locknut to 28 ft.-lbs. (38 N·m) torque.

345. ADJUST LOAD MONITOR LINKAGE. After performing steps outlined in paragraph 342, 343 and 344, proceed as follows:

Lower lift arms to full down position. Align upper edge of lift control lever (5 – Fig. 409) with Load Monitor setting notch (3) on quadrant. Position selector lever in Load Monitor (seventh) control setting (6) in selector quadrant. Top of control rod roller (4 – Fig. 406) should be 0.11 inch (2.79 mm) above lift cover surface. Use Churchill gage tool FT.8528 or Nuday tool 4655 (2) to check setting.

Loosen locknut (6 – Fig. 397) and adjust turnbuckle (5) until control valve pin (1) extends 0.19 inch (4.83 mm) beyond control valve bushing (3). Use Churchill gage tool FT.8527 or Nuday tool 4654

Fig. 408 – Rotate screw (S) to adjust Load Monitor fork on Models 5610, 6610, 6710, 7610 and 7710 as outlined in text.

Fig. 408A – Refer to paragraph 340 and adjust Load Monitor as outlined by turning screw (W).

(2) to check dimension, then tighten locknut (6) to 10 ft.-lbs. (14 N·m) torque.

Reinstall control valve front plate and tighten retaining cap screws to 34 ft.-lbs. (46 N·m) torque. Remove clamp and Churchill tool FT.8529 or Nuday tool 4656 (3–Fig. 402) from lift cover.

346. **ADJUST LOAD MONITOR FORK.** Refer to paragraph 338 and adjust Load Monitor fork as outlined.

Models 6710-7710 With Load Monitor

347. **ADJUST DRAFT CONTROL MAIN SPRING.** Refer to paragraph 323 and adjust draft control main spring as outlined.

348. **LIFT COVER LINKAGE.** With lift cover and cylinder assembly removed from tractor as outlined in paragraph 356, proceed as follows:

Measure length of strut (4–Fig. 401) between center of mounting holes, recommended length is 8.92 inch (227 mm). Adjust length as needed. Loosen locknut (2) and rotate eccentric shaft (3) upward until shaft is closest to roof of lift cover, then tighten nut. Position and clamp Churchill tool FT.8529 or Nuday tool 4656 (3–Fig. 402) on lift cover and place strut (1) on tool.

NOTE: Strut must not move from position while the following adjustments are being performed. Adjustments must be performed in sequence outlined as follows:

349. **ADJUST POSITION CONTROL INTERNAL LINKAGE.** After performing steps outlined in paragraph 348, proceed as follows:

Place lift arms in fully raised position. Align notch on body (3–Fig. 400) with position control notch on support assembly (8). Loosen securing cap screw (9). Ensure control rod roller is in contact with selector linkage assembly, then rotate actuator (5) until bottom of control rod roller (4–Fig. 404) is 0.72 inch (18.3 mm) above lift cover surface. Use Churchill gage tool FT.8528 or Nuday tool 4655 (2) to check setting. Tighten actuator securing cap screw to 13 ft.-lbs. (18 N·m) torque.

Lower lift arms to full down position. Remove cap screws securing control valve front plate (30–Fig. 426).

NOTE: Do not allow drop poppet valve (26) and spring (27) to fall free from lift cylinder housing.

Loosen locknut (6–Fig. 397) and adjust turnbuckle (5) until control valve pin (1) extends 0.19 inch (4.83 mm) beyond control valve bushing (3). Use Churchill gage tool FT.8527 or Nuday tool 4654 (2) to check dimension, then tighten

locknut (6) to 10 ft.-lbs. (14 N·m) torque. Scribe an identification mark on actuator (5–Fig. 400) and support assembly to identify setting.

350. **ADJUST DRAFT CONTROL INTERNAL LINKAGE.** After performing steps outlined in paragraph 348 and 349, proceed as follows:

Lower lift arms to full down position. Align notch on body (3–Fig. 400) with draft control notch on support assembly (7). Loosen securing cap screw (9) and rotate actuator (5) until draft control mark is aligned with scribe mark on support assembly and bottom of control rod roller (4–Fig. 405) is 2.10 inch (53.3 mm) above lift cover surface. Use Churchill gage tool FT.8528 or Nuday tool 4655 (2) to check setting. Tighten actuator securing cap screw to 13 ft.-lbs. (18 N·m) torque. Loosen locknut (2–Fig. 401) and rotate eccentric shaft (3) until control valve pin (1–Fig. 397) extends 0.19 inch (4.83 mm) beyond control valve bushing (3). Use Churchill gage tool FT.8527 or Nuday tool 4654 (2) to check dimension, then tighten locknut to 28 ft.-lbs. (38 N·m) torque.

351. **ADJUST LOAD MONITOR LINKAGE.** After performing steps outlined in paragraph 348, 349 and 350, proceed as follows:

Lower lift arms to full down position. Align notch on body (3–Fig. 400) with Load Monitor notch on support assembly (7). Loosen securing cap screw (9) and rotate actuator (5) until Load Monitor mark is aligned with scribe mark on support assembly and top of control rod roller (4–Fig. 406) should be 0.11 inch (2.79 mm) above lift cover surface. Use Churchill gage tool FT.8528 or Nuday tool 4655 (2) to check setting.

Loosen locknut (6–Fig. 397) and adjust turnbuckle (5) until control valve pin (1) extends 0.19 inch (4.83 mm) beyond control valve bushing (3). Use Churchill gage tool FT.8527 or Nuday tool 4654 (2) to check dimension, then tighten locknut (6) to 10 ft.-lbs. (14 N·m) torque.

Reinstall control valve front plate and tighten retaining cap screws to 34 ft.-lbs. (46 N·m) torque. Remove clamp and withdraw Churchill tool FT.8529 or Nuday tool 4656 (3–Fig. 402) from lift cover.

352. **ADJUST SELECTOR LEVER.** Adjust selector control rod length to where selector lever engages position control setting in quadrant when actuator arm on lift cover is placed in position control setting.

353. **ADJUST LOAD MONITOR FORK.** Refer to paragraph 338 and adjust Load Monitor fork as outlined.

LIFT COVER AND CYLINDER

Models 5000-5600-6600-6700-7000-7600-7700 Without Load Monitor

354. **R&R LIFT COVER AND CYLINDER.** To remove lift cover and cylinder assembly, first be sure all oil is exhausted from lift cylinder by placing selector lever in draft control position, pushing selector valve inward, moving lift control lever to bottom of quadrant and allowing lift arms to move to lowest position.

On all 6700 and 7700 models and all other models equipped with a Ford cab, refer to paragraphs 254 and 255 to tilt cab or platform forward for access to lift cover.

Thoroughly clean lift cover assembly. Remove top link and clevis pin securing draft control main spring yoke to rocker and swing rocker away from yoke. Remove clevis pins from lift arms to disconnect lift linkage. On Models 6700 and 7700, detach cross-shaped coupler on right side of lift cover and withdraw lift lever shaft. Remove seat assembly on models so equipped.

NOTE: Identify bolts by position as different lengths are used.

Attach a hoist to lift cover and raise lift cover assembly straight up until return oil tube and filter is clear of rear axle housing, then remove tube and filter assembly from lift cylinder. Move cover away from trctor and place it on supports with lower side (cylinder and linkage) up.

To reinstall lift cover assembly, proceed as follows: Position new lift cover gasket on rear axle center housing. Attach hoist to lift cover and suspend assembly over rear axle center housing. Install return oil tube and filter assembly, tightening retaining bolt finger tight. Lower assembly to about five inches (127 mm) from center housing and position return oil tube and filter for maximum clearance. Raise lift cover far enough to tighten tube retaining cap screw to a torque of 14-17 ft.-lbs. (19-23 N·m). Take care not to move tube from position while tightening bolt. Lower assembly to rear axle housing, install retaining cap screws and tighten to a torque of 55-75 ft.-lbs. (75-102 N·m). Adjust main draft control spring as outlined in paragraph 294 and reconnect linkage. Reinstall seat assembly or reposition cab or platform.

Models 5000-5600-6600-6700-7000-7600-7700 With Load Monitor

355. **R&R LIFT COVER AND CYLINDER.** To remove lift cover and cyl-

inder assembly, first be sure all oil is exhausted from lift cylinder by placing selector lever in draft control position, pushing selector valve inward, moving lift control lever to bottom of quadrant and allowing lift arms to move to lowest position.

On all 6700 and 7700 models and all other models equipped with a Ford cab, refer to paragraphs 254 and 255 to tilt cab or platform forward for access to lift cover.

Thoroughly clean lift cover assembly. Remove top link and clevis pin securing draft control mainspring yoke to rocker and swing rocker away from lift yoke. Remove clevis pins from lift arms to disconnect lift linkage. On Models 6700 and 7700, detach cross-shaped coupler on right side of lift cover and withdraw lift lever shaft. Remove seat assembly on models so equipped. Clean area around oil filter manifold, refer to Fig. 410 and to Fig. 371 (Models 5600, 6600, 6700, 7600 and 7700), and remove cap nut which is located in front of oil filter. Then, use a hex wrench to remove sleeve (S) from manifold.

NOTE: Failure to remove sleeve will result in damage to sleeve and/or port of lift cylinder.

Remove cap screws retaining lift cover to rear axle center housing.

NOTE: Identify bolts as to position as different lengths are used.

Attach a suitable hoist, raise lift cover approximately 3½ inches (89 mm) above rear axle center housing and disconnect Load Monitor strut from selector linkage assembly. Complete removal of cover from tractor and place it on supports with lower side (cylinder and linkage) upwards. Be sure assembly is not resting on any controls.

To reinstall lift cover assembly, proceed as follows: Position new lift cover gasket on rear axle center housing and install a new "O" ring in the bore near

right front corner of lift cover mounting surface. Position lift cover over rear axle center housing, lower cover until it is approximately 3½ inches (89 mm) above axle housing, then connect Load Monitor strut to selector linkage assembly. Carefully lower cover on rear axle housing, install cover retaining cap screws but delay tightening until sleeve is installed in oil filter manifold. Install sleeve in oil filter manifold, then install cap nut on sleeve. Now tighten cover retaining cap screws to 95-115 ft.-lbs. (129-156 N·m).

Adjust main draft control spring as outlined in paragraph 297, if necessary, then reconnect linkage and reinstall seat assembly or reposition cab or platform.

All Models 5610-6610-6710-7610-7710

356. R&R LIFT COVER AND CYLINDER. Lower lift arms to complete down position. Remove top link and detach lift linkage from lift arms by removing cotter pin and clevis pin from each arm. Withdraw spring clips, then remove clevis pin securing draft control main spring yoke to rocker and rotate rocker away from yoke. Disconnect and remove remote control valves, support brackets and lines.

NOTE: On models equipped with trailer brake, remove two cap screws retaining brake coupling support bracket and pipe.

Remove two cap screws retaining oil cooler line to front of lift cover, then remove remaining cap screw retaining auxiliary services selector valve actuating lever and withdraw lever assembly.

NOTE: On models equipped with an auxiliary pump, disconnect lines from priority valve pack. Use care when remote valve feed line is disconnected as check valve spring and stop may be discharged.

Cover exposed line openings and ports. Remove cap screw retaining rear

transmission dipstick support bracket to lift cover.

NOTE: On models equipped with trailer brake, disconnect feed tube at trailer brake valve. Remove cap screw securing trailer brake reservoir support bracket to lift cover, then place reservoir clear of lift cover. Remove cap screws retaining trailer brake valve to support bracket. Lift support bracket clear of cover.

Remove any other hoses from lift cover that will prevent its removal. On Models 5610, 6610 and 7610 without a cab, remove cap screws securing seat and withdraw seat. Tilt cab or platform forward for access to lift cover as outlined in paragraph 254 or 255. On Models 6710 and 7710, detach cross-shaped coupler on right side of lift cover and withdraw lift lever shaft. Thoroughly clean lift cover assembly. Remove fifteen cap screws securing lift cover to center housing. Attach a hoist to lift cover and raise lift cover assembly straight up.

NOTE: On models equipped with Load Monitor, raise lift cover approximately 3½ inches (89 mm) above rear axle center housing and disconnect Load Monitor strut from selector linkage assembly.

Complete removal of cover from tractor.

NOTE: Use care not to damage lift pressure tube.

Place lift cover on supports with lower side (cylinder and linkage) upwards. Be sure assembly is not resting on any controls.

To reinstall lift cover assembly, proceed in reverse order of removal. Install new "O" rings and gasket, position lift pressure tube in top face of center housing. Tighten lift cover retaining cap screws to 112 ft.-lbs. (152 N·m).

Models 5000-5600-6600-6700-7000-7600-7700

357. R&R LIFT CYLINDER. With the lift cover removed as outlined in paragraph 354 or 355, proceed as follows:

Remove piston connecting rod as shown in Fig. 411 and control valve turnbuckle assembly as shown in Fig. 412. Move selector lever to gain adequate clearance to unscrew override plunger as shown in Fig. 413, then disconnect position control link and remove override plunger and flow control override spool as shown in Fig. 414.

NOTE: The spring located in bore behind override spool should also be removed at this time.

Pull selector valve to "out" position. Remove screw from knob (23 – Fig. 415).

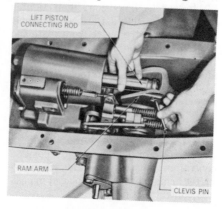

Fig. 410 — Sleeve (S) located under cap nut must be removed prior to removing lift cover assembly. Refer to text.

Fig. 411 — On models without Load Monitor remove lift cylinder connecting rod as shown. On models with Load Monitor the connecting rod is retained by two retaining pins.

Pull knob from valve stem (21), then remove stem from valve spool and bore in lift cover. Remove flow control valve knob (31), stem (28) and retainer (30) by unscrewing retainer from lift cover. Remove cap screws retaining accessory cover (16) and remove cover. Remove the four additional cap screws retaining cylinder to lift cover and remove cylinder assembly as shown in Fig. 416. Refer to paragraph 358 for cylinder overhaul.

When reinstalling lift cylinder, place new sealing "O" rings in upper face as shown in Fig. 417, then install cylinder to lift cover tightening retaining cap screws to a torque of 110-130 ft.-lbs. (150-177 N·m). Models 5600, 6600, 6700, 7600 and 7700 require 165-200 ft.-lbs. (224-272 N·m). Install accessory cover with new "O" rings and tighten retaining cap screws to a torque of 25-30 ft.-lbs. (34-41 N·m). Models 5600, 6600, 6700, 7600 and 7700 call for 42-56 ft.-lbs. (57-76 N·m). Complete reinstallation procedure by reversing removal procedure and reinstall lift cover assembly as outlined in paragraph 354 or 355.

NOTE: If installing new override valve, be sure to install valve having correct color code; refer to paragraph 365.

358. OVERHAUL LIFT CYLINDER. To service lift cylinder components, refer to the following appropriate paragraphs 359 through 367. An exploded view of the lift cylinder assembly is shown in Fig. 418.

359. LIFT CYLINDER SAFETY VALVE. To protect lift cylinder from excessive hydraulic pressure due to shock loads imposed by rear mounted implements, a safety relief valve (29 –

Fig. 418) is threaded into lift cylinder. Safety valve will open when pressure within cylinder reaches 2750 to 2850 psi (18.97-19.66 MPa). Models 5600, 6600, 6700, 7600 and 7700 have a safety valve opening pressure of 2850-2950 psi (19.66-20.35 MPa).

As opening pressure of safety valve is above hydraulic system relief pressure, valve can be bench tested only by suitable connectors to a hand pump and a high pressure hydraulic gage. If test equipment is not available and condition of safety valve is questionable, renew valve.

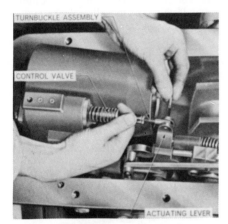

Fig. 412 — Removing turnbuckle assembly from actuating lever and control valve spool.

Fig. 416 — Removing lift cylinder assembly from lift cover.

Fig. 413 — Unscrewing flow control valve override plunger assembly.

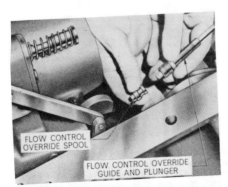

Fig. 414 — Flow control valve override plunger assembly and override control spool removed from lift cylinder.

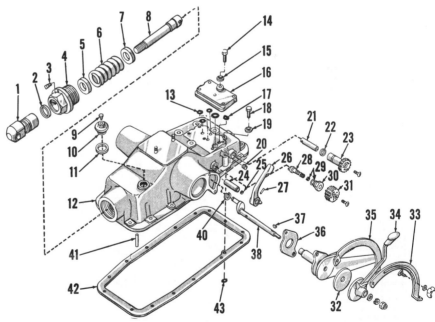

Fig. 415 — Exploded view of Model 5000 hydraulic lift cover assembly without Load Monitor. Pin (41) (models without Load Monitor) prevents draft control plunger (8) from turning with yoke (1) and can be driven from lift cover casting after removing filler plug (10).

1. Yoke	12. Lift cover	23. Selector valve knob	33. Quadrant
2. Seal ring	13. "O" rings	24. Sleeve	34. Lift control lever
3. Set screw	14. Plug	25. "O" ring	35. Support
4. Retainer nut	15. Seal ring	26. Selector lever	36. Gasket
5. Rear spring seat	16. Accessory cover	27. Pin	37. Woodruff key
6. Main spring	17. "O" rings	28. Flow control valve stem	38. Control lever shaft
7. Front spring seat	18. Plug	29. Seal rings	40. Snap ring
8. Plunger	19. Seal ring	30. Valve stem retainer	41. Pin
9. Breather	20. Plug	31. Valve knob	42. Gasket
10. Filler pump	21. Selector valve rod	32. Friction disc	43. "O" ring
11. Gasket	22. Seal ring		

NOTE: DO NOT attempt to disassemble or adjust the safety valve assembly.

Install safety valve using a new seal (28) and tighten valve assembly to a torque of 75-90 ft.-lbs. (102-122 N·m). Excessive tightening torque may distort valve body resulting in faulty valve operation.

360. PISTON AND SEALS. To remove piston, first remove cylinder safety valve as in paragraph 359, then insert a small rod through safety valve opening

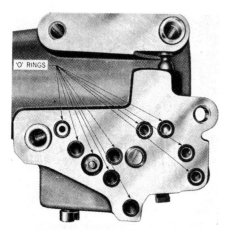

Fig. 417 — View showing "O" rings inserted in recesses in top face of lift cylinder assembly.

and push piston from rear end of cylinder.

Inspect piston (26 – Fig. 418) and cylinder bore for scoring or excessive wear and inspect cylinder casting for cracks. If cylinder bore is excessively scored, cylinder assembly must be renewed. Renew piston if scored.

To install piston seals, soak leather back-up ring in water for about two minutes, then install back-up ring in groove on piston with rough side of leather toward front (closed) end of piston. Lubricate "O" ring in hydraulic fluid and install it in groove at front side of back-up ring. After allowing back-up ring to shrink to original size, lubricate piston and cylinder and install piston in cylinder bore.

NOTE: A nylon back-up ring may be used. If so, dip it in hydraulic oil prior to installation.

361. SELECTOR VALVE. To remove selector valve (36 – Fig. 418), push valve inward and place a suitable size "O" ring on inner end of valve as shown in Fig. 421. The detent ball (34 – Fig. 418) will then ride over the inner end of valve as valve is withdrawn from bore in cylinder casting. Catch detent ball and spring (35) as valve is removed.

Carefully inspect selector valve spool and valve bore in cylinder casting. Re-

new cylinder if bore is excessively scored, worn or is otherwise damaged to extent that leakage would occur between valve bore and lands on valve spool. Renew valve spool if spool lands are scored or excessively worn, or if valve fits loosely in bore.

The selector valve spool can be renewed by selective fit only. To check fit of new valve, lubricate valve and insert in normal position in valve bore. A slight drag should be felt when valve is moved in its normal range of travel. If valve sticks or binds, select a smaller diameter valve; if valve fits loosely, select a larger valve. The selector valve is available in

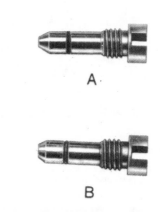

Fig. 419 — Comparison view of old-style (A) and up-dated (B) flow control valve used in hydraulic lift cylinders of Models 5600, 6600, 6700, 7600 and 7700. New valve and spring may be used to service prior models.

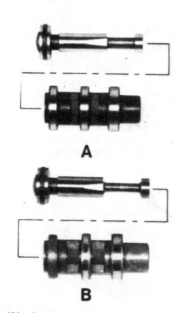

Fig. 420 — Comparison views of previous unloading valve and rear bushing (A) and current (B) type used in lift cylinders of Models 5600, 6600, 6700, 7600 and 7700. Newer style unloading valve and rear bushing may be fitted to prior models but DO NOT interchange new unloading valve plug (31 — Fig. 418) in older models.

Fig. 418 — Exploded view of lift cylinder assembly used on Models 5600, 6600, 6700, 7600 and 7700. It is typical of that used on earlier models. Sleeve (21A) is not used on earlier models. Service details in text.

1. Flow control valve
2. "O" ring
3. Retainer
4. "O" ring
5. Valve seat
6. Valve spring
7. "O" ring
8. Check valve retainer
9. Valve spring
10. Check valve ball
11. Valve seat
12. "O" ring
13. Retaining ring
14. By-pass valve
15. Valve spring
16. Hollow dowels (2)
17. Valve seal
18. Unloading valve
19. Rear bushing
20. Retainer plate
21. Control valve spring
21A. Control valve sleeve
22. Control valve
23. Valve bushing
24. Back-up ring
25. "O" ring
26. Lift piston
27. Lift cylinder
28. Valve seal
29. Safety valve
30. Front bushing
31. Unloading valve plug
32. Retainer plate
33. Plug
34. Detent ball
35. Spring
36. Lift selector valve
37. Pipe plug (¼-in.)
38. Pipe plug (⅛ socket)
39. Pipe plug (1/16-in.)

40. Ball (2)*
41. Block plug*
42. "O" ring (1)*
43. "O" ring (1)
44. "O" ring (2)
45. "O" ring (7)

46. "O" ring (1)
47. "O" ring (2)
48. "O" ring (1)
49. "O" ring (3)*

*Models w/o hydraulics

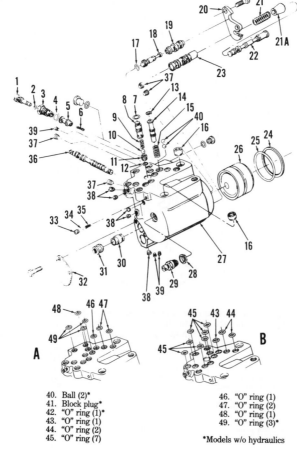

five different size ranges and each size range is color coded as follows:

Size Range	Color Code
0.6232-0.6235 in.	Blue/White
(15.829-15.837 mm)	
0.6235-0.6238 in.	White
(15.837-15.845 mm)	
0.6238-0.6241 in.	Blue
(15.845-15.852 mm)	
0.6241-0.6244 in.	Yellow
(15.852-15.860 mm)	
0.6244-0.6247	Green
(15.860-15.867 mm)	

NOTE: As color code indicates a size range only, a valve spool of one color code may fit correctly while other valves of same color code may fit too tight or too loose.

To install selector valve spool, first insert detent spring and detent ball in their bore and depress detent ball with small diameter rod as shown in Fig. 422; then install valve spool in bore to one of the detent positions.

362. CHECK VALVE AND SEAT. With lift cylinder removed as outlined in paragraph 357, check valve retainer (8 – Fig. 418) can be removed by hooking a piece of wire around groove in outer end of retainer and pulling retainer from bore. Remove spring (9) and check valve ball (10), then remove check valve seat (11) and "O" ring (12) with wire hook.

Carefully inspect spring, check valve ball and seat for wear, cracks, distortion or other damage and renew if any defect is noted. Install check valve with new "O" ring as follows:

Install new "O" ring (12) on check valve seat, lubricate "O" ring and push seat into bore with chamfered inside diameter out. Insert check ball and spring in bore and install new "O" ring (7) on retainer. Lubricate "O" ring and push retainer into bore over check valve spring.

363. FLOW CONTROL BY-PASS VALVE. With lift cylinder assembly removed as outlined in paragraph 357, remove flow control by-pass valve retaining ring (13 – Fig. 418); spring (15) should then push valve from bore. If valve is stuck in bore, it can usually be removed by pushing valve down against spring pressure, then quickly releasing it so it will pop out of bore.

Inspect valve spring for distortion, cracks or other damage and renew if defect is noted. Carefully inspect valve spool and bore in cylinder casting for scoring, excessive wear or other defect; renew cylinder if bore is excessively scored. Renew valve spool if spool lands are scored or excessively worn, or if valve fits loosely in bore, renew valve.

The flow control by-pass valve can be renewed by selective fit only. To check

fit of new valve, place spring in bore, lubricate valve and insert in bore on top of spring. The valve should fit snugly, but when valve is depressed and released, spring should quickly return valve. If valve sticks or binds, select a smaller valve; if valve fits loosely, select a larger diameter valve. The flow control by-pass valve is available in five different size ranges and each size range is color coded as follows:

Size Range	Color Code
0.6670-0.6672 in.	Blue/White
(16.942-16.947 mm)	
0.6672-0.6674 in	White
(16.947-16.952 mm)	
0.6674-0.6676 in.	Blue
(16.952-16.957 mm)	
0.6676-0.6678 in.	Yellow
(16.957-16.962 mm)	
0.6678-0.6680 in.	Green
(16.962-16.967 mm)	

NOTE: As the color code indicates a size range only, a valve spool of one color code may fit correctly while other valves of same color code may fit too tight or too loose.

To install flow control by-pass valve, insert spring and spool in bore with small diameter tip of valve out, then hold valve depressed in bore while installing retaining clip.

364. FLOW CONTROL VALVE. With lift cylinder remove as outlined in paragraph 357, remove flow control valve (1 – Fig. 418) by unscrewing valve from retainer (3) and then remove "O" ring (2) from valve stem. Unscrew retainer from cylinder casting and remove "O" ring (4) from retainer. Remove flow control valve seat (5) and flow control spring (6) with a wire hook.

Carefully inspect valve seat and valve seat bore for scoring or excessive wear. Cylinder must be renewed if bore is excessively scored or otherwise damaged. Renew spring if cracked, distorted or if free length is less than that of a new spring.

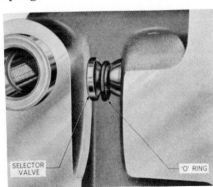

Fig. 421 – Place an "O" ring on inner end of selector valve to allow valve to be withdrawn past detent ball (34 – Fig. 418).

Flow control valve seat can be renewed by selective fit only. To check fit of new valve seat, place spring in bore, lubricate seat and insert in bore on top of spring. Seat should fit snugly in bore, but when seat is depressed against spring pressure and released, spring should quickly return seat. If seat sticks or binds, select a seat of smaller diameter; if seat fits loosely, select a larger diameter seat. Flow control valve seat is available in five different size ranges and each size range is color coded as follows:

Size Range	Color Code
0.7482-0.7485 in.	Blue/White
(19.004-19.012 mm)	
0.7485-0.7488 in.	White
(19.012-19.019 mm)	
0.7488-0.7491 in.	Blue
(19.019-19.027 mm)	
0.7491-0.7494 in.	Yellow
(19.027-19.035 mm)	
0.7494-0.7497 in.	Green
(19.035-19.042 mm)	

NOTE: As the color code indicates a size range only, a seat of one color code may fit correctly while other seats of same color code may fit too tight or too loose.

To install flow control valve, install new "O" rings on valve and seat, lubricate "O" rings and thread valve into retainer. Place spring in bore, lubricate seat and insert seat in bore with end having land flush with seat end out. Thread retainer and valve assembly into cylinder and tighten retainer snugly.

365. FLOW CONTROL OVERRIDE VALVE. Flow control override valve is removed during lift cylinder removal from cover as outlined in paragraph 357. Refer to exploded view of hydraulic lift shaft and linkage in Fig. 435 or 436.

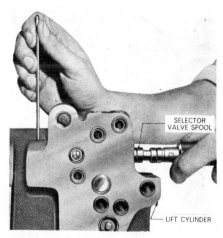

Fig. 422 – Hold detent ball down with small rod while inserting selector valve spool in lift cylinder.

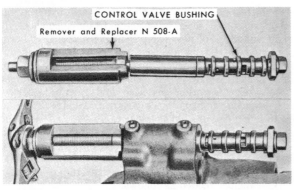

CONTROL VALVE BUSHING

Remover and Replacer N 508-A

Fig. 423 — Views showing use of valve bushing removal and installation tool (Nuday tool N-508-A is shown; tool 2191 is similar). Earlier model control valve bushing is shown.

Size Range	Color Code
1.0000-1.0002 in.	Blue/White
(25.400-25.405 mm)	
1.0002-1.0004 in.	White
(25.405-25.410 mm)	
1.0004-1.0006 in.	Blue
(25.410-25.415 mm)	
1.0006-1.0008 in.	Yellow
(25.415-25.420 mm)	
1.0008-1.0010 in.	Green
(25.420-25.425 mm)	
1.0010-1.0012 in.	Orange
(25.425-25.430 mm)	
1.0012-1.0014 in.	Green/White
(25.432-25.435 mm)	
1.0014-1.0016 in.	Red/White
(25.435-25.440 mm)	

Carefully inspect override valve (66 and 36) and valve bore in lift cylinder for scoring or excessive wear. Lift cylinder must be renewed if bore is excessively scored or otherwise damaged which would result in improper valve operation. Renew spring (65 or 37) if cracked or distorted, or if free length is less than that of a new spring. Override valve can be renewed by selective fit only. To check fit of new valve, place spring in bore, lubricate valve and insert in bore on top of spring. Valve should fit snugly in bore, but without sticking or binding. When valve is depressed against spring and released, spring should quickly return valve. If valve sticks or binds, select a smaller diameter valve; if valve fits loosely, select a larger valve. Override valve is available in five different size ranges and each size range is color coded as follows:

Size Range	Color Code
0.4995-0.4998 in.	Blue/White
(12.687-12.695 mm)	
0.4998-0.5001 in.	White
(12.695-12.702 mm)	
0.5001-0.5004 in.	Blue
(12.702-12.710 mm)	
0.5004-0.5007 in.	Yellow
(12.710-12.717 mm)	
0.5007-0.5010 in.	Green
(12.717-12.725 mm)	

NOTE: As the color code indicates a size range only, a valve of one color code may fit correctly while other valves of same color code may fit too tight or too loose.

Install override valve, spring and plunger during cylinder installation procedure, refer to paragraph 357.

366. CONTROL VALVE AND BUSHING. To service control valve, first remove lift cylinder as outlined in paragraph 357 and proceed as follows:

Unbolt and remove retainer plate (20 – Fig. 418) and remove valve (22) and spring (21) from plate, then slide spring from valve. On late models remove sleeve from valve. Then, remove baffle plate (32) from front end of cylinder and remove bushing (23) by carefully press-ing bushing out towards front (closed) end of lift cylinder.

NOTE: Special care should be used in selecting a sleeve to push bushing from lift cylinder casting to be sure that the bore in lift cylinder is not damaged in any way; if bore is scored during bushing removal, cylinder must be renewed. When available, use of special bushing removal and installation tool (Nuday tool N-508-A) is recommended. Refer to Fig. 423.

Inspect lands on control valve spool for erosion or scoring and renew valve spool and bushing if either are damaged in any way, or if troubleshooting checks indicated leakage at control valve. Neither valve nor bushing should be renewed without renewing mating part.

Control valve bushing is available in eight different outside diameter size ranges; bushing is color coded to indicate size range and lift cylinder casting is also color coded near bushing bore. Always renew bushing with one of the same size range. Bushing outside diameter size ranges and color codes are as follows:

When installing new bushing, be sure that bore in lift cylinder and bushing are clean and free of nicks or burrs. Lubricate both bushing and bushing bore and insert end of bushing having oil passage hole nearest to end in the rear (open) end of cylinder. Press bushing forward until front face of bushing is flush with front machined face of lift cylinder casting.

Control valve spools are available in five different size ranges. The correct size valve can be determined by selective fit only **after** bushing has been pressed into cylinder casting. To check valve spool fit, lubricate valve spool and bushing and insert spool in bushing from open (rear) end of cylinder. A drag should be felt on valve when moving it in bushing through normal range of travel. If valve moves freely through bushing, select a larger diameter valve spool; if valve sticks or binds, select a smaller diameter valve spool. Valve spool finally selected should be the largest diameter that will allow valve to slide through

Fig. 424 — Exploded view of unloading valve and bushings and control valve and bushing showing relative valve and bushing location when installed in lift cylinder.

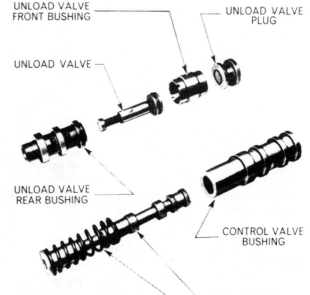

UNLOAD VALVE FRONT BUSHING

UNLOAD VALVE PLUG

UNLOAD VALVE

UNLOAD VALVE REAR BUSHING

CONTROL VALVE BUSHING

CONTROL VALVE SPRING

CONTROL VALVE

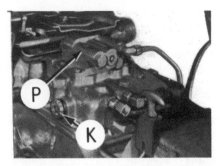

Fig. 424A — View showing location of selector valve knob (K) and priority valve (P).

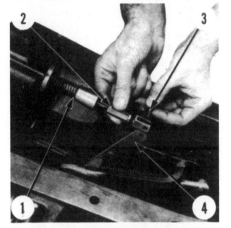

Fig. 425 — Remove clevis pin (3) to separate control valve turnbuckle assembly (2) from actuating lever (4). Control valve assembly (1).

Fig. 426 — Exploded view of lift cylinder assembly used on Models 5610, 6610, 6710, 7610 and 7710.

1. Housing
2. Rear cover plate
3. Pin
4. Control valve
5. Spring
6. Sleeve
7. Cap screw
8. Selector valve
9. Ring dowel
10. "O" ring
11. "O" ring
12. "O" ring
13. Bushing
14. Bushing
15. Spring
16. Ball
17. Spring
18. "O" ring
19. Set screw
20. Ring
21. Drop restrictor valve
22. Spring
23. Spacer
24. Retainer
25. Diffuser
26. Drop poppet valve
27. Spring
28. "O" ring
29. Plug
30. Front plate
31. "O" ring
32. Safety valve
33. Piston
34. "O" ring
35. Back-up ring

bushing from its own weight. The different valve size ranges are color coded as follows:

Size Range	Color Code
0.5917-0.5919 in.	White
(15.028-15.034 mm)	
0.5919-0.5921 in.	Blue
(15.034-15.039 mm)	
0.5921-0.5923 in.	Yellow
(15.039-15.044 mm)	
0.5925-0.5926 in.	Green
(15.049-15.050 mm)	
0.5927-0.5928 in.	Orange
(15.055-15.057 mm)	

NOTE: As the color code indicates a size range only, a valve of one color code may fit correctly while other valves of same color code may fit too tight or too loose. As the color code on bushing indicates outside diameter only, do not attempt to select a control valve spool by matching spool color code to bushing color code; however, it may happen that valve spool selected and bushing will be of the same color code.

To install control valve, slide spring over valve and fit valve into notch in retainer plate. Lubricate valve and spool and insert valve. Tighten retainer plate cap screws to a torque of 25-30 ft.-lbs. (34-41 N·m). Do not reinstall front (baffle) plate until lift cylinder has been reinstalled in cover and control linkage

is adjusted as outlined in paragraphs 295 and 296; 300 through 302; 311 and 312; or 317 through 319.

367. UNLOADING VALVE, BUSHINGS AND PLUG. To remove unloading valve, first remove control valve as outlined in paragraph 366; it is not necessary to remove control valve bushing if renewal is not indicated. Then, proceed as follows:

Thread a slide hammer adapter into unloading valve plug (31 – Fig. 418) and pull plug from valve bore with slide hammer. Unloading valve (18) can then be removed by pushing valve forward with a small screwdriver inserted through rear bushing (19). Remove "O" ring (17). Remove unloading valve bushings if renewal is indicated by pressing them out towards front (closed) end of cylinder with a suitable sleeve.

NOTE: Special care should be taken in selecting sleeve used to press bushings from cylinder casting to be sure not to damage bore. When available, use of special bushing removal and installation tool (Nuday tool N-508-A) is recommended.

Unloading valve bushings (19 and 30) are available in eight different outside diameter size ranges and should be renewed using parts of the same size range. It is usually not necessary to renew unloading valve plug (31) unless damaged during removal procedure. Plug is also available in eight different size ranges and if plug, or a new plug of the same size range fits loosely in bore, a plug of the next larger size range should be installed. Unloading valve bushing

and plug outside diameter size ranges and color codes for the different size ranges are the same as listed for control valve bushing in paragraph 366.

When installing new bushings, be sure that bore in lift cylinder is clean and free of nicks or burrs. Lubricate bushings and bushing bore. The front (large ID) bushing has one notch in one end and two notches in the other end; insert this bushing in rear (open) end of cylinder casting with end having single notch forward. Place rear bushing against rear (double notch) end of front bushing with long nose to rear. Refer to Fig. 424 for relative bushing position if necessary. Press both bushings forward into bore until rear side of land on rear bushing is flush with machined surface at rear side of lift cylinder.

Renew unloading valve (18 – Fig. 418) if scored, excessively worn or otherwise damaged. Valve is available in one size only. Lubricate valve and insert it in bushings without "O" ring (17). Valve should be a free sliding fit. Be sure that front and rear bushing bores are concentric by checking clearance between unloading valve and inside diameter of front bushing at several different points. Remove valve and install "O" ring, lubricate valve and "O" ring and reinstall in bushings. The "O" ring should impart a slight drag when moving valve back and forth in bushings. If valve sticks or binds, or moves freely as without "O" ring being installed, install a different "O" ring as there may be slight differences in "O" ring sizes.

CAUTION: Do not install an "O" ring of unknown quality or composition at this location. Some "O" ring materials may shrink or swell when subjected to hydraulic fluid and heat and therefore cause malfunction of unloading valve.

With unloading valve and "O" ring installed in bushings, install plug with threaded hole out and outer face of plug flush with machined front surface of lift cylinder.

Models 5610-6610-6710-7610-7710

368. **R&R LIFT CYLINDER.** With hydraulic lift cover removed as outlined in paragraph 356, proceed as follows:

Locate selector lever in position control setting. On models so equipped, remove priority valve assembly (P–Fig. 424A). Remove cotter pin securing clevis pin (3–Fig. 425), then extract clevis pin and separate control valve turnbuckle assembly (2) from actuating lever (4). Pull selector valve spool out and unscrew valve knob (K–Fig. 424A) retaining screw. Remove knob from lift cover, then extract selector valve stem through bore. Remove four cap screws securing lift cylinder to lift cover. Move lift arms to raised position, then separate lift cylinder from lift cover.

To reinstall lift cylinder assembly, proceed in reverse order of removal. Install new "O" rings and ensure two alignment ring dowels are correctly installed. Tighten lift cylinder securing cap screws to 182 ft.-lbs. (248 N·m) torque. Refer to appropriate paragraph for draft and position control linkage adjustment.

369. **OVERHAUL LIFT CYLINDER.** Remove safety valve (32–Fig. 426) and discard "O" ring (31).

NOTE: If safety valve (32) requires servicing, complete valve must be renewed.

Remove and note correct location of "O" rings (10, 11 and 12), then discard "O" rings. Remove cap screw (7) securing rear cover plate (2), then withdraw control valve assembly (4), spring (5) and cover plate (2). Separate control valve assembly from plate. Remove retainer (24), then withdraw components (20 through 23) and (25). Remove cap screws securing front plate (30). Note piston and size of "O" ring (28), then discard. Extract spring (27) and drop poppet valve (26).

Using Churchill tools–extension tool T.8510-1C, bushing remover and installer tool T.8510 and special nut tool T.8510-1C–or Nuday tool 2191 (Fig. 423), remove control valve bushing (13–Fig. 426). Remove set screw (19), spring (17) and ball (16), then withdraw selector valve (8). Push piston (33) from housing using a suitable rod inserted through safety valve bore.

Clean all components in a suitable sol-vent, then dry using a clean lint free cloth or blow dry using clean compressed air. Inspect all valves, bushings and housing bores for burrs, scoring, excessive wear or any other damage. Inspect for cracked, distorted or excessively worn springs. Ensure spring loaded pin (3–Fig. 426) located in control valve (4) moves freely and pin head is not excessively worn or damaged in any way. Check to be sure all valves slide freely in their related bushings. Ensure all oil passages and bushing ports flow freely and are clear of obstructions. Renew components as needed. Renew all "O" rings and seals.

NOTE: If cylinder housing bushing bores are heavily scored, a new lift cylinder assembly must be installed.

If control valve bushing is to be renewed, observe color marking located near control valve bushing bore. Control valve bushings are available in the following sizes:

Size Range	Color Code
1.0010-1.0012 in.	Orange
(25.425-25.431 mm)	
1.0008-1.0010 in.	Green
(25.420-25.425 mm)	
1.0006-1.0008 in.	Yellow
(25.415-25.420 mm)	
1.0004-1.0006 in.	Blue
(25.410-25.415 mm)	
1.0002-1.0004 in.	White
(25.405-25.410 mm)	

Use Churchill tools–guide and stop adapter tool T.8510-1C, extension tool T.8510-1C, bushing remover and replacer tool T.8510 and locator tool FT.8510-3–or Nuday bushing remover and replacer tool 2191 (Fig. 423) and locator tool 4661 to install control valve bushing (13–Fig. 426).

Select largest control valve spool that, when lightly lubricated and inserted into either end of bushing and rotated a complete revolution, will move freely without binding through entire length of bushing.

NOTE: Control valve is color-coded only for reference, obtain the best fit without binding regardless of color marking.

Control valve is available in five different size ranges and each size range is color-coded as follows:

Size Range	Color Code
0.5927-0.5928 in.	Orange
(15.055-15.057 mm)	
0.5925-0.5926 in.	Green
(15.050-15.052 mm)	
0.5921-0.5923 in.	Yellow
(15.039-15.044 mm)	
0.5919-0.5921 in.	Blue
(15.034-15.039 mm)	
0.5917-0.5919 in.	White
(15.029-15.034 mm)	

Complete reassembly in reverse order of disassembly. Lubricate all components during reassembly. Install new "O" rings and seals. If selector valve is to be renewed, valve is available in five different size ranges and each size range is color-coded as follows:

Size Range	Color Code
0.6244-0.6247 in.	Green
(15.8598-15.8674 mm)	
0.6241-0.6244 in.	Yellow
(15.8521-15.8598 mm)	
0.6238-0.6241 in.	Blue
(15.8445-15.8521 mm)	
0.6235-0.6238 in.	White
(15.8369-15.8445 mm)	
0.6232-0.6235 in.	Blue/White
(15.8293-15.8369 mm)	

Select valve that will give an optimum fit without binding throughout entire valve bore.

Tighten front plate or rear cover plate retaining cap screws to 34 ft.-lbs. (46 N·m) torque.

All Models

370. **R&R LOAD MONITOR.** To remove Load Monitor unit, split tractor between transmission and rear axle center housing as outlined in paragraphs 254, 255 or 256, then remove hydraulic lift cover as outlined in paragraph 355 or 356. See Fig. 429. Remove pto clutch and control valve assembly as outlined in paragraph 281. Remove hydraulic pump and screen (if so equipped). See Fig. 430. Remove brake pedals from brake cross shaft (C) and pull brake cross shaft from center housing and Load Monitor fork

Fig. 429 — Typical view showing location of Load Monitor unit and the components which must be removed, or disconnected prior to removal.

B. Retaining bolt
C. Brake cross shaft
L. Load Monitor unit
P. Pto clutch
1. Fork lever
3. Linkage adjustment fork
10. Load Monitor fork
11. Hydraulic pump
22. Screen

(10). See Fig. 431 for an exploded view of Load Monitor fork components.

Remove retaining bolt (B) and pull Load Monitor unit (L–Fig. 430) from bevel pinion shaft. On tractors with Select-O-Speed transmission, save any shims which may have been installed between Load Monitor unit and pinion shaft clamp nut.

371. LOAD MONITOR DISASSEMBLY. Use Figs. 432 and 433 as a guide and proceed as follows: Place Load Monitor unit in a press with input hub upward, then using a step plate, compress unit until retaining ring can be removed. With retaining ring removed, remove unit from press, then carefully lift out input hub which will remove plate and steel balls. Also, be careful not to lose drive pins and stop pins which will be free when plate clears coupling.

Reinstall unit in press with output hub upward, then using a suitable sleeve, compress unit until retaining ring can be removed. With retaining ring removed, remove unit from press and remove spring retainer. Place unit on a flat, cloth covered surface and slowly lift upward on output hub. As bottom of hub clears inside retaining ring, small steel balls will fall inside coupling onto cloth. Springs (Belleville washers) can now be removed from output hub.

372. To reassemble Load Monitor assembly, proceed as follows: Stand output hub on its pinion shaft end. Place a wood block on each side of output hub, then place coupling cover output hub so inside retaining ring clears upper surface of output hub enough to allow installation of small steel balls in their grooves. Vary height of blocks as required. Align ball grooves of coupling with ball grooves of output hub, then install five steel balls in each of the twelve grooves (total 60). Be sure none of the steel balls are installed in stop pin grooves. Remove wood blocks and allow coupling to move down which will retain steel balls in position. Be sure to maintain this positioning of coupling and output hub until retaining ring is installed so balls will not be displaced. Turn assembly so output hub is upward, then install the nine springs (Belleville washers) alternately as shown in Fig. 432.

NOTE: First and last spring will have cup facing up.

Place spring retainer over springs. Place unit in a press, then using a suitable sleeve, compress unit only enough to install retaining ring.

Remove unit from press, position it with input hub end upward and install thrust bearing on output hub. Put retaining washer in place, position input hub on output hub and be sure retaining washer fits into recess in input hub. Align stop pin grooves in input hub with stop pin grooves in coupling so remaining grooves are also aligned and install the three stop pins. Place the ten steel balls in their sockets (ramps), place plate over balls and install the twelve drive pins. Place unit in a press, use step plate and compress unit only enough to install retaining ring.

373. To install Load Monitor on tractors with manual transmission, place unit on bevel pinion shaft, install retain-

Fig. 430 — With pto clutch assembly and hydraulic pump removed, Load Monitor unit can now be removed.

B. Retaining bolt
C. Brake cross shaft
1. Fork lever

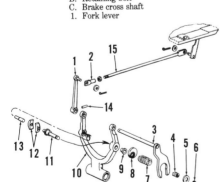

Fig. 431 — Exploded view showing Load Monitor fork and linkage assembly.

1. Lever
2. Linkage end
3. Adjustment fork
4. Inner cam spring
5. Spring seat
6. Spring retainer
7. Outer cam spring
8. Adjustment cam
9. Shaft seat
10. Fork
11. Adjustment shaft
12. Thrust blocks
13. Thrust block pin
14. Roll pin
15. Linkage strut

Fig. 432 — Cross-sectional view of the Load Monitor unit showing the relationship of component parts.

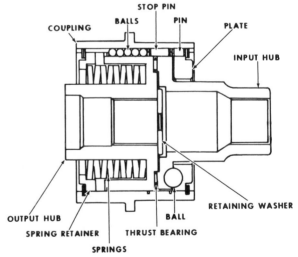

Fig. 433 — Exploded view of Load Monitor assembly to show parts order and arrangement. Refer to text.

1. Spring retainer (2)
2. Lift plate
3. Drive pin (12)
4. Ball, 0.625-in. (15.9 mm)-10
5. Input hub
6. Stop pin (2)
7. Coupling
8. Ball retainer ring
9. Pinion bolt (½-13)
10. Lockwasher
11. Retainer washer
12. Thrust bearing
13. Output hub

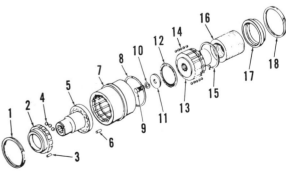

14. Balls, ¼-in. (6.35 mm)-60
15. Ball retainer ring
16. Spring
17. Spring seat
18. Retaining ring

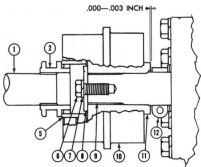

Fig. 434 — When installing Load Monitor unit on tractor with Select-O-Speed transmission, maintain clearance as shown between output hub (11) and pinion shaft clamp nut (12).

1. Transmission output shaft	8. Retaining washer
2. Disconnect coupling	9. Pinion shaft
5. Input hub	10. Coupling
6. Retaining bolt	11. Output hub
7. Lockwasher	12. Clamp nut

ing bolt and tighten bolt to a torque of 55-75 ft.-lbs. (75-102 N·m). To install Load Monitor on tractors with Select-O-Speed transmission, place unit on bevel pinion shaft so output hub protrudes slightly beyond end of bevel pinion shaft, then install retaining bolt and tighten to a torque of 55-75 ft.-lbs. (75-102 N·m). Now, refer to Fig. 434 and check distance between Load Monitor output hub and pinion shaft clamp nut. This distance should be 0.000-0.003 inch (0.00-0.08 mm). If clearance is not as stated,

vary shims on pinion shaft as required. Shims are available in thicknesses of 0.003, 0.004, 0.012, 0.032, 0.052 and 0.089 inch (0.08, 0.10, 0.30, 0.81, 1.32 and 2.26 mm).

Complete reassembly by reversing disassembly procedure.

OVERHAUL LIFT CONTROL LINKAGE, LIFT SHAFT AND COVER

374. **LIFT CONTROL LINKAGE.** With lift cylinder assembly removed from lift cover as outlined in paragraph 357 or 368, refer to Fig. 435 and paragraph 375 for information on models without Load Monitor except "10" series models. Refer to Fig. 436 and paragraph 376 for information on all models with Load Monitor (including "10" series) and "10" series models without Load Monitor.

375. MODELS WITHOUT LOAD MONITOR EXCEPT "10" SERIES. To overhaul lift control linkage, lift shaft and cover, proceed as follows:

Unbolt and remove control lever quadrant (33 – Fig. 415) from support (35), then remove control lever retaining nut, double-coil spring washer, flat washer, control lever (34), Woodruff key (37) and friction disc (32). Then, unbolt and remove support and control lever shaft

(38). Drive roll pin (27) from selector lever (26) and remove lever.

Refer to Fig. 435 and proceed as follows: Remove clevis pin retaining draft control plunger rod (51) to main draft control spring plunger, remove snap ring retaining connector (53) to actuating lever (56) and lift assembly from cover. Remove snap ring from inner end of control lever shaft (38) and remove control valve actuating lever (56). Remove clevis pin from position control link (60) and selector lever shaft (63), then remove control lever shaft and position control assembly. Drive roll pin from override valve actuator arm (69) and control lever shaft, then remove actuator arm and position control assembly from shaft. Remove selector lever shaft (63) and remove "O" ring (25 – Fig. 415) from groove in shaft.

Compress position control spring (59 – Fig. 435), remove snap ring from front end of position control rod (58), then remove rod and spring. Remove position control adjusting screw (64). Compress draft control override spring (52) and remove snap ring from front end of draft control plunger rod (51), then remove rod and spring from connector (53).

Carefully inspect all parts and renew any that are bent or excessively worn. Renew pins in position control arm and actuating arm if pins are bent or worn and arms are otherwise serviceable. Re-

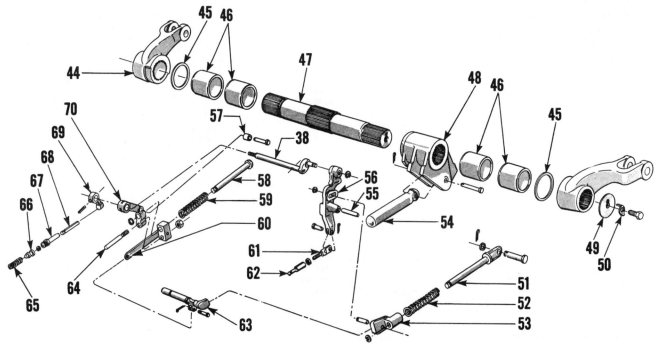

Fig. 435 — Exploded view of lift shaft and hydraulic control linkage assemblies used on early models without Load Monitor. Refer to Fig. 436 for early models with Load Monitor.

38. Control lever shaft	48. Ram arm	53. Connector	58. Position control rod	63. Selector arm	67. Guide
44. R.H. lift arm	49. Retainer plates	54. Piston connecting rod	59. Spring	64. Pin	68. Plunger
45. Dust seal	50. Locks	55. Pin	60. Link	65. Spring	69. Valve actuator arm
46. Bushings	51. Draft control plunger	56. Actuating lever	61. Link	66. Flow control override	70. Position control arm
47. Lift shaft	52. Override spring	57. Roller	62. Turnbuckle	valve	

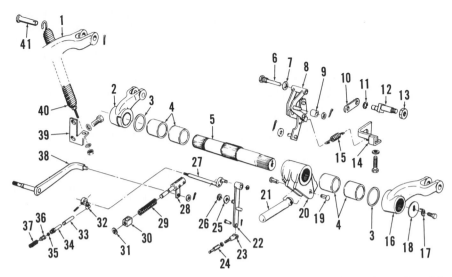

Fig. 436 — Exploded view of lift shaft and hydraulic control linkage assemblies used on early models with Load Monitor and all "10" series tractors. Also refer to Fig. 429.

1. Lift arm, L.H.
2. Lift arm, R.H.
3. Dust seal
4. Bushings
5. Lift shaft
6. Pivot shaft
7. Seal
8. Selector linkage assy.
9. Roller
10. Link
11. Retaining ring
12. Draft control adjustment screw
13. Nut
14. Spring retainer
15. Tension spring
16. Lift arm, L.H.
17. Locks
18. Retainer washers
19. Retaining pins
20. Ram arm
21. Connecting rod
22. Control valve actuating lever
23. Clevis
24. Turnbuckle
25. Spacer
26. Snap ring
27. Control lever shaft
28. Selector control rod & roller
29. Spring
30. Connector
31. Snap ring
32. Override valve actuator
33. Plunger
34. Plunger guide
35. Retaining ring
36. Override valve spool
37. Spring
38. Selector arm
39. Spring bracket
40. Lift arm return spring
41. Clevis spring

new springs if cracked or distorted or if free length is less than that of new springs.

Reassemble and reinstall control linkage by reversing disassembly and removal procedures. Adjust linkage as outlined in paragraphs 294, 295 and 296 or paragraphs 310 through 314.

376. ALL MODELS WITH LOAD MONITOR AND "10" SERIES WITHOUT LOAD MONITOR. To overhaul lift control linkage, lift shaft and cover, proceed as follows:

Remove control lever retaining nut, then remove spring washer, flat washer, friction disc and control lever from control lever shaft (27 – Fig. 436). Disconnect system selector lever clevis (external) from selector actuator arm, remove selector actuator arm anchor nut and remove actuator arm. Unbolt quadrant and slide quadrant from control lever shaft. Remove tension spring (15), then disconnect control rod and roller assembly (28) from selector arm (38). Disconnect control valve clevis (23) from control valve actuating lever (22) and pull turnbuckle (24) from control valve. Remove snap ring from pin of control lever shaft (27) and remove control valve actuating lever (22). Disconnect control rod and roller assembly from control valve actuating lever by compressing spring (29) and removing retaining ring (31). Remove nut (13) from draft control adjustment screw (12), remove (exter-

nally) pivot shaft (6), then disengage link (10) and lift out selector linkage assembly (8). Draft control adjustment screw (12) can now be removed by sliding it out of draft control plunger.

Carefully inspect all parts and renew any that are bent or excessively worn. Renew any springs that show signs of fracture or distortion and do not have a free length equal to that of new springs.

Reassemble and reinstall control linkage by reversing disassembly and removal procedure. Adjust linkage as outlined in appropriate paragraphs.

377. LIFT SHAFT. With control linkage removed as outlined in paragraph 375 or 376, use Figs. 435 and 436 as a reference and remove lift shaft and ram arm as follows:

Adequately support lift cover. Straighten locking tabs on retainers (50 or 17) and remove cap screws, retainers and flat washers (49 or 18), then remove lift arms from lift shaft (47 or 5). Bump lift shaft either way out of cover and ram arm (48 or 20). Bushings (46 or 4) will be driven from side of cover that lift shaft is removed from. Remove ram arm. Use a suitable driver or lift shaft to remove the two bushings remaining in cover.

Renew parts as necessary and reassemble as follows: Install two shaft bushings in one side of cover so outer bushing is flush with cover. Hold ram arm in position and insert shaft through

opposite side of cover from which bushings are installed, align master spline on shaft with master spline in ram arm and push shaft through cover and arm until centered. Place one of the bushings on end of shaft and bump it into lift cover, then place remaining bushing on lift shaft and bump it inward until flush with cover. Place new "O" rings on lift shaft and install lift arms with master spline on shaft and in the arms aligned. Place washer on end of shaft so hole in washer is aligned with hole in shaft, then install retainer and cap screw. Tighten cap screws evenly at each end of shaft so lift arms will just fall of their own weight, then bend locking tabs of retainers against cap screw heads.

378. MAIN DRAFT CONTROL PLUNGER. Following procedure is typical of all models. With lift shaft and ram arm removed as outlined in paragraph 377, proceed as follows:

Unscrew yoke (1 – Fig. 415) from outer end of plunger. Loosen set screw (3) and unscrew control spring retainer nut (4). Remove flat washer (5) and main control spring (6). On tractors without Load Monitor, remove filler plug (10) from top of cover and drive the roll pin (41) from lower side of cover, then push main draft control plunger (8) inward and remove it from lift cover.

Install a new seal (2) on yoke, lubricate all parts with grease and reassemble by reversing disassembly procedure. Adjust yoke and spring retainer as outlined in appropriate paragraph.

MAIN HYDRAULIC PUMP

On all models, the main hydraulic pump is mounted on the lower, right side of the rear axle center housing. An idler gear is used to transmit power from the drive gear located on the independent pto drive clutch hub to the hydraulic pump drive gear. Some "10" series tractors may be equipped with an optional auxiliary pump as noted in following section. Refer to paragraph 293 for main pump pressure testing procedures.

Models 5000-5600-6600-6700-7000-7600-7700

379. REMOVE AND REINSTALL. To remove hydraulic pump assembly, first drain rear axle center housing. On 6700 and 7700 models and models equipped with a Ford cab, disconnect brake control rods. On all other models, remove step plate (foot rest) from right side of tractor. Remove cap screws and lockwashers that retain pump, lift pump slightly away from rear axle center housing and disconnect flexible hydraulic pressure hose from pump.

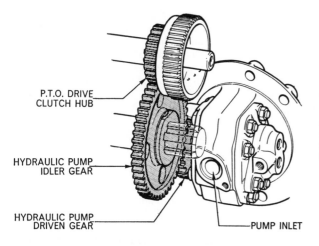

P.T.O. DRIVE
CLUTCH HUB

HYDRAULIC PUMP
IDLER GEAR

HYDRAULIC PUMP
DRIVEN GEAR

PUMP INLET

Fig. 437 — View of gear type hydraulic pump typical of that used in Model 5000 and 7000 tractors. Pump is driven from pto clutch hub through an idler gear mounted on transmission output shaft bearing retainer.

Carefully clean and inspect all parts. Renew pump body if scored or if gear track is worn more than 0.0025 inch (0.064 mm) deep in inlet side of body. The maximum runout across gear face to tooth edge (measured at 90 degrees to gear center line) should not exceed 0.001 inch (0.254 mm) and bearing journal diameter on either side of each gear should be within 0.001 inch (0.0254 mm) of the other side. If necessary to renew a pump gear, mating gear must also be renewed if face width of new gear is more than 0.002 inch (0.0508 mm) wider than mating gear. Light score marks on gear bearing faces can be removed with "O" grade emery paper on a flat (lapping) plate. Remove light score marks on gear journals with "O" grade emery paper.

The pump, intake tube and strainer can then be withdrawn from rear axle center housing. Unbolt and remove intake pipe from pump assembly. Refer to paragraph 380 for pump overhaul.

To reinstall pump, first install suction tube with new "O" ring between tube flange and pump body. Install a new "O" ring in pressure port recess in rear axle center housing and place new mounting gasket on pump. Insert suction tube and screen through opening in rear axle center housing and reconnect independent pto pressure hose. Move pump into mounting position, install retaining cap screws and lockwashers and tighten cap screws to a torque of 35-47 ft.-lbs. (48-64 N·m). Reinstall right step plate or connect brake control rods and refill axle center housing with proper fluid; refer to paragraph 286.

380. **OVERHAUL PUMP.** With hydraulic pump removed as outlined in paragraph 379, proceed as follows: Remove relief valve by unscrewing valve body (2 or 30–Fig. 438). Disassemble valve by unscrewing valve seat (9 or 37) and removing valve parts. Be careful not to lose any shims (4, 7 or 32).

Straighten locking tab on washer (24), then remove nut (23), washer, driving gear (25) and Woodruff key (41) from pump drive gear (42). Remove pump through-bolts and note location of the two bolts which act as dowel bolts and have the letter "D" stamped on head. Separate rear (small) pump body (52) from rear cover plate (46) and remove gear bearing (49), pump drive gear (50) and driven gear (51). Remove rear pump drive gear connector (43) and seal ring(s) (47 or 48) from rear cover and separate rear cover and front cover (29) from pump body (11). Carefully remove drive gear (42), driven gear (40) and bearings (39 and 44) from pump body as an assembly as shown in Fig. 444. Remove seal rings (45–Fig. 438) from pump covers. Separate gear bearings and pump gears.

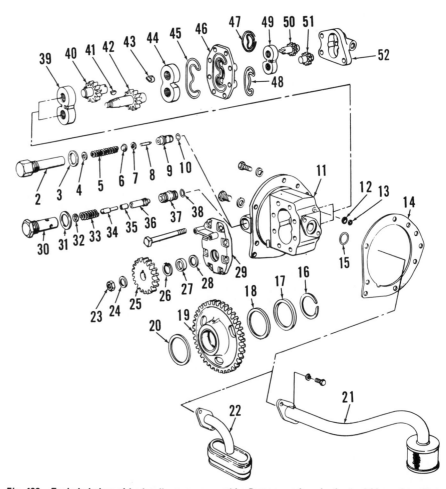

Fig. 438 — Exploded view of hydraulic pump assembly. Pumps produced prior to 4-68 used a relief valve comprised of items (2 through 10). Pumps produced 4-68 and later used a relief valve comprised of items (30 through 38). Both relief valves can be disassembled by unscrewing valve seat (9 or 37).

2. Valve body	27. Seal	40. Driven gear
3. Seal ring	28. Seal washer	41. Woodruff key
4. Shims	29. Front cover	42. Drive gear
5. Spring	30. Valve body	43. Pto pump coupling
6. Spring seat	31. Seal ring	44. Bearing assy.
7. Shims	32. Shims	45. Seal ring
8. Relief valve	33. Spring	46. Rear cover
9. Valve seat	34. Relief valve	47. Seal rings (late)
10. "O" ring	35. Seal	48. Seal ring (early)
11. Pump body	36. Relief valve	49. Bearing
12. Seal	37. Valve seat	50. Drive gear
13. "O" ring	38. "O" ring	51. Driven gear
14. Gasket	39. Bearing assy.	52. Rear pump body
15. "O" ring		
16. Retaining ring		
17. Retaining ring		
18. Thrust washer		
19. Idler gear		
20. Retaining ring		
21. Screen & pipe (5000)		
22. Screen & pipe (7000)		
23. Nut		
24. Tab washer		
25. Drive gear		
26. Snap ring		

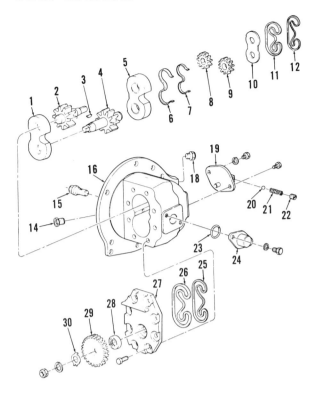

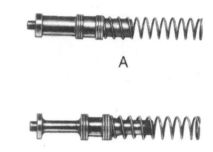

Fig. 442 — Comparison views of current (B) and previous (A) hydraulic by-pass valve and spring re-designed for increased flow or hydraulic oil. New parts may be used to service prior models.

Fig. 439 — Exploded view of hydraulic pump used on Models 5600, 6600, 6700, 7600 and 7700. Service procedures for parts renewal and overhaul are unchanged. Parts 1 through 5 are serviced as an assembly only. Items 20, 21 and 22 are used for circuit pressure relief on tractors with dual power only.

1. Bearing assy.
2. Front driven gear
3. Woodruff key
4. Front drive gear
5. Bearing assy.
6. Pressure loading strip
7. Backing strip
8. Rear drive gear
9. Rear driven gear
10. Pressure plate
11. Outer body seal
12. Inner body seal
14. Dual power fitting
15. Relief valve
16. Pump body
18. Test port plug
19. Rear cover assy.
20. Relief valve ball
21. Relief valve spring
22. Relief valve plug
23. "O" ring
24. Inlet flange
25. Inner seal ring
26. Outer seal ring
27. Front cover
28. Shaft seal
29. Drive gear (25T)
30. Tab washer

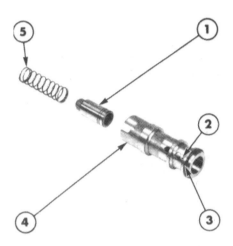

Fig. 443 — Cooler valve assembly used on Models 5600, 6600, 6700, 7600 and 7700 equipped with dual power. Refer to paragraph 289.

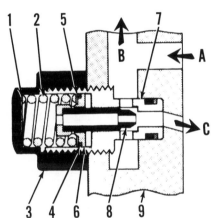

Fig. 440 — Cross-section of new type hydraulic system relief valve shown under working pressure. This valve does not interchange with previous relief valves.

1. Shims
2. Spring
3. Cap
4. Seal
5. Back-up washer
7. Valve seat
8. Valve

9. Hydraulic pump housing
   FLOW:
   A. From pump gears
   B. To hydraulic system circuits
   C. Return to reservoir

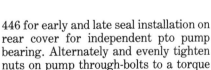

Fig. 441 — Comparison view of new (B) and previous (A) hydraulic check valve retainers. New retainer is used on 5600, 6600, 6700, 7600 and 7700 model tractors, but may be used to service prior models.

Fig. 444 — Removing the pump gears and bearings as an assembly.

The gear bearings (39 and 44) must always be renewed as a pair. On newer models (Fig. 439), pump gears and bearings, parts 1 through 5, must be replaced as an assembly.

Thoroughly clean all parts, air dry and then lubricate with hydraulic fluid prior to reassembly. Reassemble by reversing disassembly procedure using all new seals. Figure 445 can be used as an assembly guide; the large notch in gear bearings must be towards intake side of pump as shown in Fig. 444. Refer to Fig.

446 for early and late seal installation on rear cover for independent pto pump bearing. Alternately and evenly tighten nuts on pump through-bolts to a torque of 40-45 ft.-lbs. (54-61 N·m).

NOTE: Be sure the two dowel bolts are reinstalled in correct position (Fig. 446).

Install new washer (24 – Fig. 438) and bend tang of washer against gear retaining nut (23) after tightening nut.

Models 5610-6610-6710-7610-7710

381. **REMOVE AND REINSTALL.** Drain oil from rear axle center housing. On Models 5610, 6610 and 7610 without cab, detach two brake pedal return springs and remove right platform. On

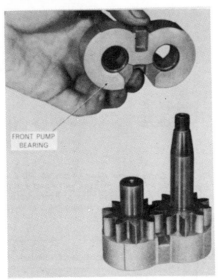

FRONT PUMP BEARING

Fig. 445 — Assembling pump gears and bearings.

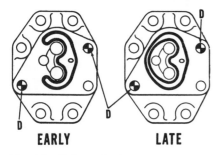

EARLY   LATE

Fig. 446 — Early production pump rear cover (46 — Fig. 438) has a single sealing ring groove. Late production cover has grooves for two sealing rings for improved sealing to rear pump body. "D" indicates dowel bolt location.

Models 5610, 6610 and 7610 with cab and Models 6710 and 7710, disconnect brake control rods. Detach brake levers from pull rods. Disconnect all fluid feed lines from hydraulic pump and position lines clear of pump. Remove cap screws securing hydraulic pump to center housing, then withdraw pump.

Reverse removal procedure to install pump. New sealing rings should be installed on hydraulic lines. Tighten bolts retaining pump to 54 ft.-lbs. (73 N·m) torque. Restore hydraulic fluid to specified level.

382. OVERHAUL PUMP. With hydraulic pump removed as outlined in paragraph 381, proceed as follows with reference to Fig. 447.

Remove filter (31). Straighten locking tab on washer (56), then remove nut (57), washer (56), driven gear (55) and Woodruff key (42) from pump drive gear (41). Remove cap screws (1) securing rear cover plate (3). Separate rear cover plate (3) from rear pump body (9) and remove bearing assembly (6), rear pump driven gear (7), rear pump drive gear (8) and connector shaft (39).

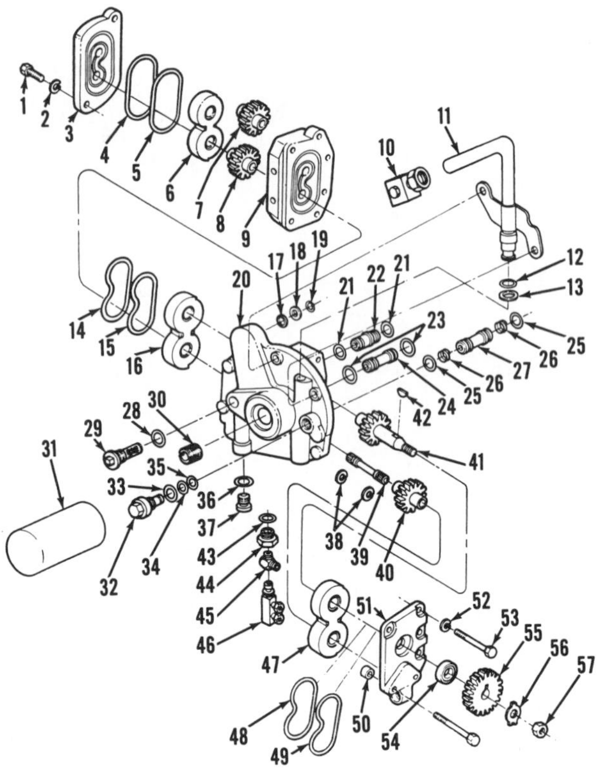

Fig. 447 — Exploded view of transmission mounted hydraulic pump used on Models 5610, 6610, 6710, 7610 and 7710.

1. Cap screw
2. Lockwasher
3. Rear cover plate
4. Outer sealing ring
5. Pressure loading ring
6. Bearing assy.
7. Rear pump driven gear
8. Rear pump drive gear
9. Rear pump body
10. Fitting
11. Tube
12. Washer
13. Seal
14. Outer sealing ring
15. Pressure loading ring
16. Bearing assy.
17. Seal
18. Washer
19. Ring
20. Housing
21. Seal
22. Inlet tube
23. Seal
24. Rear pressure tube
25. Seal
26. Pressure tube washer
27. Front pressure tube
28. Seal
29. Strainer
30. Adapter
31. Filter
32. Relief valve
33. "O" ring
34. "O" ring
35. "O" ring
36. Seal
37. Plug
38. Clip
39. Connector shaft
40. Front pump driven gear
41. Front pump drive gear
42. Woodruff key
43. "O" ring
44. Fitting
45. Elbow (without fwd)
46. Fitting (with fwd)
47. Bearing assy.

48. Pressure loading ring
49. Outer sealing ring
50. Ring dowel
51. Front end plate
52. Lockwasher
53. Through-bolt
54. Seal
55. Driven gear
56. Tab washer
57. Nut

NOTE: Observe positioning of bearing assembly (6, 16 and 47). If bearing assemblies are to be reused, they must be installed in their original positions.

Remove through-bolts (53), then complete disassembly of components with reference to Fig. 447. Remove strainer (29) and relief valve (32).

Carefully clean and inspect all parts. Renew pump body (9) and/or housing (20) if scored or if gear track is worn more than 0.004-inch (0.102 mm) deep on inlet side of body. Maximum runout across gear face to tooth edge (measured at 90 degrees to gear centerline) should not exceed 0.001 inch (0.0254 mm) and bearing journal on either side of each gear should be within 0.0005 inch (0.0127 mm) of the other side. There must be no more than 0.0002 inch (0.005 mm) difference in widths of drive gear and driven gear. Light score marks on gear bearing faces can be removed with "O" grade emery paper on a flat (lapping) plate. Remove light score marks on gear journals with "O" grade emery paper.

Thoroughly clean parts and allow to air dry. Renew all seals and "O" rings for reassembly. Lubricate all components with hydraulic fluid prior to assembly. Reassemble by reversing disassembly

procedure. Install bearings (6, 16 and 47) so notches in bearing face are towards gears and relieved area on outside of bearing is toward outlet side of body. Pack high temperature grease between lips of shaft seal (54). Tighten through-bolts (53) and cap screws (1) to 35-40 ft.-lbs. (48-54 N·m).

Install new washer (56) and bend tang of washer against gear retaining nut (57) after tightening nut.

NOTE: Prior to installation, prime pump through suction port and rotate gears by hand.

AUXILIARY HYDRAULIC PUMP

Models So Equipped

Some "10" series tractors may be equipped with an optional gear-type auxiliary pump mounted at rear, left side of engine. Drive is taken from a gear mounted on rear of engine camshaft. The auxiliary pump is used to supplement the main pump in supplying pressurized hydraulic oil to operate the remote valve circuit simultaneously with the power lift circuit. Refer to paragraph 293 for auxiliary pump pressure testing procedures.

383. REMOVE AND REINSTALL. Gear pump is mounted on rear, left side

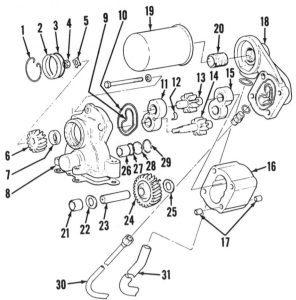

Fig. 448 – Exploded view of engine-driven hydraulic gear pump.

1. Clip
2. Plug
3. "O" ring
4. Nut
5. Tab washer
6. Gear
7. Seal
8. Housing
9. Seal ring
10. Seal ring
11. Bearings
12. Stuffer strip
13. Driven gear
14. Drive gear
15. Bearings
16. Body
17. Dowels
18. Cover
19. Filter
20. Adapter
21. Bushing
22. Washer
23. Idler shaft
24. Idler gear
25. Washer
26. Bushing
27. "O" ring
28. Plug
29. Snap ring
30. Pressure tube
31. Inlet tube

of engine and is driven by a gear attached to rear end of engine camshaft.

To remove pump, thoroughly clean area around pump and hydraulic lines. Disconnect pump outlet tube, unbolt and lift off pump and pull pump inlet tube from pump. Plug hydraulic line openings.

Reverse removal procedure to install pump. New "O" rings should be installed on hydraulic lines. Tighten pump retaining bolts to 30 ft.-lbs. (41 N·m) torque.

384. OVERHAUL PUMP. Prior to pump disassembly, mark drive housing (8 – Fig. 448), body (16) and cover (18) so they may be reassembled in original positions. Unscrew filter then remove wire retainer (1) and plug (2). Unscrew gear nut (4) and remove tab washer (5). Remove four through-bolts and using a soft mallet separate housing, body and cover. Tap on end of drive gear (14) to release taper from gear (6). Note positions of bearings and gears before removing from body. Remove snap ring (29), plug (28) and "O" ring (27). If plug will not dislodge easily, compressed air may be directed behind plug through bleed hole in pump mounting surface. Remove idler gear shaft (23), idler gear (24) and washers.

Carefully clean and inspect all parts. Renew pump body if scored or if gear track is worn more than 0.004-inch (0.102 mm) deep on inlet side of body. Maximum runout across gear face to tooth edge (measured at 90 degrees to gear centerline) should not exceed 0.001 inch (0.0254 mm) and bearing journal on either side of each gear should be within 0.0005 inch (0.0127 mm) of the other side. There must be no more than 0.0002 inch (0.005 mm) difference in widths of drive gear and driven gear. Light score

marks on gear bearing faces can be removed with "0" grade emery paper on a flat (lapping) plate. Remove light score marks on gear journals with "0" grade emery paper.

Thoroughly clean parts and allow to air dry. Lubricate components with hydraulic fluid prior to assembly. Reassemble by reversing disassembly procedure. Install bearings (11 and 15) so notches in bearing face are towards gears and relieved area on outside of bearing is toward outlet side of body. Pack high temperature grease between lips of shaft seal (7). Tighten through-bolts to 37 ft.-lbs. (50 N·m).

PRIORITY VALVE

Models With Auxiliary Hydraulic Pump

385. OPERATING PRINCIPLES. Priority valve (P – Fig. 424A) is mounted on top of the hydraulic lift cover. The priority valve regulates oil flow in the remote control valve circuit and combines auxiliary pump oil flow with main pump oil flow to maintain a constant pressure in the main hydraulic system. A pressure relief valve incorporated in the combining valve opens at 2700-2750 psi (18.63-18.97 MPa) to protect auxiliary pump and remote valve circuit.

386. R&R PRIORITY VALVE. Disconnect oil lines from connections on valve and plug or cap openings to prevent contamination. Remove mounting cap screws and withdraw valve assembly from lift cover.

Reinstall valve assembly in reverse order of removal. Tighten securing cap screws to 60 ft.-lbs. (82 N·m) torque.

387. OVERHAUL VALVE. Note and remove "O" rings located in bottom-side counterbores. Remove check valve stop (3 – Fig. 449) and withdraw spring (4) and check ball (5). Unscrew hex head retainer located at front end of flow control valve (2), then withdraw spring (1). Unscrew plug located at rear of flow control valve bore. Using care, push flow control valve out front of bore. Unscrew plug located at front end of unload valve (11), then withdraw spring (14). Stand valve assembly on end to allow unload valve (11) to slide from bore or as needed, use a pair of needle nose pliers to withdraw valve. Unscrew hex head retainer located at front end of combining (sequencing) valve (12), then withdraw spring (13) and valve (12). As equipped, unscrew plug located at front end of auxiliary pump circuit check valve bore, then remove plug located at rear of bore. Extract stop (7), spring (8) and check ball (9).

Clean all parts in a suitable cleaning solvent and blow dry with clean compressed air. Place components on a clean, lint-free cloth. Inspect valves for scoring, burrs, freedom of movement in bore, excessive wear or any other damage. Inspect check balls for corrosion, excessive wear or any other damage. Inspect springs for cracks, weakness, distortions or any other damage. Examine valve housing for cracks or restricted passages and bores for scoring or excessive wear. Renew parts as needed.

Reassemble in reverse order of disassembly. Renew all "O" rings. Lubricate valve components with hydraulic oil prior to reassembly. Securely tighten

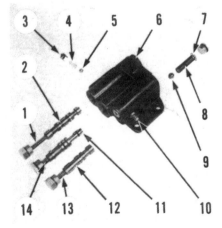

Fig. 449 – View showing priority valve assembly used on Models 5610, 6610, 6710, 7610 and 7710.

1. Flow control valve spring
2. Flow control valve
3. Check valve stop
4. Check valve spring
5. Check valve ball
6. Housing
7. Check valve stop
8. Check valve spring
9. Check valve ball
10. Pilot line connection
11. Unload valve
12. Combining (sequencing) valve
13. Combining (sequencing) valve spring
14. Unload valve spring

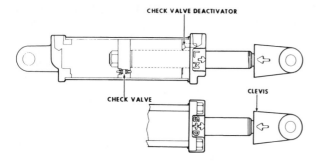

CHECK VALVE DEACTIVATOR

CHECK VALVE

CLEVIS

Fig. 450 — Remote cylinder operation can be selected by rotating cylinder rod clevis to the positions shown. Also refer to text.

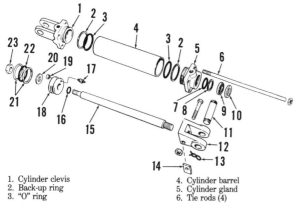

Fig. 451 — Exploded view of Load Monitor remote control cylinder. Note check valve (17).

7. "O" ring
8. Back-up ring
9. "V" block seal
10. Wiper
11. Clevis pin
12. Rod clevis
13. Spring pin
14. Clevis key
15. Cylinder rod
16. "O" ring
17. Check (poppet) valve
18. Piston
19. Check valve seat
20. Washer
21. Back-up rings
22. "O" ring
23. Nut

1. Cylinder clevis
2. Back-up ring
3. "O" ring
4. Cylinder barrel
5. Cylinder gland
6. Tie rods (4)

end of its stroke because the tractor lift cylinder will reach the end of its stroke first and in doing so, will contact the stop pin causing the control valve spool to return to neutral and stop oil flow.

To correct this out-of-phase condition, proceed as follows: Push the auxiliary service control valve fully inward and move the hydraulic lift control lever to bottom of quadrant. When lift links are fully lowered, pull auxiliary service control valve fully out and move lift control lever to top of quadrant. When the Load Monitor remote cylinder reaches the end of its stroke, the lift arms should stop moving a moment later when tractor lift cylinder reaches the end of its stroke. The cylinders should now be in phase (synchronized).

388. OVERHAUL REMOTE CYLINDER. With cylinder removed, remove hoses if necessary, then move piston back and forth to clear cylinder of oil. Remove nuts from tie-rods (6 – Fig. 451), then pull piston, piston rod and gland (5) from barrel (tube). Pull cylinder clevis (1) from opposite end of barrel. Remove nut (23) from cylinder rod (15) and remove washer (20), piston (18) and gland (5) from cylinder rod. If necessary, cylinder rod clevis (12) can be removed from cylinder rod after removing bolt and key (14). "O" ring (7) and back-up ring (8) can be removed from gland after removal of wiper seal (10) and "V" block seal (9). Check (poppet) valve (17) and valve seat (19) can be removed from piston if necessary.

Clean and inspect all parts. Renew any which are scored or excessively worn.

Use new "O" rings, back-up rings and seals and reassemble by reversing the disassembly procedure. Be sure check valve is centered in check valve deactivator area when arrow on cylinder rod clevis (12) is aligned with the STD arrow on gland (5). Install wiper seal (10) with lip toward outside.

hex head retainers and plugs. Position "O" rings in counterbores of valve bottom-side using petroleum jelly or a suitable equivalent.

LOAD MONITOR REMOTE CYLINDER

A special Load Monitor remote cylinder is available for use in conjunction with the Load Monitor hydraulic system. This cylinder can also be used as a standard 8-inch (20.3 cm) stroke cylinder and cylinder function can be selected by rotating cylinder clevis so arrow on clevis aligns with either the LM (Load Monitor) arrow on gland (cylinder) or the STD (Standard) arrow. These markings are 180 degrees apart. See Fig. 450.

When set for Load Monitor operation, the check valve unseats when cylinder reaches end of stroke and diverts oil flow to the tractor lift cylinder to obtain full lift of lift links.

To connect Load Monitor remote cylinder, attach piston end hose to auxiliary service (lift) port and rod end hose to return (drop) port. Mount cylinder on implement the same as any standard 8-inch (20.3 cm) stroke cylinder. Pull auxiliary service control valve to its fully out position to pressurize auxiliary service port.

NOTE: If the Load Monitor remote cylinder is hooked up when the main hydraulic system lift arms are partially raised, the remote cylinder will be unable to reach

REMOTE CONTROL VALVES

Models 5000, 5600, 6600, 6700, 7000, 7600 and 7700 use single and double spool remote control valves. Models 5610, 6610, 6710, 7610 and 7710 use closed-center remote control valves.

SINGLE AND DOUBLE SPOOL REMOTE CONTROL VALVES

Single spool or double spool remote control valves, with or without detents, have been used on the Model 5000 and 7000 tractors. Refer to Fig. 452 for an

exploded view of a single spool, no detent valve. Refer to Fig. 453 for an exploded view of a double spool, no detent valve used prior to production date 5-70 (8-70 double spool).

For exploded views of valves used from production date 5-70 (8-70 double spool) to production date 12-70, refer to Figs. 456 and 457.

For exploded views of valves used from production date 12-70 and after, refer to Figs. 459 and 460. This later design valve is also used on Models 5600, 6600, 6700, 7600 and 7700.

In all cases, valve body and valve spool are a select fit and are not available separately. When disassembling double spool valves, keep spools identified with their correct bores.

Valves can be identified by their exterior configuration.

Remote Control Valves Without Spool Detent (Prior to 12-70)

389. Exploded views of single spool and double spool remote control valves without spool detent are shown in Figs.

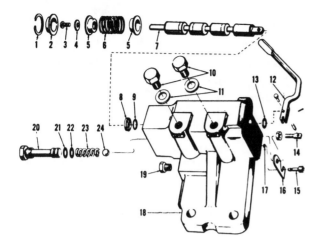

Fig. 452 — Exploded view of single spool remote control valve without spool detent mechanism.

1. Snap ring
2. Plug
3. Screw
4. Washer
5. Spring cups
6. Centering spring
7. Valve spool
8. Seal retainer
9. "O" ring
10. Port plugs
11. Seal rings
12. Control lever
13. "O" ring
14. Pivot screw
15. Switch valve
16. Retainer plate
17. "O" ring
18. Valve body
19. Plug
20. Check valve plug
21. "O" ring
22. "O" ring
23. Spring
24. Check valve ball

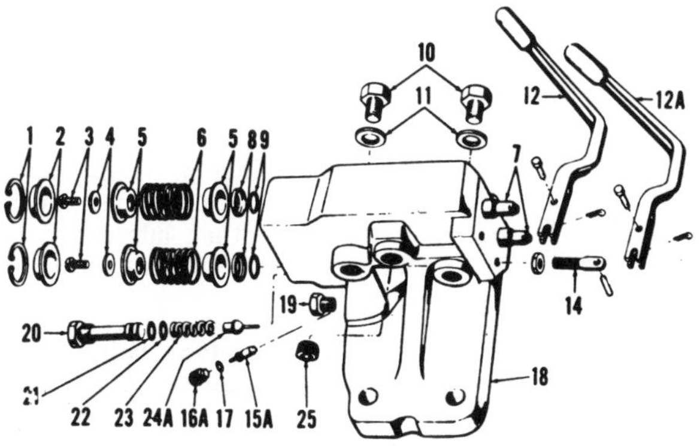

Fig. 453 — Exploded view of two spool remote control valve without spool detent mechanism. Valve spools (7) are not shown removed; refer to (7 — Fig. 452).

9. "O" rings
10. Port plugs
11. Seal rings
12. Long control lever
12A. Short control lever
14. Pivot screw
15A. By-pass valve
16A. Plug
17. "O" ring
18. Valve body
19. Plug
20. Check valve plug
21. "O" ring
22. "O" ring
23. Spring
24A. Check valve
25. Plug

1. Snap rings
2. Plugs
3. Screws
4. Washers
5. Spring cups
6. Centering springs
7. Valve spools
8. Seal retainers

Remove control levers by removing clevis pins retaining levers to valve spools, then pull levers from pivot pins in adjusting screws (14 — Fig. 452 or 453). Remove snap rings (1) that retain centering spring caps (2), then push or tap on rear (lever) end of valve spools to remove spools (7), centering spring (6) assemblies and caps from front end of valve body.

NOTE : On double spool control valves, be sure to tag valve spools as they are removed as each valve is a select fit and not interchangeable.

Remove screws (3) from ends of valve spools to remove centering springs (6), spring cups (5) and washers (4). Using a slide hammer with small internal pulling attachment (Nuday tool 7600-E or equivalent), remove seal retainers (8) from valve bores. Remove the "O" rings (9 and 13) from valve spool bores.

Thoroughly clean and inspect all parts. If control valve spools or bores are seriously damaged or scored, the complete valve assembly must be re-newed as spools and body are not available separately. Renew centering springs, retainers or washers if cracked or distorted. Centering spring free length (new) is 1.103 inch (28.07 mm); renew spring if free length varies materially from this dimension.

To reassemble, proceed as follows: Lubricate control valve spools and insert them into front (centering spring) end of valve body. Push spools rearward so rear ends are even with "O" ring groove in valve body. Lubricate "O" rings (9 — Fig. 453) and using rear ends of valve spools as "back stops," insert "O" rings in grooves in valve bores. Push spools on through "O" rings so they extend about ¾-inch (1.9 cm) from rear of valve body. Lubricate front "O" rings and push them into bore around front ends of valves, then install retainers (8) using special driver (Nuday tool N-651 or equivalent). Reinstall centering springs and cups, flat washers, retaining screws, caps, snap rings and control valve levers.

452 and 453. Either valve can be used with single or double acting remote cylinders. On single spool valve, open (turn out) switch valve (15 — Fig. 452) to oper-ate a single acting remote cylinder from "LIFT" port of valve. On double spool valve, back Allen head screw (16A — Fig. 453) out two or three turns to operate a single acting cylinder from the "No. 1 LIFT" port. Valve levers must be held in raising or lowering position as valve spools do not have detents. If valve is held open after cylinder reaches end of stroke, tractor hydraulic system relief valve will open at approximately 2500 psi (17.25 MPa).

390. REMOVE AND REINSTALL. To remove remote control valve, first disconnect any remote hoses, place lift system selector lever in draft control position and move 3-point hitch control lever to bottom of quadrant. Stand on lift arms to force all oil from 3-point hitch cylinder. Then, unbolt and remove remote control valve assembly from lift cover.

When reinstalling remote control valve, always renew sealing "O" rings between valve body and lift cover or manifold. Refer to Figs. 454 and 455.

One "O" ring (5 — Fig. 455) is placed in counterbore on lift cover and is omitted from counterbore in valve face (1B — Fig. 454). Tighten retaining cap screws to a torque of 40-45 ft.-lbs. (54-61 N·m).

391. OVERHAUL VALVE ASSEM-BLY. With remote control valve assem-bly removed from tractor as outlined in paragraph 390, refer to the following paragraphs for overhaul procedure:

392. CONTROL VALVE SPOOLS, CENTERING SPRINGS AND SEALS.

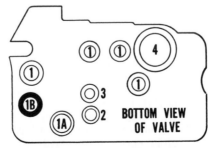

Fig. 454 — View showing bottom of remote con-trol valve assembly. "O" ring (1B) is omitted; refer to Fig. 455. On remote control valves without spool detent (Figs. 452 and 453), "O" ring is not used at location (1A).

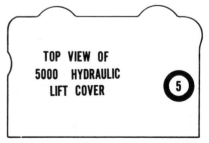

Fig. 455 — When installing remote control valves on all models, install "O" ring on hydraulic lift cover at (5) instead of "O" ring (1B) on valve; refer to Fig. 454.

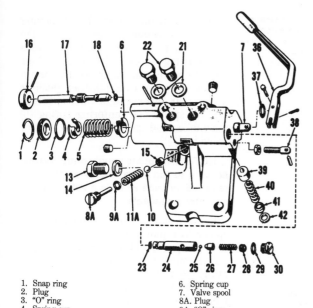

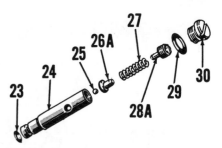

Fig. 456 — Exploded view of single spool remote control valve with spool detent (39) mechanism. Valve spool (7) is not shown removed from valve body; refer to Fig. 457. Refer to Fig. 458 for late production detent regulating valve assembly.

11A. Spring	
13. Plug	
14. Seal ring	
15. Check valve seat	
16. Knob	
17. Float valve	
18. "O" ring	
21. Seal rings	
22. Port plugs	
23. "O" ring	
24. Valve body	
25. Valve ball	
26. Ball seat	
27. Spring	
28. Adjusting plug	
29. "O" ring	
30. Plug	
39. Spool detent	
40. Spring	
41. Retainer	
42. Retaining ring	

1. Snap ring
2. Plug
3. "O" ring
4. Spring cup
5. Centering spring
6. Spring cup
7. Valve spool
8A. Plug
9A. "O" ring
10. Check valve ball

Fig. 458 — Exploded view of late production detent regulating valve assembly for valves shown in Figs. 456 and 457.

23. "O" ring	27. Spring
24. Valve body	28A. Adjusting plug
25. Valve ball	29. "O" ring
26A. Spring retainer	30. Plug

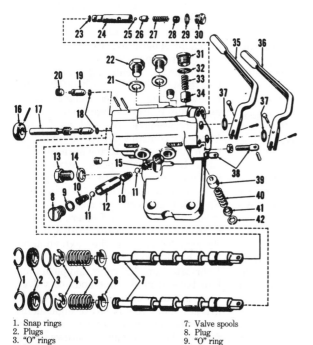

Fig. 457 — Exploded view of two spool remote control valve assembly with spool detent (34 and 39) mechanism. Refer to Fig. 458 for view of late production detent regulating valve assembly.

13. Plug
14. Seal ring
15. Check valve seat
16. Knob
17. Float valve
18. "O" rings
19. By-pass valve
20. Plug
21. Seal rings
22. Port plugs
23. "O" ring
24. Valve body
25. Valve ball
26. Ball seat
27. Spring
28. Adjusting plug
29. "O" ring
30. Plug
31. Plug
32. "O" ring
33. "O" ring
34. Detent piston
35. Long valve lever
36. Short valve lever
37. "O" rings
38. Pivot screws
39. Detent piston
40. Spring
41. Retainer
42. Retaining ring

1. Snap rings
2. Plugs
3. "O" rings
4. Spring cups
5. Centering springs
6. Spring cups
7. Valve spools
8. Plug
9. "O" ring
10. Springs
11. Check valve balls
12. Check valve seat

393. SINGLE SPOOL SWITCH VALVE. Loosen locknut on lever adjusting screw (14 – Fig. 452) and swing retainer (16) out of way. Then, using an Allen wrench, unscrew switch valve (15) from valve body. Remove and discard "O" ring (17). Renew switch valve if excessively pitted or distorted. Install and lubricate new "O" ring, then screw valve into valve body. Move retainer into position and tighten adjusting screw locknut while holding adjusting screw square with control lever.

394. DOUBLE SPOOL SWITCH VALVE. Using an Allen wrench, remove set screw (16A – Fig. 453). Then, using needle nose pliers, pull valve (15A) from valve body; remove and discard "O" ring (17). Renew valve if pitted or distorted. Install new "O" ring in groove on valve, lubricate valve and "O" ring and push into place in valve body. Reinstall Allen head plug.

395. CHECK VALVE. To remove check valve ball (24 – Fig. 452) on single

spool valve or check valve poppet (24A – Fig. 453) on double spool valve, unscrew plug (20) and remove spring (23) and ball or poppet. Remove sealing "O" rings (21 and 22) from plug. Renew ball or poppet valve if damaged in any way and renew spring if free length is not equal to that of new spring.

Install new "O" rings on check valve plug. Drop ball or poppet valve into bore, insert spring, then lubricate "O" rings and screw plug into valve body. Tighten plug securely.

Remote Control Valves With Spool Detents (Prior to 12-70)

396. Exploded views of single and double spool remote control valves with spool detents are shown in Fig. 456 and 457. Either valve may be used to operate single or double acting remote cylinders. To operate single acting remote cylinder from single spool valve or from top spool of double spool valve, turn knob (16) counterclockwise until switch valve (17) is screwed out against stop pin. To operate single acting remote cylinder from lower spool of double spool valve, screw Allen head screw (20) counterclockwise until against stop pin which will allow switch valve (19) to lift from seat. Unless float action is desired, switch valve (17 or 19) must be closed to operate a double acting remote cylinder. Detent pistons hold control valve spools in raising or lowering position until remote cylinder reaches end of stroke; then, increase in hydraulic pressure will open detent regulating valve and pressurized hydraulic fluid will lift detent pistons away from spools allowing spool centering spring to return spools to neutral position. Control levers can be returned to neutral position manually before remote cylinder reaches end of stroke, or can be held in raising or lowering position after detent release pressure is reached. Remote control cylinder and valve are protected by the 2500 psi (17.25 MPa) tractor hydraulic system relief valve when control valve levers are manually held in raising or lowering position or if detent regulating valve is adjusted to pressure exceeding that of relief valve.

397. ADJUST DETENT REGULATING VALVE. Detent regulating valve should be adjusted to release spool detent pistons at a pressure slightly higher than that of normal cylinder operating pressure. When so adjusted, control valve will be returned to neutral position without an excessive pressure buildup when remote cylinder reaches end of stroke.

To adjust detent regulating valve, refer to Fig. 457 for double spool valve or to Fig. 456 for single spool valve and proceed as follows: Remove regulating valve plug (30), then using an Allen wrench, turn socket head adjusting screw (28) into valve body (24) to increase detent release pressure or out to decrease release pressure. Reinstall plug with new sealing "O" ring (29) after completing the required adjustment. Late style detent regulating valve assembly is shown in Fig. 458.

398. REMOVE AND REINSTALL. To remove and reinstall remote control valve with spool detents, refer to paragraph 390 which outlines procedure for removing and reinstalling similar remote control valve without spool detents.

399. OVERHAUL. First, remove remote control valve from tractor as outlined in paragraph 390, then proceed as outlined in following paragraphs 400 through 403.

400. CONTROL VALVE SPOOLS, SPOOL CENTERING SPRINGS, SPOOL DETENT PISTONS AND SEALS. Refer to exploded view of double spool control valve in Fig. 457 and proceed as follows:

Remove clevis pins retaining control levers (35 and/or 36) to valve spools (7) and remove levers from pivot pins in adjusting screws (38). Remove detent plug (31) and "O" ring (32) and/or retaining ring (42) and retainer (41), then withdraw detent spring (33 and/or 40) and piston (34 and/or 39). Remove snap rings (1) from front end of valve bores and tap or push valve spools forward out of valve body; bore plugs (2) will be pushed out by valve spools. Remove "O" rings (3 and 37) from bore plugs and from rear of valve bores in body. Compress centering spring (5) and remove outer spring seats (4), then release spring pressure and remove springs and inner spring seats (6) from valve spools.

Thoroughly clean and inspect all parts. If control valve spools or bores in valve body are seriously damaged, complete valve assembly must be renewed as valve spools and body are not

available separately. Renew centering springs, retainers (spring seats) or bore caps if cracked or distorted. Centering spring free length (new) is 1.103 inch (28.07 mm); renew spring if free length varies materially from this dimension.

To reassemble, proceed as follows: Place rear spring seat (6) on valve spools, slide centering spring over end of spool and compress spring far enough to install front spring seat (4); then, release spring. Lubricate valve spools and insert them into valve bores from front end of valve body. Hold spools so rear (lever) ends are even with the "O" ring grooves in valve bores, then lubricate and install "O" rings. Push valves on through "O" rings until spring seats contact shoulders in valve bores. Install new "O" rings on bore plugs, then install plugs and retaining snap rings. Lubricate and install detent pistons, springs and spring retaining plug with new "O" ring or retainer and retaining ring. Reinstall control levers on pivot pins and connect to valve spools with clevis pins and cotter pins.

401. DETENT REGULATING VALVE. Refer to Fig. 456 for single spool valve and to Fig. 457 for double spool control valve. Then, proceed as follows:

Unscrew regulating valve plug (30) and remove plug and "O" ring (29). Insert an Allen wrench in socket head plug (28) and withdraw valve assembly from bore in valve body by prying and pulling on wrench. If valve body (24) is stuck tightly, unscrew plug, extract spring (27) and pull seat (24) from valve with hooked wire. Be careful not to lose valve ball retainer (26) or valve ball (25).

NOTE: Early production regulating valve is shown in Figs. 456 and 457, refer to Fig. 458 for exploded view of later production regulating valve.

Regulating valve spring fits inside early valve ball retainer and outside of later valve ball retainer and spring guide (26A). Later adjusting plug (28A) also has a spring guide and is used as service replacement for earlier plug (28–Fig. 456 and 457).

Inspect all parts and renew any that are worn or damaged. If necessary to renew valve seat (24), complete regulating valve assembly must be renewed. Install valve with new "O" ring (23) and lubricate "O" ring before inserting valve in bore. Refer to paragraph 397 for valve adjustment. Install regulating valve retaining plug with new "O" ring (29) when valve is correctly adjusted.

NOTE: It will be necessary to temporarily install plug while checking detent release pressure.

402. SWITCH (FLOAT) VALVES. Refer to Fig. 456 for single spool valve and to Fig. 457 for double spool valve.

To remove switch valve (17) on single spool valve or for top spool for double spool valve, drive retaining roll pin from valve body and turn knob (16) to unscrew valve completely. Remove "O" ring (18) from valve stem.

To remove switch valve (19) for lower spool of double spool valve, drive retaining roll pin from valve body and unscrew plug (20) with Allen wrench. Then, using needle nose pliers, pull valve from valve body and remove "O" ring from valve stem.

Install new sealing "O" ring on valve stem, lubricate "O" ring and push or thread valve stem into valve body. Install plug to retain switch valve for lower spool in double spool valve assembly. Install retaining roll pins in control valve body.

403. CHECK VALVES AND SEATS. On single spool valve assembly, refer to Fig. 456 and remove spring guide plug (8A), spring (11A) and check valve ball (10). Using a wire hook or other suitable tool, extract valve seat (15).

On double spool valve assembly, refer to Fig. 457 and remove plug (8), spring (10), check valve ball (11) and using a wire hook or other suitable tool, remove seat (12). Then remove second spring and check valve ball and extract inner seat (15) with suitable tool.

Renew check valve balls and seats if pitted, chipped or worn. Renew any spring if cracked, distorted or if free length is less than that of new spring. Reinstall by reversing removal procedure and using a new "O" ring (9) on retaining plug (8 or 8A).

Remote Control Valves, With Or Without Spool Detents (12-70 And After)

404. Exploded views of single and double spool remote control valves with spool detents are shown in Figs. 459 and 460. Except for detent assemblies (items 24 through 28), remote control valves are the same. Either control valve can be used to operate single or double acting cylinders. Refer to Fig. 461 and Table A.

Disassembly of detent and non-detent valves will be the same except double spool valves have twice as many parts and non-detent valves do not include detent assemblies.

405. ADJUST DETENT REGULATING VALVE. Detent regulating valve (29–Figs. 459 and 460) should be adjusted to release spool detent poppets at a pressure slightly higher than that of

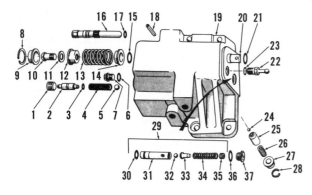

Fig. 459 — Exploded view of single spool remote control valve after production date of 12-70. Valve shown has detent assembly (items 24 through 28). Non-detent valves, except for detent assembly, remain the same. Detent valve can be made to function as a non-detent valve by removing only detent valve spring (26).

1. Plug
2. Check valve plug
3. Seal
4. Spring
5. Check valve ball
6. Seal
7. Plug
8. Retaining ring
9. End cap
10. Screw

11. Washer
12. Spring retainer
13. Centering spring
14. Spring retainer
15. "O" ring
16. Float valve stem
17. Seal
18. Spring pin
19. Valve body

20. Valve spool
21. "O" ring
22. Handle pivot
23. Seal
24. Detent ball
25. Poppet
26. Spring
27. Spring retainer
28. Snap ring

29. Detent regulating valve
30. Seal
31. Valve body
32. Steel ball
33. Spring retainer
34. Spring
35. Adjusting plug
36. "O" ring
37. Plug

Fig. 460 — Exploded view of double spool remote control valve after production date of 12-70. Except for detent assembly (items 24 through 28), non-detent valves remain the same. The inboard spool, without detents, is always double acting and cannot be converted to single acting operation. Refer to Fig. 459 for legend.

Refer to paragraph 406 to adjust linkage during reinstallation on 6700 and 7700 models.

408. **OVERHAUL.** Remove remote control valve as outlined in paragraph 328, then proceed as follows: Remove pin (bolt) securing handle to valve spool and disengage handle from spool and handle pivot (22 – Fig. 459 or 460). Remove snap ring (28), spring retainer (27), detent spring (26), poppet (25) and detent ball (24).

NOTE: On double spool detent valves, keep poppets identified with their bores as poppets differ slightly in that one poppet has a small bleed hole.

Remove retaining ring (8) and end cap (9), then push on handle end of spool and remove spool and centering spring assembly from front of valve body (19). Remove screw (10), washer (11), spring retainer (12), centering spring (13) and spring retainer (14) from spool. Remove "O" rings (15 and 21). Remove plug (37) and "O" ring (36), pull regulating valve (29) from body, then remove adjusting plug (35), regulating valve spring (34), spring retainer (33) and ball (32) from regulating valve body (31). Remove seal (30) from valve body.

Remove socket head plug (1), then remove check valve plug (2) with seal (3),

normal cylinder operating pressure. When so adjusted, control valve spool will be returned to neutral position without an excessive pressure buildup when remote cylinder reaches end of stroke.

To adjust detent regulating valve, refer to Fig. 459 or 460 and proceed as follows: Remove valve plug (37), then using an Allen wrench, turn socket head adjusting plug (35) into valve body (31) to increase detent release pressure, or out to decrease release pressure. Reinstall plug with new "O" ring (36) after completing adjustment.

406. **ADJUST REMOTE CONTROL VALVE LINKAGE (MODELS 6700 AND 7700).** Rotate turnbuckle on control rod so with remote control valve spool in neutral position there is 2½-2 5/8 inches (63.5-66.5 mm) between rear edge of control lever and rear of control lever slot in console. On models equipped with double spool control valves, adjustment applies to linkage for both valves.

407. **REMOVE AND REINSTALL.** To remove and reinstall remote control valve, refer to paragraph 390 which outlines the basic procedure for all remote control valves. Detach remote control valve linkage on Models 6700 and 7700 and other models equipped with a cab.

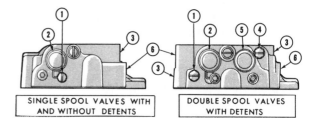

Fig. 461 — View showing front side of single and double spool remote control valves.

1. By-pass valve
2. Outboard spool
3. Drop port
4. Float valve
5. Inboard spool
6. Lift port

SINGLE SPOOL VALVES WITH AND WITHOUT DETENTS

DOUBLE SPOOL VALVES WITH DETENTS

Control Valve Type	Cylinder Application(s)	Float Valve	Bypass Valve	Single-Spool Valves or Outboard Spool of Double-Spool Valves		Inboard Spool of Double-Spool Valves	
				Lift Port	Drop Port	Lift Port	Drop Port
Single-Spool Without Detents	One Double-Acting	N/A	Closed	Lift Hose	Drop Hose	N/A	N/A
	One Single-Acting	N/A	Open	Lift Hose	Plug	N/A	N/A
Single-Spool With Detents	One Double-Acting	N/A	Closed	Lift Hose	Drop Hose	N/A	N/A
	One Single-Acting	N/A	Open	Lift Hose	Plug	N/A	N/A
Double-Spool Without Detents	Two Double-Acting	N/A	Closed	Lift Hose	Drop Hose	Lift Hose	Drop Hose
	One Double-Acting One Single-Acting	N/A	Open	Lift Hose	Plug	Lift Hose	Drop Hose
Double-Spool With Detents	Two Double-Acting	Closed	Closed	Lift Hose	Drop Hose	Lift Hose	Drop Hose
	One Double-Acting One Single-Acting	Open	Closed	Lift Hose	Plug	Lift Hose	Drop Hose
	Two Single-Acting	Open	Open	Lift Hose	Plug	Lift Hose	Plug

Table A — Refer to Fig. 461 for float valve and by-pass valve locations and to above table for cylinder application, valve settings and hose installation. The inboard spool of double spool control valve without detents is always double acting and cannot be converted to a single acting operation. N/A — Does Not Apply.

spring (4) and check valve ball (5) from valve body.

Drive out roll pin (18), then unscrew float valve (16) and remove seal (17).

Plug (7), seal (6), handle pivot (22) and seal (23) can be removed, if necessary.

Thoroughly clean and inspect all parts. If control valve spools or spool bores in valve body are excessively scored or worn, the complete valve assembly must be renewed as valve spools and body are not available separately. Centering spring free length is 2.022 inches (51.36 mm) for detent type control valves, or 2.157 inches (54.79 mm) for non-detent type control valves. Renew centering springs if they differ substantially from given dimensions, or if they show signs of fractures or distortion. Also inspect check valve spring (4) and regulator spring (34) for signs of fracture or distortion and renew as necessary.

To reassemble, use all new "O" rings and seals, lubricate parts and reassemble by reversing the disassembly procedure.

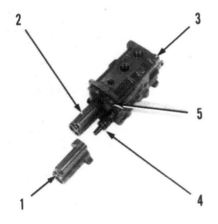

Fig. 462 – View showing detent regulating valve (1), detent housing (2), valve body (3), spool and centering spring assembly (4) and flow control knob (5) on closed-center remote control valves.

CLOSED-CENTER REMOTE CONTROL VALVES

Closed-center remote control valves are used on Models 5610, 6610, 6710, 7610 and 7710. Valves are available in options of single or double spool and may be installed in combinations of one, two, three or four spools. Double spool valves contain twice as many components as a single spool valve and in addition use a shuttle check valve located in the internal pilot line passage connecting the two spools. A flow control restrictor is used on each remote control valve spool and may be adjusted to a minimum of 3 gpm flow and a maximum of 15 gpm flow. When flow control restrictor is set to less than maximum output, excess fluid flow from pump is returned to the sump by the combining valve located in the priority valve assembly.

Closed-center valves may be used to operate either single or double acting cylinders. When used to operate single acting cylinders, a single hose is connected to lift port of quick release coupling. The cylinder is extended by moving control lever to raise position and retracted by moving lever to float position.

409. ADJUST REMOTE CONTROL VALVE LINKAGE. Loosen locknut securing trunnion and rotate trunnion so with remote control valve spool in neutral position, there is a distance of 1.5 inch (38.1 mm) from center of actuating rod attaching hole to valve body. Control valve handle should be vertically positioned (within 2 degrees) when valve spool is in neutral position.

410. ADJUST DETENT REGULATING VALVE. Detent regulating valve should be adjusted to release control valve spool at a pressure slightly higher

than that of normal cylinder operating pressure. When so adjusted, control valve spool will be returned to neutral position without an excessive pressure buildup when remote cylinder reaches end of stroke.

To adjust, loosen locknut and turn adjusting screw in to increase pressure or out to decrease pressure required to return control valve spool to neutral.

411. REMOVE AND REINSTALL. Clean area around remote control valve spool(s), then detach control linkage from valve spool(s). Tag and identify valve hoses, then remove hoses and cap valve body ports and hose openings. Remove cap screws securing remote control valve and quick-release coupling assembly to mounting bracket, then withdraw assembly. Separate remote control valve from quick-release coupling.

To reinstall, reverse removal procedure. Tighten securing cap screws to 41 ft.-lbs. (56 N·m) torque. Refer to paragraph 409 for control valve linkage adjustment.

412. OVERHAUL REMOTE CONTROL VALVE. Remove cap screws retaining detent housing (2–Fig. 462) to valve body (3), then withdraw detent housing. Extract valve spool and centering spring (4) as an assembly from detent housing end of valve body.

NOTE: Slide valve spool and centering spring assembly slowly from bore without applying undue pressure as damage to spool could result.

Remove plug at end of flow control spool bore and withdraw spring (1–Fig. 463) and spool (2). Unscrew flow control knob (5–Fig. 462) retaining screw, then remove knob. Remove circlip located behind flow control knob,

Fig. 463 – View showing flow control spool (2) and spring (1). Refer to text.

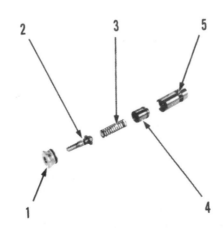

Fig. 464 – Retainer (1), shaft (2), spring (3), load check valve (4) and restrictor (5) are secured into position by a circlip located behind flow control knob.

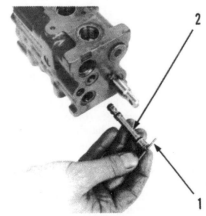

Fig. 465 – After removing securing circlip, extract shuttle check valve (2) by inserting a suitable tool (1) through hole in valve end.

Fig. 466 — View showing shuttle check valve seat (1) and ball (2) located in valve body side of double spool valves.

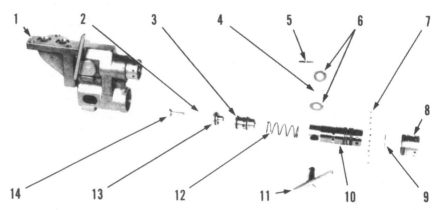

Fig. 467 — View showing quick-release coupling assembly used on closed-center remote control valves.

1. Housing	5. Roll pin	9. Clip
2. Clip	6. Washers	10. Body
3. Check valve assy.	7. Ball bearings	11. Lever assy.
4. Spring washer	8. Sleeve	12. Spring
		13. Guide
		14. Plunger

then withdraw retainer (1 – Fig. 464), shaft (2), spring (3), load check valve (4) and restrictor (5). Remove circlip retaining shuttle check valve (2 – Fig. 465), then insert a suitable tool (1) through hole in valve end and extract check valve.

Note that on double spool valves, two shuttle check valves as shown in Fig. 465 are used. A third shuttle check valve seat (1 – Fig. 466) and ball (2) are located in valve body side.

Separate removed components as needed.

Thoroughly clean and inspect all parts. If valve spools or spool bores in valve body are excessively scored or worn, complete valve assembly must be renewed as valve spools and body are not available separately. Inspect springs for signs of fracture or distortion and renew as necessary. Renew all other parts as needed if excessive wear or other damage is noted.

To reassemble, renew all "O" rings and seals, lubricate parts and reassemble by reversing the disassembly procedure. Tighten detent housing retaining cap screws and centering spring retaining coupling to 5 ft.-lbs. (7 N·m) torque. Tighten screw retaining flow control knob to 1.25 ft.-lbs. (1.7 N·m). On double spool valves, tighten shuttle valve located in valve body side to 10 ft.-lbs. (14 N·m).

413. OVERHAUL QUICK RELEASE COUPLING. Unscrew sleeve (8 – Fig. 467) from housing (1). Remove roll pin (5) and washers (6) from handle shaft, then withdraw handle (11). Complete disassembly of components with reference to Fig. 467.

Thoroughly clean components, then inspect all components for excessive wear or any other damage and renew as needed. Reassembly is reverse order of disassembly. Renew all "O" rings and lubricate components during reassembly.

CAB

Models 5600-5610-6600-6610-7600-7610

414. REMOVE AND REINSTALL. Remove access panel in cab transmission tunnel. If so equipped, detach auxiliary service control knob and remote control valve levers. Remove hydraulic lever control quadrant while noting two rear bolts are dowel type. Remove quadrant boot retainer and push boot down through opening. Move hydraulic control levers to vertical position. Detach steering wheel. Remove transmission gear shift knobs, boot and transmission access panel. Remove steering column access panel. Unscrew square head pinch bolt and remove throttle control lever. Unscrew and remove steering column boot retainer plate and boot. As equipped, detach lower end of Dual Power control rod from pivot assembly. Disconnect main wiring harness connector adjacent to steering gear. If so equipped, remove transmission handbrake lever assembly.

On models equipped with air conditioning, disconnect quick-release connectors located above rear axle. Disconnect all cables, hoses, wires, rods and tubing which will interfere with separation of cab from tractor. Loosen securing locknut and unscrew differential lock pedal. Remove scuff plates, mat and access panels in each door opening, to uncover front cab mounting bolts. Attach lifting apparatus to cab, unscrew front and rear cab mounting bolts and lift cab off tractor.

Reinstall in reverse order of removal. Tighten front cab mounting bolts to 180-220 ft.-lbs. (245-300 N·m) torque and rear cab mounting bolts to 252-308 ft.-lbs. (343-419 N·m) torque. If loosened, tighten cab mounting pad-to-axle housing bolts to 180-220 ft.-lbs. (245-300 N·m) torque.

Models 6700-6710-7700-7710

415. REMOVE AND REINSTALL. Remove gear shift knobs, then lift floor mat to expose gearshift boot. Remove shift lever boot. As needed, detach Dual Power control link clevis from pedal, then remove pedal. If so equipped, remove transmission handbrake lever assembly. Remove scuff plates, mat and access panels in each door opening, to uncover front cab mounting bolts.

On models equipped with air conditioning, disconnect quick-release connectors located above rear axle. Loosen securing locknut and unscrew differential lock pedal. Unscrew pinch bolts at gear lever pivots and extract main pivot pins, then withdraw levers. Disconnect main wiring harness connector adjacent to steering gear and steering hoses at pipe fittings. Disconnect all cables, hoses, wiring, rods and tubing which will interfere with separation of cab from tractor. Attach lifting apparatus to cab, unscrew front and rear cab mounting bolts and lift cab off tractor.

Reinstall in reverse order of removal. Tighten front cab mounting bolts to 180-220 ft.-lbs. (245-300 N·m) torque and rear cab mounting bolts to 252-308 ft.-lbs. (343-419 N·m) torque. If loosened, tighten cab mounting pad-to-axle housing bolts to 180-220 ft.-lbs. (245-300 N·m) torque.

NOTES

NOTES